STUDY GUIDE FOR

MICROBIOLOGY
An Introduction

FIFTH EDITION

Tortora • Funke • Case

Berdell R. Funke
North Dakota State University

The Benjamin/Cummings Publishing Company, Inc.
Redwood City, California • Menlo Park, California
Reading, Massachusetts • New York • Don Mills, Ontario
Wokingham, U.K. • Amsterdam • Bonn • Sydney
Singapore • Tokyo • Madrid • San Juan

Sponsoring Editor: Anne Scanlan-Rohrer

Assistant Editor: Leslie With

Editorial Assistant: Karmen Butterer

Production Supervisor: Larry Olsen

Outside Production Service: Adrienne Armstrong

Copyeditor: Alan Titche

Cover Designer: Yvo Riezebos

Compositor: Fog Press

Cover Photograph: Copyright © CNRI/Science Photo Library/Photo Researchers

ISBN 0-8053-8508-8

1 2 3 4 5 6 7 8 9 10—CRS—98 97 96 95 94

Library of Congress Cataloging-in-Publication Data

Funke, Berdell R.
    Study guide for Microbiology : an introduction / Berdell R. Funke.
    — 5th ed. / Tortora, Funke, Case.
        p.   cm.
    ISBN 0-8053-8508-8
    1. Microbiology—Outlines, syllabi, etc. 2. Microbiology—
Examinations, questions, etc. 3. Microbiology—Examinations,
questions, etc. 4. Molecular biology—Problems, exercises, etc.
I. Tortora, Gerard J. Microbiology. 5th ed. II. Title.
QR41.2.T67 1994 Suppl.
576—dc20
                                                            94-38835
                                                            CIP

The Benjamin/Cummings Publishing Company, Inc.
390 Bridge Parkway
Redwood City, California 94065

Contents

Preface

To the Student

Welcome to microbiology. As with any subject, you'll find some parts more interesting than others. But you will encounter few subjects in college that touch on daily living in so many ways. You'll find out things that are quite certain to interest you: how you become immune to diseases, how penicillin works, why oranges get moldy, and other "facts of life." On the other hand, now and then you may feel like the little girl thanking her aunt for a book on penguins. "Thank you very much," she wrote, "for the book on penguins. It told me a lot about penguins, even more than I wanted to know."

The Importance of Tutoring Yourself

I teach microbiology regularly to students required to take it for nursing, pharmacy, home economics, agriculture, civil engineering, and a number of other fields, including mortuary science. Not long ago, a mortuary science student attempted the course and failed miserably. In none of three exams did he reach a score of 50%. At the end of the following term he told me, woefully, that he had just repeated the course and had missed a grade of C by two points. To transfer the credit to the school of mortuary science he needed at least a C.

It happens that our school has a policy of providing tutors for those who need them. I turned his case over to a bacteriology graduate student who tutored in his spare time. On the first exam this formerly hapless undergraduate scored a 92 . . . and never looked back. Had three terms of exposure to my lectures finally enlightened him? Probably not; the tutor reported that in preparing for the first exam, the student denied having ever heard of the procaryotic cell.

What, then, made the difference? The tutor did, of course. But a tutor can't really put anything into your head that you can't put there all by yourself. What a tutor does is force you to think about the material instead of just staring at it. The tutor exposes your areas of ignorance, but instead of letting it go at that, forces you to pave over these areas with information.

You can do the same thing the tutor does—tutor yourself. The suggestions that follow and the use of this study guide will help you.

Lectures and Note Taking

Instructors vary in their approaches to microbiology courses. Some give lectures that largely paraphrase the text, and exams are based mainly on the text. Others scarcely use the book at all in their lectures and base the exams solely on the lectures; successful students must be careful note takers. (I have even heard of some cases in which there seems to be no apparent relationship between the information available to the student and the exams. These cases, of course, can only be left to the chaplain.)

It is important to take notes in an organized fashion such as this:

Staining of Bacteria
  Gram Stain
    Insert all material about Gram staining. Take down all terms
    likely to be associated with Gram staining.

  Acid-Fast Stain
    Insert all material about the acid-fast stain. Take down all terms
    likely to be associated with acid-fast staining.

If you look back at notes like this, you can see that you were hearing about the staining of bacteria and that there were basically two types of bacterial stains discussed. Note that you also clearly kept the information about the Gram stain separate from information about the acid-fast stain. Never let your notes run together so that one topic merges with another. On objective tests this failure to separate discussions can be fatal. Make sure you have definitions straight. Never leave your notes or your mind a blank on something as obvious as the definition of a term.

Studying

I always invite students with grade problems in the course to come see me. I ask them to bring their notes and a copy of the offending exam, corrected from a key. Very often certain patterns show up. One is that the notes have no "study marks"; that is, no underlining or circling. Another is that the student often misses a cluster of questions pertaining to one group of topics—for example, the characteristics of different antibiotics. In other words, the student has encountered a list of antibiotics and failed to differentiate one from another. This is why it is important to organize information by general headings and a system of indentation as I have described. I also notice occasionally that a term, such as the word *autotroph*, stands alone in the notes without definition or explanation. Surely the student could have filled in such an omission by consulting the book or the instructor. In my presentation of the material I am not trying to keep things secret, nor, presumably, is your instructor.

My recommendations are these: First, *organize your notes* so that it is obvious where a discussion of a topic, such as the disease diphtheria, starts, and where it ends. The most obvious way of doing this is to underline the main topic and inset all the information about that topic. (If you have trouble writing rapidly enough, or you have trouble with English as a language, ask the instructor if you can record the lectures.)

Second, *never confuse staring at a page with studying a page.* As you read, do your eyes ever travel down half a page while your mind fails to take in a single word written on it? A person has a limited capability for concentration, and by staying up all night before an exam, he or she is likely to exceed it. Do your studying in short bursts within the limits of your full attention span; start early. In order to ensure concentration, go through the notes or the text with a pencil or highlighting marker. Look deliberately at the material, and try to think like the person who goes through it to create an exam. You will find that your mind and the instructor's mind will often travel the same path. If the topic is penicillin, for example, mark the important facts: that it is produced by a mold, that its mode of action affects cell wall synthesis, that it has a β-lactam ring in its structure, that it affects mainly gram-positive bacteria, and so on. When you have done this, you have actually *thought* about the material instead of just staring at it. This method will also teach you how to take notes. You'll begin to recognize material that is likely to be on an exam and get it down, leaving to casual memory the instructor's anecdote about his memories of childhood chickenpox.

Study Guide Contents

This study guide provides a chapter-by-chapter summary that contains the important terms and concepts likely to be on examinations. This chapter summary is usually organized by the headings used in the text. Important terms are printed in boldface and defined, and important figures and tables from the text are included. Following each chapter synopsis is a sample exam, which is an extensive self-testing section containing matching questions and fill-in-the-blanks. An answer key is provided. (You may wish to supplement these questions with the Study Questions at the end of each chapter in the textbook.) Obviously, not all questions can be anticipated, and a good instructor will usually introduce material other than that found in the text. But going through these sample exams will work as a self-tutoring device to direct attention to any areas of ignorance.

If you are going to improve your time in running the mile or the amount of weights you can lift, you must do more than read a book on techniques. You must apply them to build up your endurance or your muscles. The same applies to study habits. A guide cannot do your work for you; it can only serve as a tutor. You must do the work yourself.

Credits

Chapter 2 Label the Art: Courtesy of Richard Liebaert, from *Student Study Guide for Biology: Concepts and Connections*, Redwood City, CA: The Benjamin/Cummings Publishing Company, Inc., 1994, p. 11.

Figure 4.1: © S.C. Holt, University of Texas Health Center/Biological Photo Service.

Figure 9.1: Adapted from N. Campbell, *Biology*, 3rd Ed. (Redwood City, CA: Benjamin/Cummings, 1993), Figure 19.2, p. 392. © 1993 Benjamin/Cummings Publishing Company.

Figure 16.1, Label the Art: G.J. Tortora, S.R. Grabowski, *Principles of Anatomy and Physiology*, 7th Ed. (New York: HarperCollins, 1993), Figure 22.1, p. 684. © 1993 HarperCollins College Publishers. Reprinted by permission of the publisher.

Figure 17.4, Label the Art I: Adapted from N. Campbell, *Biology*, 3rd Ed. (Redwood City, CA: Benjamin/Cummings, 1993), Figures 39.13, 39.14, pp. 862, 865. © 1993 Benjamin/Cummings Publishing Company.

Chapter 19 Label the Art: Hoth, Jr., Meyers, Stein, "Current Status of HIV Therapy," *Hospital Practice*, Vol. 27, No. 9, p. 154. Illustrations © Alan D. Iselin. Reprinted with permission.

Chapter 23 Label the Art: Adapted from N. Campbell, *Biology*, 3rd Ed. (Redwood City, CA: Benjamin/Cummings, 1993), Figure 38.5a, p. 823. © 1993 Benjamin/Cummings Publishing Company.

Figure 24.2: Adapted from N. Campbell, *Biology*, 3rd Ed. (Redwood City, CA: Benjamin/Cummings, 1993), Figure 38.22, p. 840. © 1993 Benjamin/Cummings Publishing Company.

Figure 25.13: © Science VU/Visuals Unlimited.

Chapter 27 Label the Art: Adapted from N. Campbell, L. Mitchell, J. Reece, *Biology: Concepts and Connections* (Redwood City, CA: Benjamin/Cummings, 1994), Figure 36.14, p. 710. © 1994 Benjamin/Cummings Publishing Company.

Figure 27.2: Hugo Spencer/Photo Researchers, Inc.

Answers: Figure a: © Cabisco/Visuals Unlimited. Figure b: © Barry Dowsett/SPL/Photo Researchers, Inc. Figure c: © E.C.S. Chan/Visuals Unlimited. Figure d: © Science Source/Photo Researchers, Inc.

CHAPTER 1

The Microbial World and You

Learning Objectives

After completing this chapter, you should be able to:

- Describe several ways in which microbes affect our lives.
- Explain the importance of observations made by Hooke and van Leeuwenhoek.
- Compare the theories of spontaneous generation and biogenesis.
- Identify the contributions to microbiology made by Needham, Spallanzani, Virchow, and Pasteur.
- Describe the importance of Koch's postulates.
- Describe how Pasteur's work influenced Lister and Koch.
- Identify the contributions to microbiology made by Ehrlich, Fleming, and Dubos.
- Define immunology, virology, and microbial genetics.
- Recognize the system of scientific nomenclature that uses genus and specific epithet names.
- List the five kingdoms of living organisms and the major members of each kingdom.
- Differentiate among the major groups of organisms studied in microbiology.
- List at least four beneficial activities of microorganisms.
- List two examples of biotechnology that use genetic engineering and two examples that do not.
- Define normal microbiota.

Microbes in Our Lives

Microbes include bacteria, fungi (yeasts and molds), protozoa, viruses, and the microscopic forms of algae. Most of these microorganisms are not harmful and indeed play a vital role in maintaining our global environment. Only a minority are **pathogenic** (disease producing). They are part of the food chain in oceans, lakes, and rivers; they break down wastes, incorporate nitrogen gas from the air into organic compounds, and participate in photosynthesis, which generates food and oxygen.

A Brief History of Microbiology

The First Observations

Anton van Leeuwenhoek was the first to report on the observation of microorganisms seen with magnifying lenses, beginning in 1674. He made detailed drawings of "animalcules" that have since been identified as representing bacteria and protozoa. About this time, **Robert Hooke** observed with a microscope the pores in slices of plants. He called them "cells." His discovery, reported in 1665, marked the beginning of the **cell theory**—that all living things are composed of cells.

The Debate over Spontaneous Generation

Until the second half of the nineteenth century, it was generally believed that life could arise spontaneously from nonliving matter, a process known as **spontaneous generation.** An early opponent of spontaneous generation, **Francesco Redi,** demonstrated in 1668 that maggots, the larvae of flies, do not arise spontaneously from decaying meat. He filled three jars with decaying meat and sealed them; three similar jars were left open. Maggots appeared only in the open vessels, which flies had entered. In subsequent experiments, in which he covered the first three jars with gauze rather than sealing their tops, the results were the same. These experiments demonstrated that access to air was not a factor.

Many, however, still believed that the simpler organisms observed by Leeuwenhoek might undergo spontaneous generation. In 1745, **John Needham** found that heated nutrient fluids poured into covered flasks were soon teeming with microorganisms. He took this as evidence of spontaneous generation. Twenty years later, **Lazzaro Spallanzani** showed that Needham's microorganisms had entered the fluid after boiling. Heating such fluids in a sealed flask, he showed, prevented the growth Needham had observed. Objections still remained; Needham felt that the heating had destroyed some vital force necessary for spontaneous generation. The concept of **biogenesis,** that living cells can arise only from other living cells, was introduced in 1858 by **Rudolf Virchow.**

In 1861, **Louis Pasteur** designed the experiments that finally ended the debate about spontaneous generation. He showed that flasks left open to the air after boiling would soon be contaminated, but if they were sealed, they remained free of microorganisms. He also used flasks whose long necks he bent into S-shaped curves. Air, with its presumed vital force, could enter these flasks, but airborne microorganisms were trapped in the tubes. The flask contents remained sterile. Pasteur showed that microorganisms are present throughout the environment and that they can be destroyed. He also devised methods of blocking the access of airborne microorganisms to nutrient environments; these methods were the basis of **aseptic** (germ-free) **techniques,** which are among the first things that a beginning microbiologist learns.

Fermentation and Pasteurization

At this time Pasteur was asked to investigate the problem of spoilage of beer and wine. Pasteur showed that, contrary to the belief that air acted on the sugars to convert them to alcohol, microorganisms called yeasts were responsible—and in the absence of air. This process is called **fermentation** and is used in making wine or beer. Spoilage occurs later when bacteria, in the presence of air, change the alcoholic beverage into acetic acid (vinegar). Pasteur prevented spoilage by heating the wine or beer just enough to kill such bacteria, a process that came to be known as **pasteurization.**

The Germ Theory of Disease

This association of yeasts with fermentation was the first concept to link a microorganism's activity to physical and chemical changes in organic materials. It suggested the possibility that microorganisms might be able to cause diseases as well—the **germ theory of disease.** In 1835, **Agostino Bassi** made the first association between a microorganism and a disease by proving that a silkworm disease was caused by a fungus. In 1865, Pasteur found that another silkworm disease was caused by a protozoa. Also in the 1860s, **Joseph Lister** applied the germ theory to medicine. He used carbolic acid (phenol) on surgical dressings and wounds, and greatly reduced the numbers of infections and deaths. In the 1840s **Ignaz Semmelweiz** demonstrated that chemically disinfecting the hands of physicians minimized infections of obstetrical patients. In 1876, **Robert Koch** demonstrated that rod-shaped bacteria in the blood of cattle that had died of anthrax were the cause of death. He showed that these bacteria could be isolated and grown in pure culture, be injected into healthy animals, and cause their death by anthrax. The same bacteria could then be isolated from the dead animals. This demonstration, which proved that a specific microbe is the cause of a specific disease, followed a set of criteria known today as **Koch's postulates.**

Vaccination

In 1798, **Edward Jenner** showed that the mild disease cowpox gave immunity to smallpox. He inoculated people with cowpox material by scratching their arm with a cowpox-infected needle. This process became known as **vaccination** (*vacca* is the Latin word for cow). The protection from disease provided by vaccination is called **immunity.** Years later, around 1880, Pasteur showed why vaccinations work. He found that the bacterium for fowl cholera lost its **virulence** (ability to cause disease) after it was grown for long periods in the laboratory. However, he showed that the weakened bacteria still retained their ability to induce immunity. Apparently the cowpox virus is related closely enough to smallpox to induce effective immunity.

The Birth of Modern Chemotherapy: Dreams of a "Magic Bullet"

The treatment of disease by chemical substances is called **chemotherapy.**
When prepared from chemicals in the laboratory, these substances are
called **synthetic drugs,** and when produced naturally by bacteria and
fungi they are called **antibiotics. Paul Ehrlich** speculated about a "magic
bullet" that would destroy a pathogen without harming the infected host.
In 1910, he found *salvarsan,* an arsenic derivative, that was effective
against syphilis. *Quinine,* an extract of South American tree bark, had until then been the only other such chemical available. Spanish conquista-
dors used it to treat malaria. In the late 1930s, a survey of dye derivatives
uncovered the important group of antibacterial *sulfa drugs.* The first antibiotic was discovered by **Alexander Fleming,** who observed the inhibition
of bacterial growth by the mold *Penicillium notatum* and the inhibitor penicillin. Penicillin was mass-produced and clinically tested in the 1940s.
Since then, many antibiotics have been discovered.

Modern Developments in Microbiology

Immunology, the study of immunity, has expanded rapidly in the twenti-
eth century. Smallpox has been eliminated, and many new vaccines have
become available. A major challenge will be to defeat the AIDS virus,
which attacks the immune system. In 1933 **Rebecca Lancefield** proposed
an immunologically based classification system for streptococci bacteria,
classifying them as serotypes (variants within a species).

   Virology, the study of viruses, really began in 1892 when **Dmitri
Iwanowski** demonstrated that tobacco plant pathogens would pass
through filters too fine for known bacteria. Much later, **Wendell Stanley**
showed that the organism, called the tobacco mosaic virus, was so simple
and homogeneous it could be crystallized.

Recombinant-DNA Technology

Recombinant-DNA technology had its origin in **microbial genetics** (how
microbes inherit traits) and **molecular biology** (how genetic information
is carried in DNA, which is then used to direct synthesis of proteins). Because of their simplicity and rapid reproduction rate, bacteria are the preferred organisms in this field. Beginning in the early 1940s, **George W.
Beadle** and **Edward L. Tatum** demonstrated the relationship between
genes and enzymes. DNA was established as the hereditary material by
Oswald Avery, Colin MacLeod, and **Maclyn McCarty. Joshua Lederberg**
and **Edward L. Tatum** discovered bacterial genetic transfer by conjugation. In 1958, **James Watson** and **Francis Crick** proposed the structure of
DNA. In the 1960s **François Jacob** and **Jacques Monod** discovered messenger RNA, important in protein synthesis, and later made major discoveries about the regulation of gene function in bacteria. **Paul Berg** showed
that fragments of animal DNA—genes—could be attached to bacterial
DNA, the first examples of **recombinant DNA.** These genetically altered
bacteria can be used to make large quantities of a desired protein.

Naming and Classifying Microorganisms

The system of naming **(nomenclature)** we now use was established by **Carolus Linnaeus.** Scientific nomenclature assigns each organism two names. The **genus,** the first name, is always capitalized, and the **specific epithet (species),** which follows, is not capitalized. The scientific names of organisms are always either underlined or italicized. At first, organisms were grouped into either the animal kingdom or the plant kingdom. **H. R. Whittaker,** in 1969, devised a **five-kingdom classification** system, which groups organisms into **Procaryotae (Monera), Protista, Fungi, Plantae,** and **Animalia,** based on cellular organization and nutritional patterns.

The Diversity of Microorganisms

Bacteria

Bacteria are simple, one-celled organisms whose genetic material is not enclosed in a special nuclear membrane. For this reason, bacteria are called **procaryotes** (prenucleus); they make up the kingdom Whittaker calls Monera. Bacterial cells generally have one of three shapes: **bacillus** (rodlike), **coccus** (spherical or ovoid), and **spiral** (curved or corkscrew). Individual bacteria may form pairs, chains, or other groupings, which are usually the same within a species. Bacteria are enclosed in cell walls largely made of peptidoglycan (cellulose is the main substance of plant cell walls). Bacteria generally reproduce by **binary fission** into two equal daughter cells. Many move by appendages called flagella, and although most use organic material for nutrition, some use inorganic substances or carry out photosynthesis.

Fungi

Fungi are **eucaryotes;** they contain DNA within a distinct nucleus surrounded by a nuclear membrane. They may be unicellular or multicellular. Their cell walls are composed primarily of *chitin.* **Yeasts** are unicellular nonfilamentous fungi larger than bacteria. **Molds** form *mycelia* of long filaments (*hyphae*).

Protozoa

Protozoa are unicellular, eucaryotic microbes, members of the kingdom Protista. They are classified by their means of locomotion, such as *pseudopods, cilia,* or *flagella.*

Algae

Algae are photosynthetic eucaryotes, mostly of the kingdom Protista, and are usually unicellular. They need light and air for growth.

Viruses

Viruses are very small and are not cellular. They have a core of either DNA or RNA, surrounded by a protein coat. They may have a lipid envelope layer as well. They reproduce only inside the cells of a host organism.

Multicellular Animal Parasites

Flatworms and **roundworms,** collectively called **helminths,** are not strictly microorganisms. A part of their life cycles involves microscopic forms, however, and identifying them requires many of the techniques used in identifying traditional microorganisms.

Microbes and Human Welfare

Recycling Vital Elements

Microbes recycle vital elements such as nitrogen, carbon, oxygen, sulfur, and phosphorus. *Cyanobacteria* (called blue-green algae) and certain soil bacteria may use atmospheric nitrogen directly (*nitrogen fixation*). The nitrogen fixed from the air is incorporated into living organisms and eventually returned as gaseous nitrogen, making up the **nitrogen cycle.** **Martinus Beijerinck** and **Sergei Winogradsky** first showed how bacteria helped recycle vital elements. In the **carbon cycle,** carbon dioxide is removed from the air by plants and algae, which convert it to food. In the **oxygen cycle,** oxygen is recycled to the air during photosynthesis. Microbes are used in treatment of sewage. Microbes are useful in treating oil spills, toxic waste sites, and so on, a process called **bioremediation.** Bacteria such as *Bacillus thuringiensis* are used in control of insect pests.

Modern Biotechnology and Genetic Engineering

Practical applications of microbiology are called **biotechnology.** The use of recombinant DNA technology has led to the advent of **genetic engineering,** which now produces insulin, interferon, clotting substances, vaccines, and other substances. Eventually it may become common to replace missing or defective genes in human cells, a process called *gene therapy.* Agricultural applications, including drought resistance and resistance to insects and microbial diseases, may also result from genetic engineering.

Microbes and Human Disease

The relationship between microbes and disease will remain of great interest to us all. Our normal microbiota (microorganisms) do not disturb us and often are helpful. However, many diseases are caused by microorganisms. The study of the body's resistance to microbial infection and disease is a continuing part of microbiological research.

Self-Tests

In the matching section, there is only one answer to each question; however, the lettered options (a, b, c, etc.) may be used more than once or not at all.

I. Matching

e 1. In 1668, demonstrated that maggots appeared only in decaying meat that had been exposed to flies.

n 2. Introduced the concept that living cells arise from other living cells.

k 3. Introduced the technique of vaccination for smallpox.

m 4. First to use the microscope to observe "cells."

f 5. Made an association between silkworm disease and a fungus.

g 6. A surgeon who used carbolic acid to control wound infections.

i 7. First to speculate about the possibility of a "magic bullet" that would destroy a pathogen without harming the host.

j 8. Discovered penicillin.

h 9. Using anthrax as a model, demonstrated that a specific microorganism is the cause of a specific disease.

l 10. Originated our system of scientific nomenclature.

a. Anton van Leeuwenhoek
b. John Needham
c. Lazzaro Spallanzani
d. Louis Pasteur
e. Francesco Redi
f. Agostino Bassi
g. Joseph Lister
h. Robert Koch
i. Paul Ehrlich
j. Alexander Fleming
k. Edward Jenner
l. Carolus Linnaeus
m. Robert Hooke
n. Rudolph Virchow

II. Matching

1. Assigned a microbial cause to fermentation.

2. First to crystallize a virus.

a 3. Showed that mild heating of spirits kills spoilage bacteria without damage to the beverage.

4. Developed a classification system that groups organisms into five kingdoms.

5. Devised a classification system for the streptococci based on an immunological system of serotypes.

6. Demonstrated that infections in obstetrical wards could be minimized by disinfecting hands of physicians.

7. Participated in determining the structure of DNA.

8. First demonstrated that genetic information could be exchanged between bacteria by conjugation.

a. Louis Pasteur

b. Wendell Stanley

c. H. R. Whittaker

d. Francis Crick

e. Paul Berg

f. Ignaz Semmelweiz

g. Rebecca Lancefield

h. George Beadle

i. Joshua Lederberg

III. Matching

1. Grouped as Procaryotae, or Monera.

d 2. Noncellular; reproduce only inside cells of host organism.

f 3. Helminths.

c 4. Yeasts.

5. Procaryotes.

6. Unicellular, eucaryotic microorganisms; members of kingdom Protista.

a. Protozoa

b. Elephants

c. Fungi

d. Bacteria

e. Viruses

f. Multicellular animal parasites

IV. Fill-in-the-Blanks

1. Bacteria generally reproduce by a process called ___binary fission___ into two equal daughter cells.

2. Microorganisms classified as photosynthetic eucaryotes are ___algae___ .

3. Techniques that keep areas free of unwanted microorganisms are referred to as ___aseptic___ (germ-free) techniques.

4. The set of criteria that prove that a specific microorganism is the cause of a specific disease is known today as _____ .

5. The concept that living cells can arise only from other living cells is called *biogenesis*

6. One objection proponents of spontaneous generation made to experiments in which nutrient fluids were heated in sealed containers was that heating destroyed some *vital force* in the air.

7. According to the rules applied to the scientific naming of a biological organism, the *Genus Species* name is always capitalized.

8. The protection from a disease that is provided by vaccination is termed *immunity*.

9. Monera (also called blue-green algae), which may fix nitrogen from the air, are called
    _____ .

10. The general name for a rod-shaped bacterium is *Bacillus*.

11. The general name for a spherical or ovoid bacterium is *coccus*.

12. Protists that are classified by their means of locomotion are *Protoza*.

13. The process by which yeasts change sugars into alcohol is termed *fermentation*

14. Molds form mycelia of long filaments called *hyphae*.

15. The treatment of a disease with chemical substances is called *Chemotherapy*

16. Paul Ehrlich discovered an arsenic derivative, *Salvarsan*, that was effective against syphilis.

17. Antimicrobial chemicals produced naturally by bacteria and fungi are called *antibodics*.

18. The use of microbes to help clean up an oil spill is called *Biomeditation*

V. Critical Thinking

1. Discuss some contributions microorganisms make that help maintain a balanced environment.

2. What are the advantages of using microorganisms to control insect pests?

3. Discuss three ways that microorganisms have been harnessed to improve the environment.

4. Although penicillin was first discovered in 1928, it wasn't clinically tested until the 1940s. What kinds of issues and problems must have been addressed before making penicillin available for treatment of infectious diseases?

5. Discuss the similarities and differences between the syphilis epidemic of the 1940s and the current AIDS epidemic.

6. List three characteristics unique to procaryotes.

7. Explain how the breakthroughs in genetics that occurred in the 1940s, 1950s, and 1960s led to recombinant DNA and genetic engineering.

8. Discuss the relationship between humans and their normal microbiota. What can cause this usually healthy relationship to change, allowing the same microbes to produce disease?

9. How did Bassi's and Pasteur's work on silkworm disease help to reinforce the germ theory of disease?

10. How is sourdough bread different from conventional bread?

CHAPTER 2

Chemical Principles

Learning Objectives

After completing this chapter, you should be able to:

- Discuss the structure of an atom and its relation to the chemical properties of elements.
- Define ionic bond, covalent bond, hydrogen bond, molecular weight, and mole.
- List several properties of water that are important to living systems.
- Define acid, base, salt, and pH.
- Distinguish between organic and inorganic compounds.
- Identify the building blocks of carbohydrates, simple lipids, phospholipids, proteins, and nucleic acids.
- Identify the role of ATP in cellular activities.

Structure of Atoms

All matter is made up of small units called **atoms.** Atoms in certain combinations form **molecules.** All atoms have a centrally located **nucleus** and particles called **electrons,** which have a negative (–) charge, moving around the nucleus. The nucleus of an atom is made up of positively (+) charged particles called **protons** and uncharged particles called **neutrons.** Both neutrons and protons have about the same weight, 1840 times that of an electron. Because the total positive charge on the nucleus equals the total negative charge of the electrons, each atom is an electrically neutral unit. Different kinds of atoms are listed by **atomic number,** the number of protons in the nucleus. The **atomic weight** is the total number of protons and neutrons in an atom.

Chemical Elements

All atoms with the same atomic number are classified as the same **chemical element.** Examples of elements are hydrogen (H), carbon (C), sodium (Na), nitrogen (N), and sulfur (S). About 26 of the 92 naturally occurring elements are commonly found in living things. Most elements have **isotopes,** or forms of the element, having the same number of protons in the nucleus but different weights because of differing numbers of neutrons. For example, oxygen isotopes have the same atomic number, 8, but different atomic weights: $^{16}_{8}O$, $^{17}_{8}O$, $^{18}_{8}O$.

Electrons are arranged in **electron shells,** which are regions corresponding to different **energy levels** (Table 2.1). The arrangement is called an **electronic configuration.** The chemical properties of atoms are largely a function of the number of electrons in the outermost electron shell. When the outer shell of an atom is filled, as in helium, it is stable, inert; it does not react with other atoms. Partially filled electron shells make for atoms that tend to react with other atoms to become more stable.

How Atoms Form Molecules: Chemical Bonds

The **valence,** or combining capacity of an atom, is dependent on the number of extra electrons or missing electrons in its outermost electron shell. Hydrogen has a valence of one and carbon has a valence of four. This means that hydrogen can form only one chemical bond and carbon four. When atoms gain stability by completing the full complement of electrons, they form **molecules.** A molecule containing at least two kinds of atoms, such as water (H_2O), is a **compound.**

Ionic and Covalent Bonds

The attractive forces holding molecules together are chemical bonds. In general, atoms form bonds by (1) gaining or losing electrons from the outer electron shell to form **ionic bonds,** or (2) sharing outer electrons to form **covalent bonds.** Sodium chloride (NaCl) is an example of a substance formed by ionic bonding. Sodium (Na) has a single electron in the outer shell, which it tends to lose; it is an *electron donor.* The atom is positively charged. This atom reacts strongly with chlorine, which is lacking one electron of outer-shell capacity. Chlorine has a negative charge because it usually fills this vacancy with electrons from other atoms; that is, it is an *electron acceptor.* **Cations,** such as potassium (K^+) or sodium (Na^+), are positively charged **ions** (a negatively or positively charged atom or group of atoms); their outer electron shell is less than half filled, and they lose electrons. **Anions** are negatively charged ions; with their outer shells *more* than half filled, they tend to gain electrons. Examples of anions are iodide ion (I^-), chloride ion (Cl^-), and sulfide ion (S^{2-}). Covalent bonds are stronger and more common in organisms than are ionic bonds.

Hydrogen Bonds

The **hydrogen bond** consists of a hydrogen atom covalently bonded to one oxygen or nitrogen atom (these being the most commonly involved elements). In water, for example, the electrons are all closer to the oxygen nucleus than to the hydrogen nucleus. As a result, the oxygen portion of the molecule has a slight negative charge and the hydrogen portion a slight positive charge. The hydrogen bond is a result of the water molecule's slightly positive polar charge being attracted to the negative end of the other molecules. Although these bonds are relatively weak, large molecules may have hundreds of them.

TABLE 2.1 **Electronic Configurations for the Atoms of Some Elements Found in Living Organisms**

ELEMENT	FIRST ELECTRON SHELL	SECOND ELECTRON SHELL	THIRD ELECTRON SHELL	DIAGRAM	NUMBER OF VALENCE (OUTERMOST) SHELL ELECTRONS	NUMBER OF UNFILLED SPACES	MAXIMUM NUMBER OF BONDS FORMED
Hydrogen	1	—	—		1	1	1 (by either losing or gaining 1 electron)
Carbon	2	4	—		4	4	4 (by either losing or gaining 4 electrons)
Nitrogen	2	5	—		5	3	3 (by gaining or losing 3 electrons)
Oxygen	2	6	—		6	2	2 (by gaining 2 electrons)
Magnesium	2	8	2		2	6	2 (by losing 2 electrons)
Phosphorus	2	8	5		5	3	5 (by losing 5 electrons)
Sulfur	2	8	6		6	2	2 (by gaining 2 electrons)

Key: ● electron
○ unfilled space
◉ atomic nucleus

Molecular Weight and Moles

The **molecular weight** of a molecule is the sum of the atomic weights of all its atoms. A **mole** is the number of grams equal to the molecular weight.

Chemical Reactions

The making and breaking of bonds between atoms are **chemical reactions.** The process of two or more atoms, ions, or molecules combining to form new and larger molecules is called a **synthesis reaction.** The combining substances are **reactants,** and the **product** is the substance formed. All synthesis reactions are collectively called **anabolism;** an example is the formation of starch from sugar molecules. The reverse of synthesis is a **decomposition reaction;** an example is the digestion of food. Collectively, these reactions are called **catabolism. Exchange reactions** are partly synthesis and partly decomposition. In theory, all reactions are **reversible reactions,** but in practice some reverse more easily than others.

Energy of Chemical Reactions

Synthesis *requires* energy (an **endergonic reaction**) and decomposition *yields* energy (an **exergonic reaction**). Energy used by organisms for synthesis is provided by food digestion or photosynthesis.

How Chemical Reactions Occur

The minimum collision energy required for a chemical reaction to occur, according to **collision theory,** is its **activation energy.** This is the amount of energy needed to disrupt the stable electron configuration of a molecule so that a new arrangement can occur. Increases in temperature, pressure, and concentration increase collision frequency and increase **reaction rates.**

Enzymes and Chemical Reactions.   **Enzymes** are large protein molecules that have a three-dimensional **active site** tailored to interact with a specific **substrate.** They act as **catalysts** to accelerate chemical reactions. The enzyme–substrate complex that is formed lowers the activation energy required and thus enables collisions to be more effective. Enzymes speed up reactions without an increase in temperature, which is important in biological systems.

IMPORTANT BIOLOGICAL MOLECULES

Inorganic Compounds

Water

Inorganic compounds, such as water, are essential for living cells. **Water, a polar molecule,** has an unequal charge distribution that gives it four important characteristics: (1) a high boiling point and, incidentally, a solid form (ice) that is less dense than the liquid, a characteristic that allows it to float; (2) excellent solvent properties (that is, it is a good dissolving medium. Polar substances undergo **dissociation** into individual molecules in water; they dissolve because the negative part of the water molecule is attracted to the positive part of the molecules in the **solute,** or dissolving substance); (3) an important role as a reactant or product in many chemical reactions; and (4) an excellent temperature buffer capability that helps protect the cell from changes in environmental temperature.

Acids, Bases, and Salts

When inorganic salts such as NaCl are dissolved in water, they dissociate, or ionize, into ions. An **acid** dissociates into one or more hydrogen ions (or protons H^+) and one or more negative ions. In other words, acids are a proton (H^+) donor; HCl, for example, yields H^+ and Cl^-. A **base,** such as NaOH, dissociates into Na^+ and the **hydroxide ion** (OH^-). A **salt** is a substance that dissociates into ions that are neither H^+ nor OH^-. NaCl is an example; it dissociates into Na^+ and Cl^-.

Acid–Base Balance

The **pH** of a solution is the negative logarithm to the base 10 of the hydrogen ion concentration in moles per liter. Acidic solutions contain more H^+ ions than OH^- ions and have a pH lower than 7. Alkaline, or basic, solutions have more OH^- ions than H^+ ions. A pH of 7, with equal concentrations of H^+ and OH^-, is neutral. Each change of number on the logarithm pH scale represents a tenfold change in concentration. **Buffers** are compounds that keep the pH from changing drastically.

Organic Compounds

All **organic compounds** contain carbon, whose four covalent bonds allow it to combine with neighboring or other atoms to form large structures. Hydrogen fills most of the free bonds in these structures. Many other **functional groups** such as the **hydroxy** (OH) (not to be confused with hydroxide) and **amino** (NH_2) groups are also found. Such fundamental groups can be added to **carbon skeletons** to make alcohols or amino acids. When single bonds hold a carbon atom to four other atoms to form a skeleton of three-dimensional shape, it is called a **tetrahedron.** Small organic molecules can be combined into very large molecules called **macromolecules.** Generally, these are **polymers,** which are formed by repeating small molecules called **monomers.** When two monomers join together, they usually release a molecule of water, a reaction called **condensation reaction** or **dehydration synthesis.**

Carbohydrates

Carbohydrates have the general formula CH_2O, with a hydrogen–oxygen ratio of 2:1. Carbohydrates form **monosaccharides,** simple sugars of three to seven carbon atoms; **disaccharides,** two monosaccharides bonded together; and **polysaccharides,** eight or more monosaccharides such as glucose joined by dehydration. Two molecules with the same chemical formula but different structures are called **isomers.**

Lipids

Lipids also are composed of atoms of carbon, hydrogen, and oxygen, but they do not have the 2:1 hydrogen–oxygen ratio. They are diverse in structure but share the property of being soluble in nonpolar solvents such as ether and alcohol, but not in water. Examples are **fats,** which are formed from glycerol and fatty acids joined by an **ester link. Saturated fats** have no double bonds; **unsaturated fats** have several. Important in the membranes of cells are **phospholipids,** which also contain phosphorus, nitrogen, and sulfur. Lipids called **sterols** (members of the steroids) are important constituents in the plasma membranes of cells.

Proteins

Amino acids are the subunits of **proteins,** which are organic molecules containing carbon, hydrogen, oxygen, and nitrogen, as well as some sulfur. Amino acids have at least one carboxyl (—COOH) group and one amino (—NH$_2$) group attached to the same carbon atom, the alpha carbon—hence the name **alpha-amino acid.** Amino acids exist in configurations called **stereoisomers,** which are designated D or L. (Only L-isomers are found in biological proteins, except for those found in some bacterial cell walls and in some antibiotics.) Only 20 different kinds of amino acids occur naturally in proteins. Amino acids in proteins are connected by **peptide bonds.**

Proteins have four levels of organization: **primary,** the order in which amino acids are linked (their sequence); **secondary,** localized, repetitious twisting or folding of the polypeptide chain that form helixes and pleated sheets, held together by hydrogen bonding; **tertiary,** three-dimensional shapes in which sulfhydryl groups (—SH) can form covalent disulfide links (—S—S—), hydrogen bonds providing the holding force; and a **quaternary** structure, an aggregation of two or more individual polypeptide units that operate as a single functional unit. A protein that loses its characteristic shape has undergone **denaturation** and is not functional. **Conjugated proteins** are formed by combinations of amino acids and other organic or inorganic components. **Glycoproteins,** for example, contain sugars; **nucleoproteins** contain nucleic acids; and **lipoproteins** contain lipids.

Nucleic Acids

The basic units of **nucleic acids** are **nucleotides.** There are two principal nucleic acids: **deoxyribonucleic acid (DNA)** and **ribonucleic acid (RNA).** Each nucleotide unit of DNA contains three parts: a nitrogen-containing base, a five-carbon sugar **(deoxyribose or ribose),** and a phosphoric acid molecule (Figure 2.1). The nitrogen-containing base is either **adenine, guanine, cytosine,** or **thymine.** Adenine and guanine are double-ring structures, **purines.** Thymine and cytosine are single-ring structures, **pyrimidines.** Note in the figure how adenine always pairs with thymine

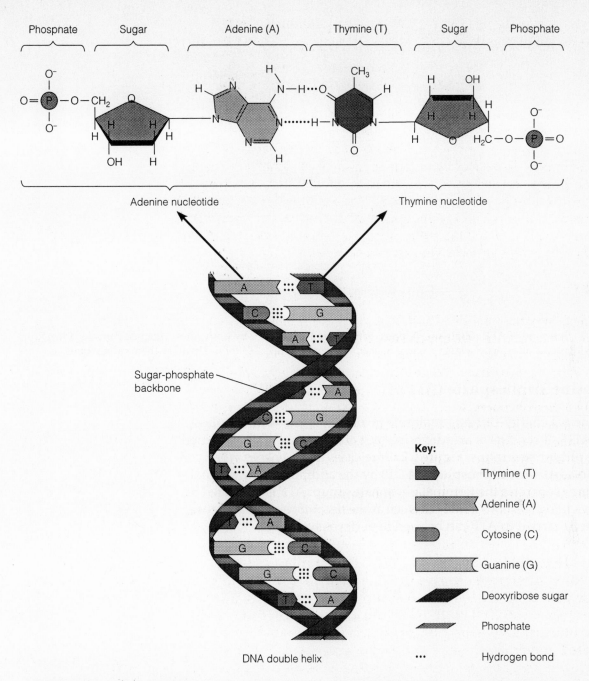

FIGURE 2.1 The structure of DNA. Nucleotides (top) are composed of a deoxyribose sugar molecule linked to a phosphate group and to a base. The two nucleotides shown here are linked by hydrogen bonds between their complementary bases. The ladderlike form of DNA's double helix (bottom) is made up of many nucleotides, with the repeating sugar-phosphate combination forming the backbone and the complementary bases the rungs.

and cytosine with guanine. This **complementary pairing** allows one strand to be reproduced from the structure of another. RNA differs from DNA in several ways. It is usually single-stranded, the five-carbon sugar is ribose, and **uracil** replaces thymine. **Nucleosides** have a purine or pyrimidine attached to a pentose sugar but have no phosphate group.

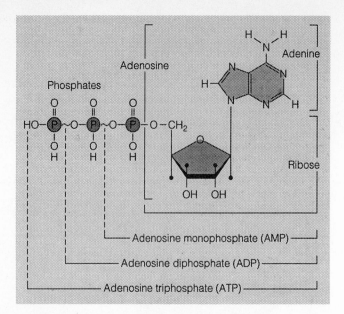

FIGURE 2.2 **Structure of ATP.** High-energy phosphate bonds are indicated by wavy lines. When ATP breaks down to ADP and inorganic phosphate, a large amount of chemical energy is released for use in other chemical reactions.

Adenosine Triphosphate (ATP)

The principal energy-carrying molecule in all cells is **adenosine triphosphate (ATP).** It consists of an adenosine unit of adenine and ribose joined to three phosphate groups (Figure 2.2). It releases much energy when converted to **adenosine diphosphate (ADP)** by the addition of a water molecule, which separates the terminal phosphate group. ATP is generated by energy-yielding reactions in the cell, such as the decomposition of glucose. The energy carried in ATP can be used to perform synthesis.

Self-Tests

In the matching section, there is only one answer to each question; however, the lettered options (a, b, c, etc.) may be used more than once or not at all.

I. Matching

1. The strongest of the three chemical bonds listed.

2. An uncharged particle in the atomic nucleus.

3. A hydrogen ion.

4. The number of protons in the nucleus.

5. Particles with a negative charge that move in shells around the nucleus.

6. A bond formed by sharing electrons in the outermost shell.

7. A weak bond formed, for example, by the slight positive charge at the hydrogen end of the water molecule reacting with the negative end of other molecules.

8. A bond formed by the gain or loss of electrons from the outer electron shell.

a. Proton

b. Atomic weight

c. Electron

d. Nucleus

e. Atomic number

f. Neutron

g. Ionic bond

h. Covalent bond

i. Hydrogen bond

II. Matching

1. The substance upon which an enzyme acts.

2. A protein that lowers the activation energy required for a reaction.

3. The sum of the atomic weights of a molecule's atoms.

4. The collective term for all decomposition reactions.

5. The number of grams equal to molecular weight.

6. The collective term for all synthesis reactions.

7. The combining capacity of an atom.

8. An ion with a positive charge.

9. One of two molecules with the same chemical formula but different structures.

a. Cation

b. Anion

c. Substrate

d. Valence

e. Mole

f. Molecular weight

g. Anabolism

h. Catabolism

i. Mole

j. Enzyme

k. Isomer

III. Matching

1. Prevents drastic change in pH.

2. Substance that dissociates into ions that are neither OH⁻ or H⁺.

3. A proton donor.

4. Dissociates into one or more negative hydroxide ions, such as OH⁻.

5. Combinations of atoms that have gained stability by completing the full complement of electrons in the outermost shell.

a. Compound

b. Molecule

c. Acid

d. Salt

e. Base

f. Hydroxide ion

g. Buffer

h. Hydroxy group

IV. Matching

1. Eight or more glucose molecules in a chain.

2. Sterol.

3. Fat.

4. Production of a molecule of water during synthesis.

5. Formed from chains of amino acids.

6. Lipoprotein.

7. Results from the release of energy by separation of the terminal phosphate group.

8. DNA.

a. Condensation

b. Monosaccharide

c. Disaccharide

d. Polysaccharide

e. Lipid

f. Protein

g. Conjugated proteins

h. Nucleic acid

i. Adenosine triphosphate

j. Adenosine diphosphate

V. Fill-in-the-Blanks

1. All atoms with the same atomic number are classified as the same _____ .

2. When discussing synthesis, the combining substances are called _____ and the substance formed is the _____ .

3. Carbon has a valence of _____ .

4. Most elements have _____ , which have the same number of protons in the nucleus but different weights due to differing numbers of neutrons.

5. Amino acids in proteins are connected by _____ .

6. The five-carbon sugar in DNA is _____ .

7. In DNA, adenine always pairs with _____ .

8. In RNA, thymine is replaced by _____ .

9. The principal energy-carrying molecule in all cells is _____ .

10. RNA differs from DNA in being usually _____ stranded.

11. In a protein, the order of the amino acid sequence is the _____ level of organization.

12. The basic units of nucleic acids are _____ .

13. Thymine and cytosine are single-ring structures called _____ .

14. An example of a nitrogen-containing base in a nucleotide is _____ . (More than one answer is acceptable.)

15. The _____ level of protein organization provides it with a three-dimensional shape.

16. About _____ different kinds of amino acids occur naturally in proteins.

17. Amino acids can exist as _____ , designated D or L.

18. Some important characteristics of water are its high _____ and its capacity as a temperature _____ .

19. A molecule with at least two kinds of atoms, such as water, is a _____ .

20. Cations are positively charged ions; their outer electron shell is _____ than half filled and they lose electrons.

21. The minimum collision energy required for a chemical reaction to occur is its _____ .

22. _____ have the general formula CH_2O, with a hydrogen–oxygen ratio of 2:1.

23. Sucrose, formed of the sugars glucose and fructose, is an example of a _____ saccharide. (Provide the prefix.)

24. _____ share the property of being soluble in nonpolar solvents such as ether and alcohol, but not in water.

25. Neutrons and protons have a weight about 1840 times that of _____ .

26. Decomposition yields energy, which is called an _____ reaction.

VI. Label the Art

It is most important to know what the subatomic particles are, where they are located in an atom, and that atoms of different elements differ because they contain different numbers of protons. Sketch each atom, labeling and coloring **protons** blue, **neutrons** green, and **electrons** red. Also use + and – signs to remind yourself of the electrical charges carried by the particles.

	Element	Symbol	Atomic Number	Mass Number	Number of Protons	Number of Neutrons	Number of Electrons
1.	Carbon-12	C	6	12	6	6	6
2.	Nitrogen-14	——	7	14	——	——	——
3.	Chlorine-35	——	——	35	17	——	——
4.	Oxygen-16	——	——	——	——	——	8
5.	Oxygen-17	——	——	——	——	——	——

VII. Critical Thinking

1. Explain how catalysts lower the activation energy of a chemical reaction? Why are catalysts important in living systems?

2. Why is carbon so important in the chemistry of life?

3. Why are lipids important components of living cells?

4. Explain the role of pH buffers in the growth of bacterial cultures in laboratory media.

5. What are the advantages and disadvantages of using bacteria as environmental pollution fighters? How are scientists trying to improve the efficiency of these microbes?

6. What are the products of the following reaction? Identify the peptide bond in the main product.

7. Discuss the four levels of protein structure. How is structure related to protein function?

8. Determine the number of moles in 240 grams of methane (CH_4).

9. Why is ATP indispensable to cells?

10. How do DNA and RNA differ?

CHAPTER 3

Observing Microorganisms Through a Microscope

Learning Objectives

After completing this chapter, you should be able to:

- List the metric units of measurement used for microorganisms, and know their U.S. equivalents.
- Diagram the path of light through a compound microscope.
- Define total magnification and resolution.
- Identify a use for darkfield, phase-contrast, differential interference contrast, and fluorescence microscopy, and compare each with brightfield illumination.
- Explain how electron microscopy differs from light microscopy.
- Differentiate between an acidic dye and a basic dye.
- Compare simple, differential, and special stains.
- List the steps in preparing a Gram stain, and describe the appearance of gram-positive and gram-negative cells after each step.
- Explain why each of the following is used: capsule stain, endospore stain, flagella stain.

Units of Measurement

Microorganisms are measured by metric units unfamiliar to many of us. The **micrometer (μm),** formerly known as the **micron,** is equal to 0.000001 (10^{-6}) meter. The prefix *micro* indicates that the unit following should be divided by one million. A **nanometer (nm),** formerly known as a **milli-micron (mμ)** is equal to 0.000000001 (10^{-9}) meter. *Nano* tells us that the unit should be divided by one billion. An **angstrom (Å)** is equal to 0.0000000001 (10^{-10}) meter.

Microscopy: The Instruments

Compound Light Microscopy

The **compound light microscope** has two sets of lenses: the **objective** and the **ocular.** Specimens magnified by the objective lens—magnified 100 times, for example—are remagnified by the ocular, usually 10 times. Thus, the total magnification is 1000 times. Most microscopes provide magnifications of 100, 400, and 1000. A magnification of 2000 times is

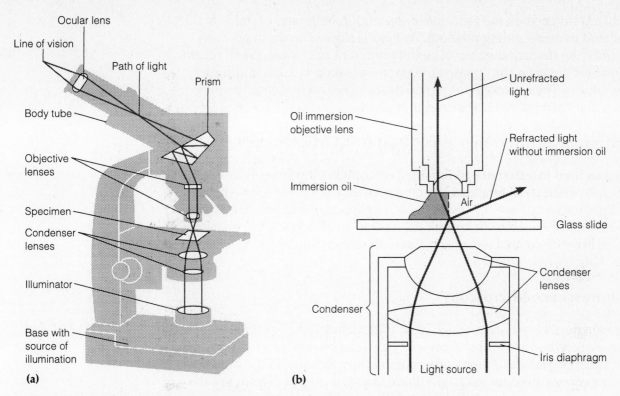

FIGURE 3.1 The compound light microscope. (a) The path of light (bottom to top). **(b)** Refraction. Because the refractive indexes of the glass microscope slide and immersion oil are the same, the light rays do not refract when passing from one to the other when the oil immersion objective lens is used. This method produces images with better resolution at magnifications greater than 100x.

about the highest obtainable. The specimen is illuminated by visible light that is passed through a **condenser,** which directs the light rays through the specimen (Figure 3.1a). **Resolution,** or **resolving power,** is the ability of a microscope to distinguish between two points. The shorter the wavelength of the illumination, the better the resolution. The white light used in a compound light microscope limits resolving power to about 0.2 μm.

For the highest magnification, it is necessary to use **oil immersion objectives.** Immersion oil has the same **refractive index** as glass; that is, the relative velocity of light passing through it is the same. Without immersion oil filling the space between the slide bearing the specimen and the objective, the image will be fuzzy with poor resolution (Figure 3.1b).

Darkfield Microscopy

Some microorganisms, such as the thin spirochete *Treponema pallidum,* which causes syphilis, are best seen with **darkfield microscopy.** In the darkfield microscope, an opaque disk blocks light from entering the objective directly. The light hits only the sides of the specimen, and scattered light enters the objective and reaches the eyes. The specimen appears white against a black background.

Phase-Contrast Microscopy

Living microorganisms do not show up well in the ordinary compound light microscope. The **phase-contrast microscope** takes advantage of subtle differences in the refractive index of different parts of the living cell and its surrounding medium. As light is slowed down in portions of differing density, they travel slightly different pathways. When recombined for viewing, the "phase differences" are seen as areas of differing brightness. The microorganism (and many of its internal structures) is seen in its natural state, alive and unstained.

Differential Interference Contrast (DIC) Microscopy

Differential interference contrast microscopy is similar to phase-contrast microscopy. It uses differences in refractive indexes but uses two beams of light instead of one. Prisms split each light beam, adding contrasting colors. Compared to standard phase-contrast microscopes, DIC images are more brightly colored, are nearly three-dimensional in appearance, and have higher resolution.

Fluorescence Microscopy

Certain *fluorochrome dyes*, which glow with visible light—yellow, for example—when illuminated by ultraviolet light, can be used to view and identify microorganisms. **Fluorescence microscopy** techniques use a special microscope with ultraviolet light illumination; this light illuminates the specimen but is not permitted to reach the eye. The stained microorganism is highly visible against a dark background in such a microscope. However, the principal use of these dyes and microscopes is in the **fluorescent-antibody** technique, or **immunofluorescence.** In this technique, the organisms are allowed to react on a slide with antibodies (highly specific proteins produced by the body's defense system). A fluorescent dye is attached to the **antibody.** The combination of the antibody, the attached dye, and the microorganism for which the antibody is specific (called an **antigen;** it stimulates the body to produce these antibodies) allows the microorganism's presence to be detected (Figure 3.2). Because the antibody is specific for a particular microorganism, this is a very useful diagnostic technique. It is often used for diagnosis of syphilis and rabies.

Electron Microscopy

The wavelengths of electrons, which travel in waves much as light does, are only about 1/100,000 as long as those of visible light, and therefore have much better resolving power. They can be focused by magnetic lenses.

Transmission Electron Microscopy. In **transmission electron microscopy,** a beam of electrons is passed through ultrathin sections of the specimen and focused on a fluorescent screen, where it is visible to the eye and can be photographed. Objects are generally magnified 10,000 times to 100,000 times, and structures, called *artifacts,* may appear as a result of the method of preparation.

Scanning Electron Microscope. In **scanning electron microscopy,** the electron beam is directed at the intact specimen from the top, rather

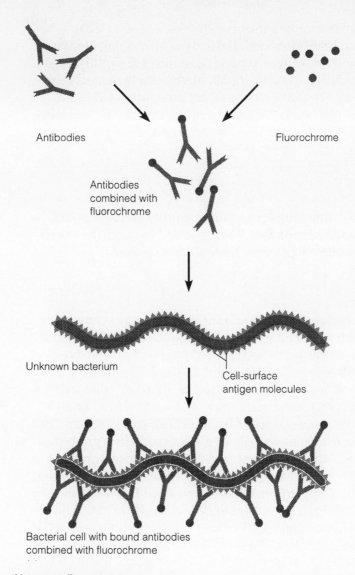

FIGURE 3.2 The principle of immunofluorescence.
A type of fluorochrome is combined with antibodies against a specific type of bacterium. When the preparation is added to bacterial cells on a microscope slide, the antibodies attach to the bacterial cells and the cells fluoresce when illuminated with ultraviolet light.

than passing through a section, and electrons leaving the surface of the specimen (secondary electrons) are viewed on a televisionlike screen. Spectacular pictures of seemingly three-dimensional, intact organisms are possible. Objects are generally magnified 1000 times to 10,000 times with a resolving power of about 20 nm.

Preparation of Specimens for Light Microscopy

Preparing Smears and Staining

Most microorganisms are viewed in stained preparations; that is, they are colored with a dye to make them visible or to emphasize certain structures. A thin film of a microbial suspension, called a **smear,** is spread on the surface of a slide. Flaming the air-dried smear coagulates the microbial proteins and **fixes** the microorganisms to the slide so they do not wash off. The smear can then be stained. **Basic dyes** have a colored ion that is positive, helping them adhere to bacteria, which are slightly negative. Examples of basic dyes are crystal violet, methylene blue, and safranin. **Acidic dyes,** having a negative color ion, are more attracted to the background than to the negatively charged bacteria; thus, a field of colorless bacteria is presented against a stained background. This is an example of **negative staining.**

Simple Stains

To visualize shapes and arrangements of cells, a **simple stain** is usually sufficient. A chemical called a **mordant** may be added to make the microorganisms stain more intensely or otherwise enhance visibility.

Differential Stains

The most useful **differential stain** is the **Gram stain,** developed by Hans Christian Gram. It divides bacteria into two large groups: **gram-positive** and **gram-negative.** In preparing a Gram stain (1) apply a purple dye, crystal violet, to a heat-fixed smear. This stains all the cells and is called the **primary stain.** After a water rinse, (2) an iodine mordant is added. When a smear stained in this manner is (3) washed with ethanol or an ethanol-acetone solution, some species of bacteria are decolorized and others are not. If the smear retains the purple dye, the organism is gram-positive. If the alcohol removes the dye, the colorless microorganisms are no longer visible. (4) Safranin, a red dye, is then applied and the decolorized, or gram-negative, bacteria appear pink. Safranin is used here as a **counterstain.** The Gram stain reflects a basic difference in the cell wall structure of bacteria. It is a first step in identification, and the susceptibility of microorganisms to antibiotics is often related to the Gram reaction.

Acid-Fast Stain.   Members of the genera *Mycobacterium* (which includes the causes of tuberculosis and leprosy) and *Nocardia* possess a cell wall with waxy components. The red dye carbolfuchsin is more soluble in these waxes than in acid-alcohol and is retained by the cell. Therefore, the **acid-fast stain,** in which carbolfuchsin is applied and gently steamed for several minutes, will stain them red. This dye is held so firmly that the cells are not decolorized by acid-alcohol, which does remove the dye from bacteria that are not acid-fast. A methylene blue counterstain will produce a slide in which acid-fast organisms are red and others are blue. The acid-fast stain is an invaluable aid in the diagnosis of tuberculosis.

Special Stains

A colloidal suspension of dark particles such as India ink can be used as a **capsule stain.** The capsule will appear around each bacterial cell as a halo from which the India ink carbon particles are excluded. Endospores do not stain by ordinary methods, but the **Schaeffer–Fulton endospore stain,** which uses malachite green as a primary stain and safranin as a counterstain, shows endospores as green within red or pink cells. Flagella are too small to be resolved by light microscopes. In a **flagella stain,** a mordant can be used to increase the diameter of the flagella until they are visible in a light microscope.

Self-Tests

In the matching section, there is only one answer to each question; however, the lettered options (a, b, c, etc.) may be used more than once or not at all.

I. Matching

1. The electrons pass through a thin section of the specimen.

2. Visible light passes through the specimen; uses separate objective and ocular lenses.

3. Details become visible because of differences in the refractive index of different parts of the cell.

4. Visible light is scattered after striking the specimen, and the specimen is visible against a darkened background.

5. A special microscope using ultraviolet illumination.

6. The electrons strike the surface of the specimen, and secondary electrons leaving the surface are viewed on a televisionlike screen.

a. Compound light microscope

b. Scanning electron microscope

c. Phase-contrast microscope

d. Transmission electron microscope

e. Fluorescence microscope

f. Darkfield microscope

II. Matching

1. Pertaining to the relative velocities of light through a substance.

2. Involves the use of antibodies and ultraviolet light.

3. One millionth of a meter.

4. One ten-billionth of a meter.

5. The ability to separate two points in a microscope field.

a. Micrometer

b. Nanometer

c. Angstrom

d. Resolving power

e. Refractive index

f. Immersion oil

g. Immunofluorescence

III. Matching

1. Adhere(s) best to bacteria, which have a negative charge, because the color molecule has a positive charge.

2. Used in diagnosis of tuberculosis.

3. Involve(s) the use of a negative stain made from India ink particles.

4. Schaeffer–Fulton stain.

5. Use(s) carbolfuchsin dye.

6. Use(s) malachite green.

7. Reflect(s) a basic difference between microbial cell walls; ethanol will not remove stain from some bacteria.

a. Basic dyes

b. Acidic dyes

c. Gram stain

d. Acid-fast stain

e. Capsule stain

f. Endospore stain

IV. Fill-in-the-Blanks

1. The prefix *nano* indicates that the unit should be divided by a _____ .

2. About the highest magnification possible in a compound light microscope is _____ .

3. Immersion oil has about the same refractive index as _____ .

4. Fluorochrome dyes glow with visible light when illuminated by _____ .

5. Electron wavelengths are only about _____ as long as visible light and therefore have much _____ resolving power.

6. Bacteria tend to have a slightly _____ electrical charge.

7. The electron microscope that tends to give a seemingly three-dimensional view of the specimen is the _____ electron microscope.

8. The thin film of a microbial suspension spread on the surface of a slide is called a _____ .

9. Flaming the slide before applying the stain is called _____ .

10. Transmission electron microscopy permits magnifications as high as about 10,000 times to _____ .

11. In the flagella stain, a _____ is used to increase the diameter of the flagella.

12. Two bacterial genera that are acid-fast are _____ and _____ .

13. A disease for which the acid-fast stain is useful in diagnosis is _____ .

14. In order to see shapes and arrangements of cells, a _____ stain is usually sufficient.

15. A chemical that makes microorganisms stain more intensely is called a _____ .

16. _____ dyes have a negative color ion.

17. Some microorganisms, such as the thin spirochete *Treponema pallidum,* are best seen with _____ microscopy.

18. Differential _____ microscopy is similar to phase contrast microscopy but uses two beams of light instead of one.

V. Critical Thinking

1. What is the limit of resolution of the compound light microscope? What factor is responsible for this limit of resolution?

2. How is the use of immersion oil with the 100X objective lens of the compound light microscope related to achieving high magnification with good resolution?

3. What is refractive index, and how is it related to the ability to view specimens with a microscope? How can the refractive index of a specimen be changed?

4. What type of microscopy would be most appropriate for viewing the following specimens or for the following situations?

    a. To detect bacteria in clinical specimens.

    b. To view objects smaller than 0.2 µm, such as viruses.

    c. To view heat-fixed, stained bacterial cells.

    d. To view microorganisms that can't be stained by standard methods.

    e. To view the internal structure of living microorganisms

5. What are basic dyes? List three basic dyes used to stain microorganisms. Why are basic dyes more often used to stain bacteria than acidic dyes?

4. Cell walls almost always contain peptidoglycan.
5. Usually divide by binary fission.

Distinguishing characteristics of **eucaryotic** (*true nucleus*) cells are:

1. Nucleus bounded by a membrane.
2. DNA of chromosomes consistently associated with proteins called histones and nonhistones.
3. Possess mitotic apparatus, mitochondria, endoplasmic reticulum, and sometimes chloroplasts.

THE PROCARYOTIC CELL

Size, Shape, and Arrangement of Bacterial Cells

Bacteria range in size from 0.2 to 2.0 μm in diameter and 2 to 8 μm in length. Basic bacterial shapes are the spherical **coccus** (meaning "berry"), the rod-shaped **bacillus** (meaning "little staff"), and the **spiral. Diplococci** form pairs; **streptococci** form chains; **tetrads** divide in two planes, forming groups of four; **sarcinae** divide in three regular planes and form cubelike packets; **staphylococci** divide in irregular, random planes and form grapelike clusters. Most bacilli are single rods, but they can appear in pairs—**diplobacilli**—or in chains—**streptobacilli. Coccobacilli** are ovals. **Vibrios**—slightly curved, commalike rods—are also included among spiral bacteria. **Spirilla** have a helical corkscrew shape and are motile by means of flagella. **Spirochetes** are shaped like spirilla but have an axial filament for motility. **Pleomorphic** bacteria have an irregular morphology; if they maintain a single shape, they are **monomorphic.**

Structures External to the Cell Wall

Glycocalyx

The general term for substances surrounding bacterial cells is **glycocalyx,** which is usually a polysaccharide, polypeptide, or both. If organized and tightly attached, it is called a **capsule** (Figure 4.1). If unorganized and loosely attached, the glycocalyx is called a **slime layer.** The glycocalyx aids in attachment to surfaces; capsules contribute to pathogenicity by protecting from phagocytosis, an important part of the body's defenses.

Flagella

Flagellar filaments are composed of a protein, **flagellin.** The base of the flagellar filament widens to a **hook.** Attached to the hook is a **basal body** (a rod with rings), which anchors the flagellum to the cell wall and plasma membrane (Figure 4.2). The basal body of gram-negative bacteria is anchored to the cell wall and plasma membrane; in gram-positive bacteria, it is anchored only at the plasma membrane. Flagella, when present, are arranged in one of four basic ways: **monotrichous** (single polar flagellum); **lophotrichous** (two or more polar flagella at one or both ends of

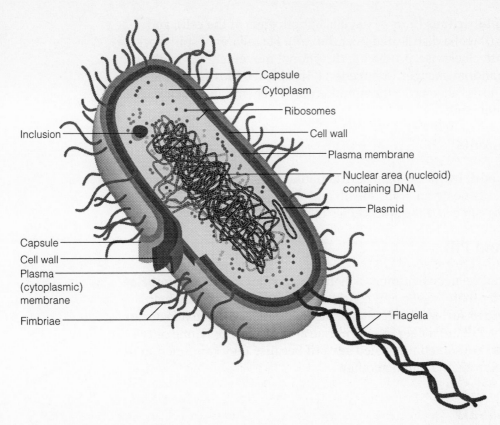

Capsule
Cytoplasm
Ribosomes
Inclusion
Cell wall
Plasma membrane
Nuclear area (nucleoid) containing DNA
Plasmid
Capsule
Cell wall
Plasma (cytoplasmic) membrane
Fimbriae
Flagella

FIGURE 4.1 Structure of a typical procaryotic (bacterial) cell. In this diagram, the cell has been cut lengthwise to reveal the internal structures.

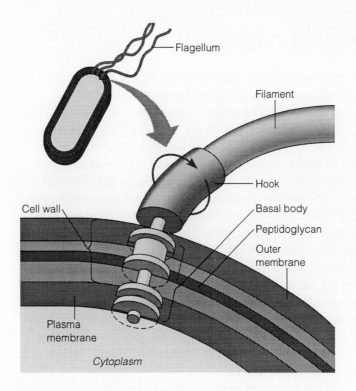

Flagellum
Filament
Hook
Cell wall
Basal body
Peptidoglycan
Outer membrane
Plasma membrane
Cytoplasm

FIGURE 4.2 One type of bacterial motility. This diagram shows a bacterium "running" and "tumbling." Note that the direction of flagellar rotation determines which of these movements occurs.

the cell); **amphitrichous** (tufts of flagella at both ends of the cell); and **peritrichous** (flagella distributed over the cell). Flagella may spin clockwise or counterclockwise, producing directional movement (a "run" or "swim") or random changes in direction ("tumbles"). Movement to or from a stimulus is called **taxis;** the stimulus may be chemicals **(chemotaxis)** or light **(phototaxis).**

Axial Filaments

Spirochetes move by means of **axial filaments,** bundles of fibrils that arise near cell poles beneath an outer sheath and wrap in spiral fashion around the cell. These can cause the spirochetes to move in a corkscrew manner.

Fimbriae and Pili

Many bacterial cells have numerous hairlike appendages called **fimbriae** that are shorter than flagella and consist of a protein, **pilin.** They help the cell adhere to surfaces such as mucous membranes—often a factor in pathogenicity. **Pili** are longer than fimbriae and number only one or two per cell. These are sometimes called **sex pili** because they can function to transfer DNA from one cell to another.

The Cell Wall

The bacterial cell wall is a semirigid structure giving the characteristic shape of the cell.

Composition and Characteristics

The cell wall of gram-positive bacteria is composed of **peptidoglycan** (murein), which consists of two sugars, N-acetylglucosamine and N-acetylmuramic acid, and also chains of amino acids. The two sugars alternate with each other, forming a carbohydrate (glycan) backbone. Peptide side chains of four amino acids attached to the N-acetylmuramic acid are crosslinked to form the macromolecule of the cell wall. Many gram-positive bacteria also contain polysaccharides called **teichoic acids.** The cell wall of acid-fast bacteria (otherwise considered gram-positive) consists of peptidoglycan and a waxy lipid, mycolic acid. The cell wall of gram-negative bacteria also contains peptidoglycans, but only thin layers. These cells have a lipoprotein, lipopolysaccharide **(LPS),** and a phospholipid **outer membrane** surrounding their peptidoglycan layers. A **periplasmic space** is found between the outer membrane and the **plasma membrane.** The outer membrane also provides resistance to phagocytosis and the action or complement (also part of host defenses). When the cell disintegrates in the host's bloodstream, the lipid portion of the LPS **(Lipid A)** is released as an **endotoxin** that can cause illness. Materials may penetrate the outer membrane through channels called **porins.**

Mycoplasma bacteria do not have cell walls. They are unique also in having sterols in their plasma membranes. Archaeobacteria do not have peptidoglycan in their walls, but have a similar substance, **pseudomurein.**

Damage to the Cell Wall

Lysozyme—an enzyme occurring in tears, mucus, and saliva—damages the cell walls of many gram-positive bacteria. A bacterium that has lost its cell wall and is surrounded only by the plasma membrane is a **protoplast.** Gram-negative cells treated with lysozyme retain much of the outer membrane layer and are called **spheroplasts.** Both are sensitive to rupture by **osmotic lysis.**

Structures Internal to the Cell Wall

Plasma (Cytoplasmic) Membrane

The **plasma (cytoplasmic) membrane** is just internal to the cell wall and encloses the cytoplasm. In procaryotes it consists primarily of phospholipids and proteins. Eucaryotic plasma membranes also contain sterols, making them more rigid. Both procaryotic and eucaryotic membranes have a two-layered structure, molecules in parallel rows, called a **phospholipid bilayer.** One end (phosphate) is water soluble and the other (hydrocarbon) is insoluble. The water-soluble ends are on the outside of the bilayer. Protein molecules are embedded in the membrane; along with phospholipids, they may move freely within the membrane. This arrangement is called the **fluid mosaic model.**

The most important function of the plasma membrane is as a selective barrier. It is **selectively permeable (semipermeable),** and certain molecules and ions pass through, whereas others do not. Several factors affect permeability. Large molecules such as proteins cannot pass; smaller molecules such as amino acids and simple sugars can pass if uncharged. (The phosphate end of the bilayer is charged.) Lipid-soluble substances, because of the phospholipid content, pass more easily. Plasma membranes contain enzymes that help break down nutrients and produce more energy. The **chromatophores** or **thylakoids,** which contain pigments for bacterial photosynthesis, are found in the plasma membranes.

Mesosomes are folds in the plasma membrane that may be only an artifact of preparation for electron microscopy.

Movement of Materials Across Membranes

Material crosses plasma membranes by passive processes such as **simple diffusion** (movement of molecules or ions from an area of higher concentration to an area of lower concentration). At equilibrium, the concentration gradient has been eliminated. **Osmosis** is the force with which a solvent (such as water) moves from a solution of lower solute concentration (such as dissolved sugar) to a solution of higher solute concentration. **Isotonic** solutions have equal solute concentrations on both sides of the membrane. **Hypotonic** solutions have a lower concentration of solutes outside the cell than inside; this is the case with most bacteria. **Hypertonic** solutions have a higher concentration of solutes outside the cell, and bacterial cells placed in such solutions lose water by osmosis and shrink, and the cytoplasm collapses within the cell wall. A third passive process is **facilitated diffusion,** in which a **carrier protein (permease)** combines with and transports a substance across the membrane, but only where a concentration gradient is present. **Active transport** requires cell energy (ATP) and also involves carrier proteins moving substances across the plasma membrane. In **group translocation,** the substance is chemically altered during transport. Once inside, the plasma membrane is impermeable. This is important for low-concentration substances.

Cytoplasm, Nuclear Area, Inclusions, and Ribosomes

The term **cytoplasm** refers to everything inside the plasma membrane. It has many **inclusions,** such as **metachromatic granules** of stored phosphate **(volutin), polysaccharide granules** of glycogen and starch, **lipid inclusions** such as *poly-β-hydroxybutyric acid,* and sulfur granules. The cytoplasm also contains many **ribosomes,** the sites of protein synthesis. **Carboxysomes** are inclusions found in bacteria that use carbon dioxide as their sole source of carbon. **Gas vacuoles** or gas vesicles help some bacteria maintain buoyancy. The **bacterial chromosome,** which contains the genetic information, is a single, long, circular molecule of DNA found in the **nuclear area,** or **nucleoid.** Small circular DNA molecules, **plasmids,** are not connected to the chromosome and replicate independently. Plasmids do not contain normally essential genes but may provide a selective advantage under abnormal conditions—antibiotic resistance, for example.

Endospores

Endospores are highly resistant bodies formed by a few bacterial species, such as *Bacillus* and *Clostridium.* **Sporulation** or **sporogenesis** is the process of their formation. First, there is an ingrowth of the plasma membrane **(spore septum).** A small portion of the cytoplasm and newly replicated bacterial chromosome is then surrounded by a membrane, the **forespore.** A thick **spore coat** of protein forms around this membrane. The endospore core is dehydrated and contains considerable *dipicolinic acid,* as well as a few essential materials necessary to return it to its vegetative state, which is accomplished through the process of **germination.**

THE EUCARYOTIC CELL

Flagella and Cilia

Eucaryotic **flagella** are relatively long; **cilia** are more numerous and are shorter. Both are involved in locomotion, and both contain small tubules of protein called **microtubules.**

The Cell Wall and Glycocalyx

Most algae and some fungi have **cell walls** containing *cellulose,* and often fungi have *chitin* as well. Yeast cell walls contain the polysaccharides *glucan* and *mannan.* No eucaryotic cell wall contains peptidoglycans.

Protozoa have a flexible outer covering called a *pellicle*. In animal cells the plasma membrane is covered by sticky carbohydrates called the **glycocalyx.**

Plasma (Cytoplasmic) Membrane

In eucaryotic cells, the **plasma membrane,** which contains sterols, may be the external cell covering. Substances cross the membrane by mechanisms similar to those in procaryotes. In addition, a process of engulfment, **endocytosis,** brings particles, even some viruses, into the cell. Examples are **phagocytosis,** used by white blood cells to ingest (destroy) bacteria (Chapter 16), and **pinocytosis** (cell drinking), by which liquids enter cells.

Cytoplasm

Cytoplasm is the matrix in which various cellular components are found. The complex internal structure of **microfilaments, intermediate filaments,** and **microtubules** is called the **cytoskeleton.**

Organelles

In eucaryotes, unlike procaryotes, many important enzymes are found in, and functions carried out by, **organelles.**

Nucleus

The **nucleus** is an oval organelle containing the DNA. It is surrounded by a **nuclear envelope. Pores** in the nuclear membrane allow the nucleus to communicate with the endoplasmic reticulum of the cytoplasm. The **nucleoplasm** is a gellike fluid in the nucleus. **Nucleoli,** which may be the center for the synthesis of ribosomal RNA, are present. DNA is combined with protein **histones,** a combination called a **nucleosome.**

Endoplasmic Reticulum and Ribosomes

The **endoplasmic reticulum (ER)** is a network of canals running throughout the cytoplasm. It may provide a surface area for chemical reactions, a pathway for transporting molecules, and a storage area for synthesized molecules; it has a role in lipid and protein synthesis. **Ribosomes** are sites of protein synthesis in the cell.

Golgi Complex

The **Golgi complex** consists of four to eight flattened sacs **(cisternae)** connected to the ER. Its function is largely the secretion (removal from the cell) of proteins, lipids, and carbohydrates. The Golgi complex also functions in carbohydrate synthesis, including glycoproteins.

Self-Tests

In the matching section, there is only one answer to each question; however, the lettered options (a, b, c, etc.) may be used more than once or not at all.

I. Matching

1. Helical; move by flagella, if present.

2. Spherical; in chains.

3. Divide in three regular planes; spheres form cubelike packets.

4. Helical; axial filaments for motility.

5. A simple, commalike curve.

6. Name means "little staff."

7. Ovals.

a. Sarcinae

b. Tetrads

c. Streptococci

d. Spirochetes

e. Vibrios

f. Bacilli

g. Cocci

h. Spirilla

i. Diplococci

j. Coccobacilli

II. Matching

1. Golgi complex.

2. Meiosis occurs in reproduction.

3. Single circular chromosome without histones.

4. Sterols generally present in cell membrane.

5. Cell wall almost always contains peptidoglycans.

6. Nucleus bounded by a membrane.

a. Eucaryotic cell

b. Procaryotic cell

III. Matching

1. Contain pigments for photosynthesis by bacteria; found in the plasma membrane.

2. Gram-negative bacterial cell walls after their treatment with lysozyme.

3. Specialized pili that assist in the transfer of genetic material between cells.

4. Numerous hairlike appendages that help attachment to mucous membranes.

5. General term for substances surrounding bacterial cells.

6. Polysaccharides found in the cell wall of many gram-positive bacteria.

a. Glycocalyx

b. Flagellin

c. Fimbriae

d. Sex pili

e. Capsules

f. Teichoic acids

g. Spheroplasts

h. Protoplasts

i. Chromatophores

j. Chloroplasts

IV. Matching

1. Metachromatic granules of stored phosphate in procaryotes.

2. Entrance of fluids into eucaryotic cells.

3. Membrane-enclosed spheres in phagocytic cells that contain powerful digestive enzymes.

4. The "powerhouses" of the cell.

5. A gellike fluid found in the eucaryotic nucleus.

6. A folded inner membrane found in the mitochondria.

a. Volutin

b. Plasmids

c. Cristae

d. Zymogens

e. Ribosomes

f. Nucleoplasm

g. Lysosomes

h. Mitochondria

i. Phagocytosis

j. Pinocytosis

V. Matching

1. Arrangement of flagella distributed over the entire cell.

2. Arrangement of tufts of flagella at both ends of the cell.

3. A widening at the base of the flagellar filament.

4. An enzyme affecting gram-positive cell walls; found in tears.

5. A compound found in bacterial endospores.

6. A compound frequently found in the cell walls of yeasts.

a. Exocytosis

b. Dipicolinic acid

c. Chitin

d. Lysozyme

e. Hook

f. Peritrichous

g. Amphitrichous

h. Lophotrichous

i. Monotrichous

j. Flagellin

VI. Matching

1. Closely involved in protein synthesis.

2. Structure(s) characteristic of both eucaryotic and procaryotic plasma membranes.

3. Found in the flagella and cilia of eucaryotic cells.

a. Phospholipid bilayer

b. Transverse septum

c. Microtubules

d. Ribosomes

VII. Fill-in-the-Blanks

1. Chemically, the capsule is a _____ , a polypeptide, or both.

2. Capsules protect pathogenic bacteria from _____ , a process by which protective host cells engulf and destroy microorganisms.

3. _____ are small circular DNA molecules that are not connected with the main chromosome.

4. In the chloroplasts of eucaryotic cells, the chlorophyll is found in stacks of thylakoid membranes called _____ .

5. _____ in the nuclear membrane allow the nucleus to communicate with the endoplasmic reticulum of the cytoplasm.

6. _____ are highly resistant bodies formed by a few bacterial species.

7. The Golgi complex consists of eight flattened sacs called _____ that are connected to the endoplasmic reticulum.

8. Archaeobacteria do not have peptidoglycan in their cell wall but have a similar substance called _____ .

9. In the sequence of spore formation, first there is an ingrowth of the plasma membrane called the spore _____ . A small portion of the cytoplasm is then surrounded by a membrane forming the _____ .

10. The semifluid center portion of the mitochondrion is called the _____ .

11. Bundles of microtubules, which probably play a role in cell division and are located within centrosomes, are called _____ .

12. The _____ complex consists of four to eight flattened sacs connected to the endoplasmic reticulum. The function is largely secretion of proteins, lipids, and carbohydrates.

13. The term _____ means a lower concentration of solutes outside the cell than inside.

14. Three examples of passive diffusion across membranes are _____ , _____ , and _____ .

15. Gram- _____ cells have a relatively smaller amount of peptidoglycan surrounded by an outer membrane of polysaccharide-phospholipid-lipoprotein than other bacterial cells.

16. The flagella of bacteria are anchored to the cell wall and plasma membrane by a _____ body.

17. The protein in the flagellar filaments of bacteria is called *flagella* .

18. The protein forming fimbriae and pili is called *pili* .

19. Bacteria with irregular morphology are termed *Pleomorphic* .

20. Another name for an extracellular polymeric substance that helps cells adhere to surfaces is _____ .

VIII. Critical Thinking

1. What is a glycocalyx? How is the presence of a glycocalyx related to bacterial virulence?

2. Draw and label the parts of the procaryotic flagellum. Discuss how attachment of the flagellum differs in gram-positive and gram-negative cells. Draw and label the four arrangements of flagella seen in bacterial cells.

3. What are teichoic acids? Explain the function(s) of teichoic acids.

4. What substances are able to most easily cross the plasma membrane?

5. Describe how a bacterial cell will respond to the following osmotic pressures: isotonic, hypotonic, hypertonic.

6. (A) Describe the process of endospore formation. (B) What bacteria can form endospores? (C) Name three genera of endospore-forming bacteria. (D) What evolutionary advantage(s) might endospore-forming bacteria have over nonendospore-forming bacteria? (E) Define sporulation, forespore, and germination.

7. How is the presence of peptidoglycan in bacterial cells clinically significant?

8. Discuss the endosymbiont hypothesis. Is there any evidence to support the endosymbiont hypothesis?

9. Discuss the relationship between the endoplasmic reticulum and the Golgi complex.

10. Discuss the differences in the ways procaryotic and eucaryotic cells are supported and how their shapes are maintained.

CHAPTER 5

Microbial Metabolism

Learning Objectives

After completing this chapter, you should be able to:

- Define metabolism and describe the fundamental differences between anabolism and catabolism.
- Describe the mechanism of enzymatic action.
- List the factors that influence enzymatic activity.
- Explain what is meant by oxidation–reduction.
- List and provide examples of three types of phosphorylation reactions that generate ATP.
- Explain the overall function of biochemical pathways.
- Describe the chemical reactions of glycolysis.
- Explain the products of the Krebs cycle.
- Describe the chemiosmotic model for ATP generation.
- Compare and contrast aerobic and anaerobic respiration.
- Describe the chemical reactions of and list some products of fermentation.
- Compare and contrast cyclic and noncyclic photophosphorylation.
- Compare and contrast the light and dark reactions of photosynthesis.
- Compare and contrast oxidative phosphorylation and photophosphorylation.
- Categorize the various nutritional patterns among organisms according to carbon source and mechanisms of carbohydrate catabolism and ATP generation.
- Describe the major types of anabolism and their relationship to catabolism.
- Define amphibolic pathways.

Metabolism is the sum of all chemical reactions within a living organism, including **anabolic** (*biosynthetic*) reactions and **catabolic** (*degradative*) reactions. Anabolism is the combination of simpler substances into complex substances and *requires* energy. Catabolism releases energy stored in organic molecules, for example, and *yields* energy. Energy liberated by catabolism is stored in energy-rich bonds of **adenosine triphosphate (ATP).**

Enzymes

Enzymes are proteins that function as **catalysts,** substances that speed up a reaction without being changed by it; that is, they lower the **activation**

energy needed. An enzyme has a three-dimensional structure, with an active site that reacts with the surface of a **substrate** (the substance acted upon). The **turnover number** is the number of molecules metabolized per enzyme molecule per second. Enzymes generally are named after the substrate they react with or the type of reaction they perform, using the suffix *-ase*. Examples are *dehydrogenase* for enzymes that remove hydrogen and *oxidase* for enzymes that add oxygen.

Enzyme Components

Many enzymes contain an **apoenzyme,** a protein that is inactive without a **cofactor,** a nonprotein component. Together, they are an activated **holoenzyme** (whole enzyme). Cofactors may serve as a bridge binding enzyme and substrate together. Cofactors may be metal ions or organic molecules called **coenzymes.** Many coenzymes are derived from vitamins; two of the most important are **NAD$^+$ (nicotinamide adenine dinucleotide)** and **NADP$^+$ (nicotinamide adenine dinucleotide phosphate).** Both are derivatives of the B vitamin nicotinic acid (niacin). They function as dehydrogenases, removing and transferring hydrogen ions and electrons. The flavin coenzymes, **FMN (flavin mononucleotide)** and **FAD (flavin adenine dinucleotide)** are derivatives of the B vitamin riboflavin. They also are dehydrogenases. **Coenzyme A** (CoA) is a derivative of the B vitamin pantothenic acid. It is important in reactions of the Krebs cycle, decarboxylation, and the synthesis and breakdown of fats.

Mechanism of Enzymatic Action

The substrate surface contacts a specific region of the enzyme molecule, the **active site,** forming an intermediate **enzyme–substrate complex.** The substrate is transformed by breakdown of the molecule or combination with another substrate molecule. The enzyme is freed unchanged.

Factors Influencing Enzymatic Activity

Temperature.   Most chemical reactions occur more rapidly as the temperature rises, but, above a certain point, denaturation of enzyme proteins results in a drastic decline in biological reaction rates. **Denaturation** usually involves breakage of the hydrogen bonds and similar weak bonds that hold the enzyme in its three-dimensional structure.

pH.   Enzymes have a **pH optimum** at which activity is maximal. Extreme pH changes cause denaturation.

Substrate Concentration.   At high substrate concentrations, the enzyme may have its active site occupied at all times by substrate or product molecules; that is, it may be **saturated.** No further increase in substrate concentrations will have an effect on the reaction.

Inhibitors.   There are two forms of enzyme inhibitors. **Competitive inhibitors** compete with the normal substrate for the active site of the enzyme. These inhibitors have a shape and chemical structure similar to the normal substrate. The action of sulfa drugs depends on competitive inhibition. **Noncompetitive inhibitors** decrease the ability of the normal substrate to combine with the enzyme. The site of a noncompetitive inhibitor's binding is an **allosteric site.** In this process, **allosteric** ("other space") **inhibition,** the inhibitor changes the shape of the active site, making it nonfunctional. Other examples are *enzyme poisons* that tie up metal ions (cofactors) and prevent enzymatic reactions, such as cyanide, which binds iron, and fluoride, which binds calcium or magnesium.

Feedback Inhibition

In some metabolic reactions, several steps are required. In many instances the final product can inhibit enzymatic activity at some step and prevent making of excessive *end-product.* This is called **feedback inhibition** (or **end-product inhibition**).

Energy Production

Nutrient molecules have energy stored in bonds that can be concentrated into the "high-energy" (or "unstable") bonds of ATP.

Oxidation–Reduction

Oxidation is the addition of oxygen or, more generally, the removal of electrons (e^-) or hydrogen ions (H^+). Because H^+ are lost, most biological reactions are called **dehydrogenation** reactions. When a compound gains electrons or hydrogen atoms, or loses oxygen, it is **reduced.** Oxidation and reduction in a cell are always coupled—one substance is oxidized and another is reduced; thus the reaction is an **oxidation–reduction** (or *redox*) reaction. NAD^+ and $NADP^+$ commonly carry the hydrogen atoms in these oxidation–reduction reactions. These reactions are usually energy producing. Highly reduced compounds such as glucose, with many hydrogen atoms, contain much potential energy.

Generation of ATP

The energy from oxidation–reduction is used to form ATP (see Figure 2.2). The addition of a phosphate group is called **phosphorylation.** In **oxidative phosphorylation,** electrons removed from organic compounds are transferred in sequence down an **electron transport chain** to an electron acceptor such as oxygen or another suitable compound, releasing energy in the process. The energy is used to make ATP from ADP by adding another phosphate. In **substrate-level phosphorylation,** no oxygen or other inorganic final electron acceptor is required. ATP is generated by the direct transfer of a high-energy phosphate from an intermediate metabolic compound to ADP. Another mechanism is **photophosphorylation,** which occurs in photosynthetic cells. Light liberates an electron from chlorophyll. The electron passes down an electron transport chain, forming ATP.

Biochemical Pathways of Energy Production

A sequence of enzymatically catalyzed chemical reactions in a cell is a **bio-chemical pathway.** Such pathways are necessary to extract energy from organic compounds; they allow energy to be released in a controlled manner instead of in a damaging burst with a large amount of heat.

Carbohydrate Catabolism

Glycolysis

The six-carbon sugar glucose plays a central role in carbohydrate metabolism (Figure 5.1). Glucose is usually broken down to **pyruvic acid** by **glycolysis** (splitting of sugar), which is the first stage of both fermentation and respiration.

   Glycolysis, also called the **Embden–Meyerhof pathway,** is a series of ten chemical reactions. The main points are that a six-carbon glucose molecule is split and forms two molecules of pyruvic acid. Two molecules of ATP were needed to start the reaction, and four molecules of ATP are formed by substrate-level phosphorylation; the net yield, therefore, is two ATP molecules. An alternate pathway is the **pentose phosphate pathway (hexose monophosphate shunt).** It produces important intermediates (pentoses) that act as precursors in the synthesis of nucleic acids, certain amino acids, and glucose from carbon dioxide by photosynthesizing organisms. Another way to oxidize glucose to pyruvic acid is the **Entner–Doudoroff pathway (EDP).** Bacteria (generally gram-negative) that utilize EDP can metabolize without either glycolysis or the pentose phosphate pathway. This pathway yields NADPH from glucose, which may be used for biosynthetic reactions.

Definition of Respiration

Respiration is an ATP-generating process in which molecules are oxidized and the final electron acceptor is almost always an inorganic molecule. In **aerobic respiration** the final electron acceptor is oxygen, and in **anaerobic respiration** it is usually an inorganic molecule other than molecular oxygen.

The Krebs Cycle

Note in Figure 5.1 that glycolysis forms pyruvic acid from carbohydrates such as glucose. Pyruvic acid loses a molecule of CO_2 **(decarboxylation)** to form an acetyl group. This then forms a complex, acetyl CoA, that can enter the **Krebs cycle.** The Krebs cycle releases the energy in acetyl CoA in a series of steps. The first step, which requires energy, yields citric acid (other names for the cycle are the *tricarboxylic acid cycle* and the *citric acid cycle*). Subsequent energy-yielding steps result in loss of CO_2 or loss of hydrogen atoms. Four molecules of CO_2 are released into the atmosphere for every two molecules of acetyl CoA that enter the cycle. Most of the energy is contained in six molecules of NADH and two molecules of $FAHD_2$.

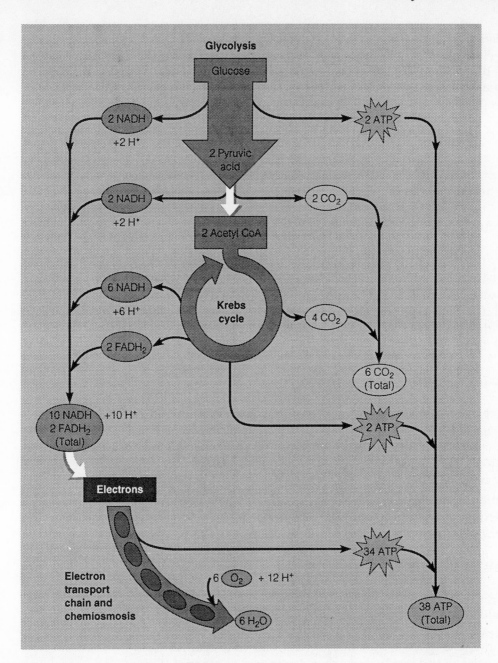

FIGURE 5.1 Summary of aerobic respiration in procaryotes. Glucose is broken down completely to carbon dioxide and water, and ATP is generated. This process has three major phases: glycolysis, the Krebs cycle, and the electron transport chain. The key event in the process is that electrons are picked up from intermediates of glycolysis and the Krebs cycle by NAD^+ or FAD and are carried by NADH or $FADH_2$ to the electron transport chain. NADH is also produced in the conversion of pyruvic acid to acetyl CoA. Most of the ATP generated by aerobic respiration is made by the chemiosmotic mechanism during the electron transport chain phase; this is called oxidative phosphorylation.

Electron Transport Chain

The energy of NADH and $FADH_2$ is generated by the electron transport chain (Figure 5.1). Note that the electrons are accepted by oxygen to form water (for each H^+ there is an equivalent electron, $2H = H^+ + 2e^-$). Electrons are released from NADH and $FADH_2$, and there is a stepwise release of energy as they pass down the chain. The energy is used to drive the chemiosmotic generation of ATP (described below). In eucaryotic cells the chain is in mitochondria; in procaryotic cells it is in the plasma membrane.

There are three classes of carrier molecules in the chain: **flavoproteins** (a flavin coenzyme is oxidized and reduced), **cytochromes** (iron is oxidized and reduced), and **ubiquinones** (coenzyme Q is a nonprotein carrier). These carriers accept and release protons (H^+) and/or electrons. At places, the electrons are pumped from one side of the membrane to another, causing a buildup of protons on one side. This buildup provides stored energy, like water behind a dam, providing energy for generation of ATP by **chemiosmosis.**

Chemiosmotic Mechanism of ATP Generation

As electrons from NADH pass down the chain, some protons are pumped across the membrane by carrier molecules called *proton pumps.* The protons accumulate on one side of the membrane, resulting in a *protein motive force* from the positive charges accumulated there. These protons diffuse back across the membrane through special channels, where *ATP synthase* synthesizes ATP.

Summary of Aerobic Respiration

Procaryotes generate 38 molecules of ATP aerobically for each molecule of glucose; eucaryotes produce only 36.

Anaerobic Respiration

Some bacteria can use oxygen substitutes such as nitrate ion (forming nitrite ion, nitrous oxide, or nitrogen gas), sulfate (forming hydrogen sulfide), or carbonate (forming methane).

Fermentation

After glucose is broken down to pyruvic acid, the pyruvic acid can undergo **fermentation.** Fermentation does not require oxygen or an electron transport chain. It uses an organic molecule as the final electron acceptor. In fermentation, the pyruvic acid accepts electrons (hydrogen) and is turned into various end-products, such as **lactic acid** or **ethanol** (Figure 5.19b in the text).

Lipid Catabolism

Some microorganisms produce extracellular enzymes, lipases, that break fats into fatty acids and glycerol. These components are then metabolized separately. Fatty acids are oxidized by **beta oxidation,** in which carbon fragments are removed two at a time to form acetyl coenzyme A, whose molecules then enter the Krebs cycle. Glycerol forms one of the intermediates of glycolysis and then is further oxidized.

Protein Catabolism

Proteases and peptidases are extracellular enzymes produced by some microbes to break down proteins, which are much too large for microbes to use unaltered, into component amino acids. Amino acids are first **deaminated** (—NH_2 removed) and **decarboxylated** (—COOH removed). They then enter the Krebs cycle in various ways.

Photosynthesis

We have just discussed organisms that obtain energy by oxidizing organic compounds. This energy source is a result of **photosynthesis,** by which electrons are taken from the hydrogen atoms of water and incorporated into sugar. Formation of ATP in this manner is called **photophosphorylation** and requires light. In **cyclic photophosphorylation,** the electrons return to chlorophyll. In **noncyclic photophosphorylation,** they are incorporated into NADPH. Photosynthesis also includes dark reactions (no light directly required) such as the **Calvin–Benson cycle.** In this cycle the CO_2 used to form sugars is "fixed."

Nutritional Patterns Among Organisms

Phototrophs use light as their primary *energy source;* **chemotrophs** extract their energy from inorganic or organic chemical compounds. The principal *carbon source* of **autotrophs** is carbon dioxide; **heterotrophs** require an organic carbon source. These terms can be combined into terms that reflect the primary energy and carbon sources.

Photoautotrophs. Photoautotrophs include photosynthetic bacteria. **Green sulfur bacteria** are anaerobes that use sulfur compounds or hydrogen gas to reduce carbon dioxide and form organic compounds. Light is the energy source. Most commonly, they produce sulfur from hydrogen sulfide. **Purple sulfur bacteria** also use sulfur compounds or hydrogen gas to reduce carbon dioxide. Neither of these photosynthetic bacteria use water to reduce carbon dioxide, as do plants, and they do *not* produce oxygen gas (they are **anoxygenic**) as a product of photosynthesis. Their photosynthetic pigment is bacteriochlorophyll, which absorbs longer wavelengths of light than chlorophyll *a*. Cyanobacteria use chlorophyll *a* in photosynthesis. They produce oxygen gas (they are **oxygenic**), just as higher plants do.

Photoheterotrophs. **Green nonsulfur** and **purple nonsulfur bacteria** are **photoheterotrophs.** They use light as an energy source but must use organic compounds instead of carbon dioxide as a carbon source. Otherwise, they are similar to the green and purple sulfur bacteria.

Chemoautotrophs. *Inorganic* compounds such as hydrogen sulfide, elemental sulfur, ammonia, nitrites, hydrogen, and iron are used by **chemoautotrophs** as sources of energy. Carbon dioxide is their principal carbon source. These compounds contain energy that may be extracted by oxidative phosphorylation reactions.

Chemoheterotrophs. With **chemoheterotrophs,** the carbon source and energy source are usually the same *organic* compound—glucose, for example. Most microorganisms are chemoheterotrophs.

Saprophytes live on dead organic matter, and **parasites** derive nutrients from a living host.

Biochemical Pathways of Energy Use (Anabolism)

Energy generated by catabolism may be used in the synthesis of new compounds for the cell. **Polysaccharides** such as glycogen are synthesized from glucose. Glucose is first joined with ATP, forming *adenosine diphosphoglucose (ADPG)*. The energy of the ATP is, in essence, used to fasten together the sequence of glucose molecules to form a polysaccharide. **Lipids** such as fats are formed by combining glycerol with fatty acids. The glycerol is derived from a glycolysis intermediate, and the fatty acids are built up from two-carbon fragments of acetyl coenzyme A. The **amino acids** required for the biosynthesis of proteins are synthesized by some bacteria; other bacteria require them to be preformed. Intermediates of carbohydrate metabolism are used in the synthesis of amino acids. DNA and RNA are made up of repeating units called **nucleotides.** These consist of a purine or pyrimidine, a five-carbon sugar, and a phosphate group. Sugars for nucleotides are derived from the pentose phosphate pathway or the Entner–Doudoroff pathway. Amino acids such as glycine and glutamine furnish the atoms from which are derived the backbone of purines and pyrimidines.

Integration of Metabolism

Anabolic and catabolic reactions are integrated through common intermediates. The Krebs cycle, for example, can operate in both anabolic and catabolic reactions. Such pathways are called **amphibolic pathways.**

Self-Tests

In the matching section, there is only one answer to each question; however, the lettered options (a, b, c, etc.) may be used more than once or not at all.

I. Matching

1. Energy-yielding series of reactions.

2. Means "whole enzyme."

3. A nonprotein component of an active enzyme.

4. A measure of the rate of activity of an enzyme.

5. A protein portion of an enzyme, inactive without a cofactor.

6. A group of enzymes that function as electron carriers in respiration and photosynthesis.

7. A mechanism by which fatty acids are degraded.

a. Catabolism

b. Anabolism

c. Turnover number

d. Apoenzyme

e. Coenzyme

f. Holoenzyme

g. Beta oxidation

h. Cytochromes

II. Matching

1. Both the carbon source and energy source are usually the same organic compound.

2. Photosynthetic, but uses organic material rather than carbon dioxide as a carbon source.

3. The photosynthetic purple nonsulfur bacteria would be classified in this nutritional group.

4. Photosynthetic bacteria that use carbon dioxide as a carbon source.

5. Changes the shape of the active site of an enzyme.

6. Very similar in shape or chemistry to the normal enzyme substrate.

a. Competitive inhibitor

b. Noncompetitive inhibitor

c. Photoautotroph

d. Chemoautotroph

e. Photoheterotroph

f. Chemoheterotroph

III. Matching

1. Hexose monophosphate shunt.

2. The final electron acceptor is oxygen.

3. Produces important intermediates that act as precursors in the synthesis of nucleic acids and so on.

4. Bacteria use oxygen substitutes such as nitrates.

5. Pyruvic acid accepts electrons and is turned into various end-products, such as lactic acid or ethanol.

6. Glucose to pyruvic acid.

a. Fermentation

b. Glycolysis

c. Pentose phosphate pathway

d. Substrate-level phosphorylation

e. Anaerobic respiration

f. Aerobic respiration

IV. Matching

1. Electrons are removed from an organic compound and are transferred by an electron transport chain to oxygen.

2. An electron is liberated from chlorophyll and passes down an electron transport chain.

a. Oxidative phosphorylation

b. Substrate-level phosphorylation

c. Photophosphorylation

V. Matching

1. A dehydrogenase coenzyme derived from nicotinic acid (niacin).

2. A dehydrogenase coenzyme derived from riboflavin.

3. In chemiosmosis, protons can diffuse across a membrane only through special channels that contain this enzyme.

4. Pyruvic acid loses carbon dioxide to form an acetyl group.

a. NAD^+

b. Decarboxylation

c. Coenzyme A

d. FMN

e. ATP synthase

f. Dehydrogenation

VI. Fill-in-the-Blanks

1. A chemoheterotroph that lives on dead organic matter is called a _____ .

2. When an organism does not produce oxygen by photosynthesis, it is termed _____ .

3. When an enzyme's active site is occupied at all times by substrate or product molecules, it is called _____ .

4. Cyanide is an example of a general type of inhibitor called _____ .

5. Sulfa drugs are an example of a type of inhibitor called _____ .

6. The removal of electrons or hydrogen ions, or the addition of oxygen, is termed _____ .

7. In _____ phosphorylation, no oxygen or other inorganic final electron acceptor is required.

8. Cyanobacteria produce _____ gas, just as do higher plants.

9. The amount of ATP yield from aerobic respiration by a procaryote is _____ .

10. The amount of ATP yield from glycolysis is _____ .

11. The removal of NH_2 from an amino acid is called _____ .

12. The removal of —COOH from an amino acid is called _____ .

13. The substance acted upon by an enzyme is called the _____ .

14. Coenzyme A is a derivative of the B vitamin _____ acid.

15. Inorganic sources of energy such as hydrogen sulfide, elemental sulfur, ammonia, nitrites, and so on are used by the _____ nutritional group of bacteria as a source of energy.

16. A sequence of enzymatically catalyzed chemical reactions in a cell is called a _____ pathway.

17. Glucose is usually broken down to pyruvic acid by _____ .

18. Another name for glycolysis is the _____ pathway.

19. In aerobic respiration, pyruvic acid is converted to acetyl _____ ; this product can then enter the Krebs cycle.

20. DNA and RNA are made up of repeating units called _____ .

VII. Label the Art

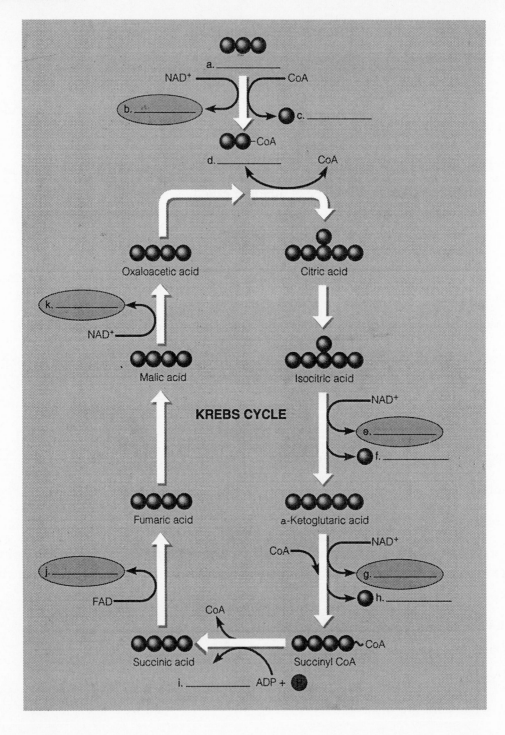

VIII. Critical Thinking

1. Why are catabolic and anabolic reactions referred to as coupled reactions?

2. Explain how competitive and noncompetitive enzyme inhibitors work.

3. How does the ultimate fate of electrons liberated differ in cyclic and noncyclic photophosphorylation?

4. What are the key features of the pentose phosphate pathway?

5. List and discuss the differences in three pathways for the oxidation of glucose to pyruvic acid.

6. What three things does a cell require to produce energy?

7. Complete the following table with examples of final electron acceptor(s) for each process.

Process	Electron acceptor
Aerobic respiration	
Anaerobic respiration	
Fermentation	

8. Complete the following table.

Nutritional category	Energy source	Carbon source
Photoautotroph		
Photoheterotroph		
Chemoautotroph		
Chemoheterotroph		

9. Define *bacteriochlorophylls, intracytoplasmic membranes,* and *chlorosomes.*

10. List and describe the three classes of carrier molecules in electron transport chains.

CHAPTER 6

Microbial Growth

Learning Objectives

After completing this chapter, you should be able to:

- Classify microbes into five groups on the basis of preferred temperature range.
- Identify how and why the pH of culture media is controlled.
- Explain the importance of osmotic pressure to microbial growth.
- Provide a use for each of the four elements (carbon, nitrogen, sulfur, and phosphorus) needed in large amounts for microbial growth.
- Explain how microbes are classified on the basis of oxygen requirements.
- Identify ways in which aerobes avoid damage by toxic forms of oxygen.
- Distinguish between chemically defined and complex media.
- Justify the use of each of the following: anaerobic techniques, living host cells, candle jars, selective and differential media, enrichment media.
- Define colony and clone.
- Describe how pure cultures can be isolated by using the streak plate method.
- Explain how microbes are preserved by deep-freezing and freeze-drying.
- Define bacterial growth, including binary fission.
- Compare the phases of microbial growth and describe their relation to generation time.
- Differentiate between direct and indirect methods of measuring cell growth.
- Explain four direct methods of measuring cell growth.
- Explain three indirect methods of measuring cell growth.

Microbial growth refers to the number of cells, not to the changes in the size of cells. **Colonies** are accumulations of cells large enough to be visible without a microscope.

3. The **hydroxyl radical** (OH·) is formed in cytoplasm by ionizing radiation and as a by-product of aerobic respiration. It is probably the most reactive form.

Organic Growth Factors. **Organic growth factors** are organic compounds such as vitamins, amino acids, and pyrimidines that are needed for life, but that a given organism is unable to synthesize.

Culture Media

Any nutrient material prepared for the growth of bacteria in a laboratory is called a **culture medium.** Microbes growing in a container of culture medium are referred to as a **culture.** To ensure that the culture will contain only the microorganisms originally added to the medium (and their offspring), the medium must initially be sterile. When a solid medium is required, a solidifying agent such as **agar** is added. Agar is a polysaccharide derived from a marine alga. Few microbes can degrade agar, so it remains a solid. It melts at about the boiling point of water but remains liquid until the temperature drops to about 40°C.

In a **chemically defined medium,** the exact chemical composition is known.

Most heterotrophic bacteria and fungi are routinely grown on **complex media,** in which the exact chemical composition varies slightly from batch to batch. Complex media are made up of nutrients such as extracts from yeasts, beef, or plants, or digest of proteins from these and other sources. In many of these media, the energy, carbon, nitrogen, and sulfur requirements of the microorganisms are largely met by partially digested protein products called peptones. Vitamins and other organic growth factors are provided by meat extracts or yeast extracts. Such extracts supplement the organic nitrogen and carbon compounds but mainly are sources of soluble vitamins and minerals. This type of medium in liquid form is **nutrient broth;** when agar is added, it is **nutrient agar.**

Anaerobic Growth Media and Methods

Obligately anaerobic bacteria often require **reducing media** for isolation. Because oxygen may be lethal, these media contain ingredients, such as sodium thioglycolate, that chemically combine with dissolved oxygen to deplete the oxygen content of the culture medium. Obligate anaerobes may be grown on the surface of solid media in anaerobic atmospheres produced in special jars in which an oxygen-free atmosphere is generated by a chemical reaction. Surface growth also can be accomplished by substituting deep vials for Petri dishes. In this process, an inert gas replaces the atmosphere in the open vial, and then the bacterial inoculum is mixed with melted nutrient medium and pipetted into the vial. The vial is tightly capped and rolled on horizontal rollers to distribute the medium and bacteria against the interior walls (**roll tubes**). Colonies appear in this film of medium and are easily counted. It is also possible to handle anaerobic organisms in anaerobic glove boxes filled with inert gases and fitted with airtight rubber glove arms and air locks.

Special Culture Techniques

A few bacteria have never been successfully grown on artificial laboratory media; the leprosy and syphilis organisms are examples. Obligate intracellular parasites such as rickettsias and chlamydias ordinarily also do not grow on artificial media. They, like viruses, require a living host cell. Carbon dioxide incubators, candle jars, and plastic bags with self-contained chemical gas generators are used to grow bacteria with special carbon dioxide concentration requirements.

Selective and Differential Media

Selective media are designed to suppress the growth of unwanted bacteria and encourage the growth of the desired microorganisms. Antibiotics, high concentrations of salt, or high acidity might be used. **Differential media** make it easier to distinguish colonies of the desired organism from other colonies growing on the same plate. The colonies have different colors or cause different changes in the surrounding medium. Sometimes selective and differential functions are combined in the one medium.

Enrichment Culture

Because bacteria may be present only in small numbers and may be missed, and because the bacterium to be isolated may be of an unusual physiological type, it is sometimes necessary to resort to enrichment cultures. For example, only a tiny number of organisms in a soil sample are capable of growing on phenol, a disinfectant. These organisms can be isolated, however, by placing the soil sample in a medium in which the only source of carbon and energy is phenol. A series of transfers between fresh culture media of the same type, each of which is allowed to incubate for a few days, gradually eliminates organic matter from the original inoculum. Finally, only organisms capable of growth on phenol will survive.

Obtaining Pure Cultures

There are several methods for isolating bacteria in **pure cultures,** which contain only one kind of organism.

Streak Plate Method. Probably the most common method of obtaining pure cultures is the **streak plate** (Figure 6.1). A sterile inoculating needle is dipped into a mixed culture and streaked in a pattern over the surface of the nutrient medium. The last cells rubbed from the needle are wide enough apart that they grow into isolated visible masses called **colonies.**

Preserving Bacterial Cultures

In **deep-freezing,** a pure culture of microbes is placed in a suspending liquid and quick-frozen at –50° to –95°C. In **lyophilization (freeze-drying),** a suspension of microbes is quickly frozen at temperatures of –54° to –72°C and the water removed by a high vacuum. The remaining powder can be

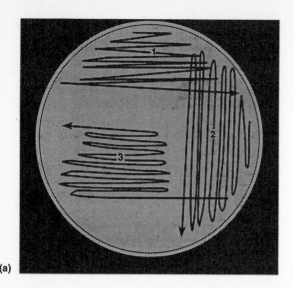

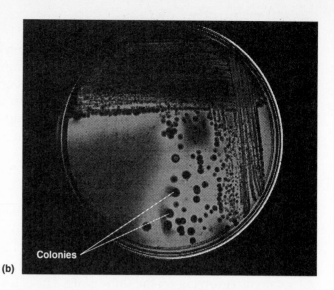

(a)   (b)

FIGURE 6.1 **Streak plate method for isolation of pure cultures of bacteria.**   (a) The direction of streaking is indicated by arrows. Streak series 1 is made from the original bacterial mixture. The inoculating loop is sterilized following each streak series. In series 2 and 3, the loop picks up bacteria from the previous series, diluting the number of cells each time. There are numerous variants of such patterns. (b) In series 3 of this example, note that well-isolated colonies of two different types of bacteria have been obtained.

stored for many years and the surviving microorganisms cultured by hydrating them with a suitable liquid nutrient medium.

Growth of Bacteria Cultures

Bacterial Division

Bacteria normally reproduce by **binary fission.** Genetic material becomes evenly distributed; then a transverse wall is formed across the center of the cell, and it separates into two cells. A few bacterial species **bud;** that is, an initial outgrowth enlarges to cell size and then separates. Some filamentous species produce **reproductive spores** or simply **fragment** into viable pieces.

Generation Time

The time required for a cell to divide or the population to double is called **generation (doubling) time.** Bacterial populations are usually graphed **logarithmically** rather than **arithmetically** to permit the handling of the immense differences in numbers (Figure 6.2).

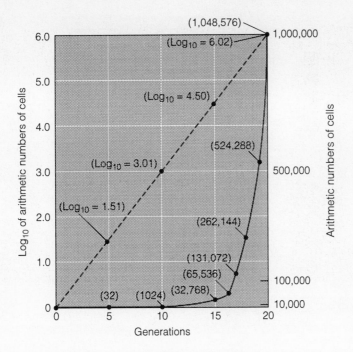

FIGURE 6.2   **Growth curve for an exponentially increasing population, plotted logarithmically (dashed line) and arithmetically (solid line).**

Phases of Growth

When bacterial population changes are graphed as a bacterial growth curve, certain phases become apparent. The **lag phase** shows little or no cell division. However, metabolic activity is intense. In the **log phase,** the cells are reproducing most actively, and their generation time reaches a minimum and remains constant; a logarithmic plot produces an ascending straight line. They are then most active metabolically and most sensitive to adverse conditions. In a *chemostat,* it is possible to keep a population in such exponential growth indefinitely. Without a chemostat, however, microbial deaths eventually balance numbers of new cells, and a **stationary phase** is reached. When the number of deaths exceeds numbers of new cells formed, the **death phase,** or **logarithmic decline,** is reached. The entire culture may die out in time.

Measurement of Microbial Growth

Plate Counts.   Dilutions of a bacterial suspension are distributed into a suitable solid nutrient medium by **serial dilution,** and the colonies appearing on the plates are counted (Figure 6.3).

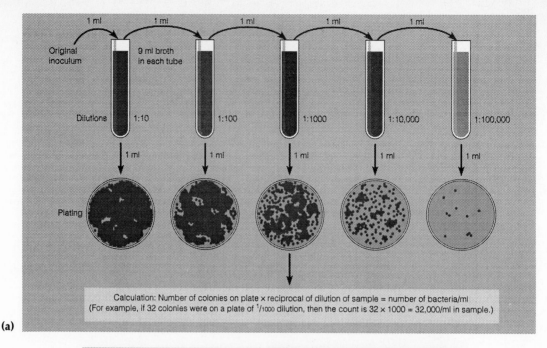

(a)

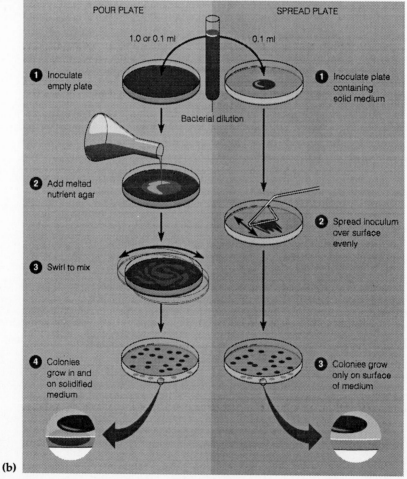

(b)

FIGURE 6.3 Plate counts and serial dilutions. **(a)** In serial dilutions, the original inoculum is diluted in a series of dilution tubes. In our example, each succeeding dilution tube will have only one-tenth the number of microbial cells as the preceding tube. Then samples of the dilution are used to inoculate Petri plates, on which colonies grow and can be counted. This count is then used to estimate the number of bacteria in the original sample. **(b)** Methods for preparation of plates for plate counts.

Filtration. Bacteria may be sieved out of a liquid suspension onto a thin membrane filter with pores too small for bacteria to pass. This filter can be transferred to a pad soaked in nutrient medium where colonies arise on the surface of the filter.

Most Probable Number (MPN). In the most probable number method, a sample is diluted out in a series of tubes of liquid medium. The greater the number of bacteria, the more dilutions it takes to dilute them out entirely and leave a tube without growth. Results of such dilutions can be compared to statistical tables, and a cell count can be estimated.

Direct Microscopic Count. In this method, a measured volume of a bacterial suspension is placed into a defined area on a special microscope slide, of which there are several designs. A microscope is used to count the cells in microscope fields. The average number per field can be multiplied by a factor that will estimate the total numbers.

Turbidity. To estimate turbidity, a beam of light is transmitted through a bacterial suspension to a photoelectric cell. The more bacteria, the less light passes. This is recorded as **absorbance** (sometimes called *optical density*, or *OD*) on the spectrophotometer or colorimeter.

Metabolic Activity. Microbial numbers can be estimated by the time required to deplete oxygen (**reduction tests**) or to produce acid or other products.

Dry Weight. Measuring weight is often the most satisfactory method for filamentous organisms such as fungi.

Self-Tests

In the matching section, there is only one answer to each question; however, the lettered options (a, b, c, etc.) may be used more than once or not at all.

I. Matching

1. Adapted to high salt concentrations, which are required for growth.

2. The general term used for organisms capable of growth at 0°C.

3. Capable of growth at high temperatures; optimum 50° to 60°C.

4. Used in media to neutralize acids.

5. A phenomenon that occurs when bacteria are placed in high salt concentration.

6. Term used in text for organisms that grow well at refrigerator temperatures; optimum growth is at Celsius temperatures in the upper teens or low twenties.

a. Buffer

b. Mesophile

c. Thermophile

d. Psychrophile

e. Psychrotroph

f. Plasmolysis

g. Extreme halophile

h. Facultative halophile

II. Matching

1. An enzyme acting upon hydrogen peroxide.

2. *Rhizobium* bacteria do this in symbiosis with leguminous plants.

3. Requires atmospheric oxygen to grow.

4. Requires atmospheric oxygen, but in lower than normal concentrations.

5. Does not use oxygen, but grows readily in its presence.

6. Does not use oxygen and usually finds it toxic.

7. Important source of energy, carbon, nitrogen, and sulfur requirements in complex media.

a. Nitrogen fixation

b. Obligate aerobe

c. Obligate anaerobe

d. Aerotolerant anaerobe

e. Catalase

f. Microaerophile

g. Peptones

h. Agar

III. Matching

1. Isolation method for getting pure cultures; uses an inoculating loop to trace a pattern of inoculum on a solid medium.

2. A device for maintaining bacteria in a logarithmic growth phase.

3. Used to increase the numbers of a small minority of microorganisms in a mixed culture to arrive at a pure culture.

4. Preservation method that uses quick-freezing and a high vacuum.

5. Accumulations of microbes large enough to see without a microscope.

a. Pour plate

b. Streak plate

c. Spread plate

d. Differential medium

e. Reducing medium

f. Enrichment culture

g. Lyophilization

h. Deep-freezing

i. Chemostat

j. Colonies

IV. Matching

1. New cell numbers balanced by death of cells.

2. No cell division, but intense metabolic activity.

3. A logarithmic plot of the population produces an ascending straight line.

a. Log phase

b. Lag phase

c. Death phase

d. Stationary phase

V. Matching

1. Used to grow obligate anaerobes.

2. Designed to suppress the growth of unwanted bacteria and to encourage growth of desired microbes.

3. Generally contain ingredients such as sodium thioglycolate that chemically combine with dissolved oxygen.

4. Nutrients are digests or extracts; exact chemical composition varies slightly from batch to batch.

a. Selective media

b. Differential media

c. Complex media

d. Reducing media

e. Chemically defined media

VI. Matching

1. Breaks down hydrogen peroxide without generation of oxygen.

2. Formed in cytoplasm by ionizing radiation.

3. An enzyme that converts hydrogen peroxide into oxygen and water.

4. The toxic form of oxygen neutralized by superoxide dismutase.

a. Hydroxyl radical

b. Peroxidase

c. Superoxide dismutase

d. Superoxide free radicals

e. Singlet oxygen

f. Catalase

VII. Fill-in-the-Blanks

1. Agar is a _____ derived from a marine alga.

2. A few bacteria and the photosynthesizing _____ are able to use gaseous nitrogen directly from the atmosphere.

3. _____ are the most common microbes; their optimum temperatures are 25° to 40°C.

4. Osmotic effects are roughly related to the _____ of molecules in a given volume of solution.

5. A complex medium in liquid form is called nutrient _____ .

6. For preservation by _____ , a pure culture of microbes is placed in a suspending liquid and quick-frozen at –50° to –95°C.

7. Bacteria usually reproduce by _____ fission.

8. Turbidity is recorded in a spectrophotometer as _____ .

9. The growth of filamentous organisms such as fungi is often best recorded by means of _____ .

10. _____ anaerobes grow more efficiently aerobically than they do anaerobically.

11. _____ halophiles do not require high salt concentrations, but they are able to grow at salt concentrations that may inhibit the growth of many other bacteria.

12. Examples of buffers are _____ salts; peptones and _____ found in complex media are also buffers.

13. Any nutrient material prepared for the growth of bacteria in a laboratory is called a _____ .

14. Agar melts at about the boiling point of water but remains liquid until the temperature drops to about _____ .

15. Dilutions of a bacterial mixture are poured into a Petri dish and mixed with melted agar. This plate-counting method is called the _____ .

16. Partially digested protein products used in complex media are called _____ .

17. In order to grow obligate intracellular parasites such as rickettsias and chlamydias, it is usually necessary to provide _____ .

18. The general term for tests that estimate microbial growth by the time required for them to deplete oxygen in the medium is _____ tests.

19. The _____ growth temperature is that at which the organism grows best.

VIII. Label the Art

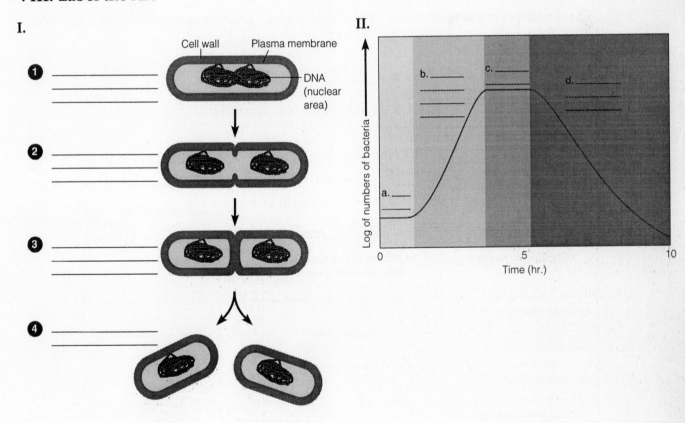

IX. Critical Thinking

1. List and briefly describe the physical requirements for bacterial growth.

2. List and briefly describe the chemical requirements for bacterial growth.

3. Discuss the method used to retard spoilage in each of the following foods.

   a. Grape jelly   d. Olives

   b. Pickles   e. Canned beans

   c. Salted fish   f. Cheese

4. What kind(s) of microorganism(s) would be most likely to cause spoilage of each of the foods listed above? (*Hint:* See Chapter 28.)

5. Explain the significance of nitrogen-fixing bacteria to agriculture.

6. Complete the following table with information about where microorganisms will grow in a tube of solid media on the basis of their relationship to oxygen.

Relation to oxygen	Where in the tube does growth occur?	Why?
Obligate aerobe		
Facultative anaerobe		
Obligate anaerobe		
Aerotolerant anaerobe		
Microaerophile		

7. Explain the function of the palladium catalyst in an anaerobic container.

8. What kind of media would be most appropriate for each of the following situations? Why?

   a. To diagnose strep throat.

   b. For the routine growth of heterotrophic bacteria.

   c. To isolate *Staphylococcus aureus*.

   d. To detect soil bacteria.

9. Draw a bacterial growth curve indicating the four phases of growth. At which phase of growth would exposure to antibiotics cause the most adverse effects on the bacterial population? Why?

10. Discuss advantages and disadvantages of using the plate count method to measure microbial populations.

CHAPTER 7

Control of Microbial Growth

Learning Objectives

After completing this chapter, you should be able to:

- Identify the contributions made by Semmelweis and Lister to the control of microbes.
- Define the following terms related to microbial control: sterilization, disinfection, antisepsis, germicide, bacteriostasis, asepsis, degerming, and sanitation.
- Explain how microbial growth is affected by the type of microbes, the microbes' physiological state, and the ambient environmental conditions.
- Describe the effects of microbial control agents on cellular structures.
- Describe the patterns of microbial death caused by treatments with microbial control agents.
- Compare the effectiveness of moist heat (boiling, autoclaving, pasteurization) and dry heat.
- Describe how filtration, cold, desiccation, and osmotic pressure suppress microbial growth.
- Explain how radiation kills cells.
- Describe the factors related to effective disinfection.
- Interpret results of use-dilution and filter paper tests.
- Identify the methods of action and preferred uses of chemical disinfectants.
- Differentiate between halogens used as antiseptics and as disinfectants.
- Identify the appropriate uses for surface-active agents.
- List the advantages of glutaraldehyde over other chemical disinfectants.
- Identify the method of sterilizing plastic labwares.

Terminology Related to the Control of Microbial Growth

Sterilization is the destruction of all forms of microbial life. **Disinfection** is the destruction of vegetative pathogens on a surface, usually with chemicals. Spores and viruses are not necessarily destroyed. **Antisepsis** is the chemical disinfection of living tissue, such as skin or mucous membrane. **Asepsis** is the absence of pathogens on an object or area, as in **aseptic surgery**. **Degerming** is the removal of transient microbes from skin by mechanical cleansing or by an antiseptic. **Sanitization** is the reduction of microbial populations on objects to safe public health levels. In general, the suffix -*cide* indicates the killer of a specified organism, as in **germicide,**

fungicide, sporicide, virucide, and so on. The suffix *-stat* used in this way indicates only that the substance inhibits.

Conditions Influencing Microbial Control

The control of microorganisms can be affected by the **temperature** of chemical solutions (generally disinfectants are more efficient when warm), by the **type of microbe** involved (pseudomonads, tuberculosis bacillus, endospores, protozoan cysts, and many viruses are relatively resistant). For example, pseudomonad resistance is related to characteristics of their **porins** (wall openings). Important factors affecting microbial control are the **physiological state of the microbe** (actively growing organisms are more susceptible) and the **environment** (organic matter such as feces or vomit may inactivate disinfectants).

Actions of Microbial Control Agents

Alteration of Membrane Permeability

The plasma membrane controls the passage of nutrients and wastes into and out of the cell. Damage to the plasma membrane causes leakage of cellular contents and interferes with cell growth.

Damage to Proteins and Nucleic Acids

Chemicals may denature proteins by reacting, for example, with disulfide bonds (or disulfide bridges), which give proteins their three-dimensional active shape. Chemicals and radiation may prevent proper replication or functioning of DNA or RNA.

Rate of Microbial Death

Bacterial populations killed by heat or chemicals tend to die at constant rates—for example, 90% every 10 minutes. Plotted logarithmically, these figures form straight descending lines.

Physical Methods of Microbial Control

Heat

Thermal death point is the lowest temperature required to kill a liquid culture of a certain species of bacteria in 10 minutes at pH 7. **Thermal death time** is the length of time required to kill all bacteria in a liquid culture at a given temperature. **Decimal reduction time** is the length of time, in minutes, required to kill 90% of the population of bacteria at a given temperature.

Moist Heat.   Boiling (100°C) kills vegetative forms of bacterial pathogens, many viruses, and fungi within 10 minutes. Endospores and some viruses survive boiling for longer times. **Steam under pressure** allows temperatures above boiling to be reached. **Autoclaves**, retorts, and pressure cookers are vessels in which high steam pressures can be contained. A typical operating condition for sterilization is 15 psi (pounds per square inch) at 121°C for 15 minutes. Moisture must touch all surfaces in order to bring about sterilization. Air must be completely exhausted from the container. An autoclave is shown in Figure 7.1.

Pasteurization is mild heating that is sufficient to kill particular spoilage or disease organisms without seriously damaging the taste of the product. **High-temperature, short-time pasteurization** uses temperatures of at least 72°C for about 15 seconds to pasteurize milk. **Equivalent treatments** are illustrated by, for example, heat of 115°C acting on an organism for 70 minutes as equivalent to heat of 125°C acting on an organism for only 7 minutes; that is, applying a higher temperature for a shorter time may kill the same number of microbes as a lower temperature for a longer time.

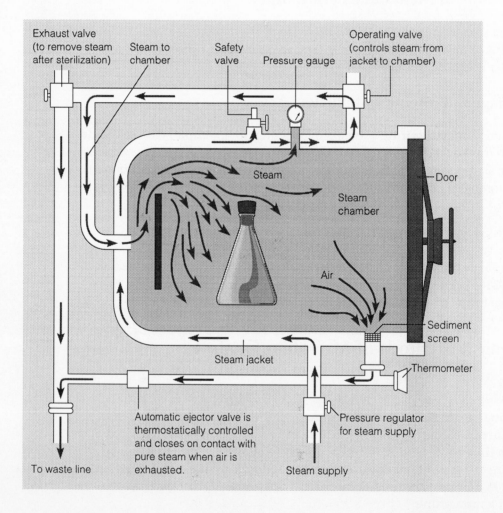

FIGURE 7.1 **Autoclave.**   The entering steam forces the air out of the bottom (black arrows). The automatic ejector valve remains open as long as an air–steam mixture is passing out of the waste line. When all the air has been ejected, the higher temperature of the pure steam closes the valve, and the pressure in the chamber increases.

Dry Heat Sterilization.   **Incineration,** as in *direct flaming,* is efficient for limited purposes. **Hot-air sterilization,** as in an oven, requires higher temperatures (such as 170°C) and longer times (such as 2 hours) to ensure sterilization. Moist heat is generally more efficient.

Filtration

Liquids sensitive to heat can be passed through a thin **membrane filter** of cellulose esters or plastic polymers that has carefully controlled pore sizes to retain microorganisms. Operating theaters and special clean rooms receive air passed through **high-efficiency particulate air filters.**

Cold

Refrigerator temperatures (0° to 7°C) slow the metabolic rate of microbes; however, psychrotrophic species still grow slowly. Some organisms grow at temperatures slightly below freezing, but microbes at the usual temperatures of freezer compartments are completely dormant.

Desiccation

Microbes require water for growth, and adequately dried **(desiccated)** foods will not support their growth.

Osmotic Pressure

High salt or sugar concentrations cause water to leave the cell; this is an example of **osmosis** (see *plasmolysis* in Chapter 6). Generally, molds and yeasts resist osmotic pressures better than bacteria.

Radiation

Ionizing radiation such as X rays, gamma rays, and high-energy electron beams carry high energy and break DNA strands. Such radiation is used to sterilize pharmaceuticals. **Nonionizing radiation** such as ultraviolet (UV) light has a longer wavelength and less energy. UV light causes bonds to form between adjacent thymines in DNA chains. The most effective wavelength of UV light is about 260 nm. Penetration is low. Sunlight has some germicidal activity, mainly due to formation of singlet oxygen (see Chapter 6) in the cytoplasm.

Chemical Methods of Microbial Control

Evaluating a Disinfectant

The relative effectiveness of disinfectants is determined by tests such as the **phenol coefficient test.** Bacteria are tested under standard conditions against concentrations of phenol and the test disinfectant.

Use-Dilution Test.   The **American Official Analytic Chemist's use-dilution test** is more elaborate. A series of tubes containing increasing concentrations of the test disinfectants are inoculated and incubated. The more the chemical can be diluted and still be effective, the higher its rating. The test bacteria are usually *Staphylococcus aureus*, *Pseudomonas aeruginosa*, and *Salmonella typhi*.

Filter Paper Method.   A disk of filter paper is soaked in a chemical agent, which is placed on an inoculated surface of an agar plate. A clear zone around the disk indicates inhibition.

Types of Disinfectants

Phenol and Phenolics.   Phenol (carbolic acid) is seldom used today. Derivatives of the phenol molecule, however, are widely used. **Phenolics** injure plasma membranes, inactivate enzymes, or denature proteins. They are stable, persistent, and are not sensitive to organic matter. **O-phenylphenol,** a *cresol*, is the main ingredient in most formulations of Lysol®. **Hexachlorophene** is effective against staphylococcal and streptococcal bacteria that cause skin infections. Problems of neurological damage to infants from excessive use have occurred, however.

Chlorhexidine.   Chlorhexidine is not a phenol, but its structure and applications resemble hexachlorophene. It is frequently used for surgical skin preparation and surgical hand scrubs.

Halogens.   Iodine is effective against all kinds of bacteria, many endospores, fungi, and some viruses. Its mechanism of activity may be its combination with the amino acid tyrosine in enzyme and cellular proteins. A **tincture** of iodine means in an aqueous-alcohol solution. An **iodophore** is a combination of iodine and an organic molecule. Iodophores do not stain and are less irritating than iodine. Examples are Isodine® and the antiseptic Betadine®. **Chlorine** is used as a gas or in combination with other chemicals. Chlorine gas is used for disinfecting municipal water supplies, swimming pools, and sewage. Sodium hypochlorite—ordinary household bleach—is a good disinfectant. Chloramines consist of chlorine and ammonia. They are more stable than most chlorines. The germicidal action of all chlorines is based on the formation of hypochlorous acid ($HOCl$) in water. One factor in its effectiveness is its neutral charge, which allows it to diffuse as rapidly as water into the cell.

Alcohols.   Both **ethanol** and **isopropanol** (rubbing alcohol) are widely used, normally at a concentration of about 70%. Concentrations of 60% to 95% are effective. They are bacteriocidal and fungicidal but are not effective against endospores or nonenveloped viruses. Alcohols enhance the effectiveness of other chemical agents.

Heavy Metals and Their Compounds.   The fact that tiny amounts of heavy metals are effective antimicrobials can be illustrated by **oligodynamic action.** A silver coin on an inoculated nutrient medium will inhibit growth for some distance. A *1% silver nitrate* solution has been used to prevent gonorrheal eye infections in newborns. Silver combines with sulfhydryl groups on proteins, denaturing them. *Mercuric chloride* is highly bacteriocidal, but is toxic and corrosive and is inactivated by organic matter. Organic mercury compounds such as Mercurochrome are less irritating and less toxic than inorganic mercuries. Copper sulfate is often used to destroy green algae in reservoirs or other waters. Zinc chloride is used in mouthwashes, and *zinc oxide* is used in paints as an antifungal.

Surface-Active Agents. **Surface-active agents,** or **surfactants,** decrease the surface tension of a liquid. Soaps and detergents are examples. They emulsify oils and are good degerming agents. Acid-anionic sanitizers are important for cleaning dairy equipment. Deodorant soaps contain compounds such as *triclocarban* that inhibit gram-positive bacteria.

Quaternary Ammonium Compounds. The **quaternary ammonium compounds** (Figure 7.2) are most effective against gram-positive bacteria, and less so against gram-negative bacteria. These *cationic* detergents have good fungicidal, amoebicidal, and virucidal (enveloped virus) activity, but they are not sporicidal. They are colorless, odorless, tasteless, nontoxic, and stable, but they are inactivated by organic matter, soaps, detergents, and surfaces such as gauze. They may even support the growth of *Pseudomonas* bacteria. They act by disrupting the plasma membranes and by denaturing enzymes. Widely used examples of quats are *benzalkonium chloride* (Zephiran®) and *cetylpyridinium chloride* (Cepacol®).

Organic Acids and Derivatives. Sorbic acid (potassium sorbate) inhibits mold spoilage in foods such as cheese. Benzoic acid (sodium benzoate) is an antifungal used in soft drinks and other acidic foods. Methylparaben and propylparaben, which are derivatives of benzoic acid, work at a neutral pH. They inhibit molds in liquid cosmetics and shampoos. Calcium propionate prevents mold growth in bread. All of these organic acids inhibit enzymatic or metabolic activity; their activity is not related to their acidity.

Aldehydes. Among the most effective antimicrobials are the **aldehydes** such as *formaldehyde*. In the form of an aqueous solution, this gas is called formalin and is used to preserve biological specimens. *Glutaraldehyde* is a less irritating form; in a 2% solution (Cidex®), it is bacteriocidal, tuberculocidal, and virucidal in 10 minutes. It is sporicidal after about 3 to 10 hours of contact. Both glutaraldehyde and formaldehyde are used for embalming.

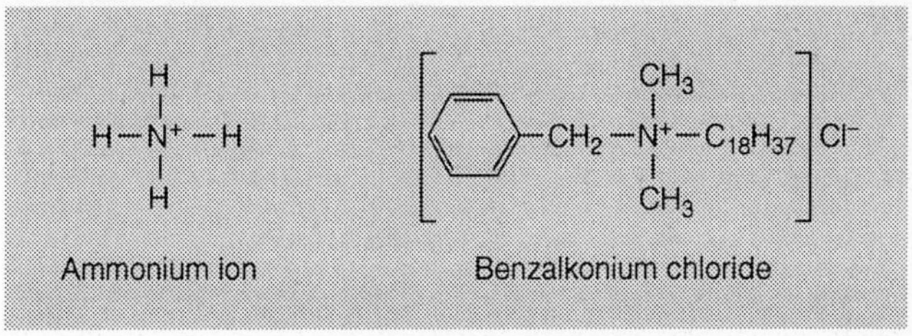

FIGURE 7.2 The ammonium ion and a quaternary ammonium compound, benzalkonium chloride (Zephiran™). Note how other groups replace the hydrogens of the ammonium ion.

Gaseous Chemosterilizers. Ethylene oxide inactivates protein by a mechanism similar to the aldehydes; that is, it combines with organic functional groups of proteins. It is a sterilant that kills all microbes and endospores and has great penetrating power. It is used to sterilize disposable medical supplies and equipment used in hospitals when heat treatment is not practical. Propylene oxide and beta-propiolactone are also important gaseous sterilants. These gases are suspected cancer-causing agents.

Oxidizing Agents. Hydrogen peroxide is a common household antiseptic. It is not a good antiseptic for open wounds, but it is useful in deep anaerobic wounds where the oxygen released by the action of the enzyme catalase in tissue is effective against anaerobes such as *Clostridium*. Benzoyl peroxide is also useful in treatment of deep wounds, although it is better known as the main ingredient in many acne medications. Ozone (O_3) is a highly reactive form of oxygen. It is generated by high-voltage electric discharges and is often used to supplement chlorination in water treatment.

Self-Tests

In the matching section, there is only one answer to each question; however, the lettered options (a, b, c, etc.) may be used more than once or not at all.

I. Matching

1. A suffix meaning "to kill."

2. Destroying *all* forms of microbial life.

3. The absence of pathogens on an object or area.

4. The reduction of microbial populations to safe public health levels.

5. The chemical disinfection of living tissue, such as skin or a mucous membrane.

6. The removal of transient microbes from skin by mechanical cleansing or by an antiseptic.

a. Disinfection

b. Sterilization

c. Antisepsis

d. Asepsis

e. Sanitization

f. Degerming

g. *-cide*

h. *-stat*

II. Matching

1. The lowest temperature required to kill a liquid culture of a certain species of bacteria in 10 minutes at pH 7.

2. The time in minutes required to kill 90% of a bacterial population.

3. Mild heating to destroy particular spoilage organisms or disease organisms in milk or similar products.

4. A test for the effectiveness of a chemical disinfectant.

5. The absence of water, resulting in a condition of dryness.

a. Thermal death time

b. Decimal reduction time

c. Thermal death point

d. Phenol coefficient

e. Pasteurization

f. Desiccation

g. Incineration

III. Matching

1. Ethylene oxide.

2. Sodium hypochlorite.

3. Copper sulfate.

4. Mercurochrome.

5. Benzalkonium chloride.

6. Acid-anionic detergents.

7. Sorbic acid.

8. Benzoyl peroxide.

9. Hexachlorophene.

10. Isopropanol.

11. Ozone.

a. Phenolic

b. Halogen

c. Alcohol

d. Heavy metal

e. Quaternary ammonium compounds

f. Surface-active agents

g. Organic acids

h. Aldehydes

i. Gaseous chemosterilizer

j. Oxidizing agent(s)

IV. Fill-in-the-Blanks

1. Ultraviolet light is an example of _____ radiation.

2. Sunlight owes its germicidal activity mainly to the formation of _____ oxygen.

3. A good example of ionizing radiation is _____ .

4. A _____ of iodine, for example, means an aqueous-alcohol solution.

5. An _____ is a combination of iodine and an organic molecule.

6. Chloramines consist of chlorine and _____ .

7. Ethanol is usually used in a concentration of about _____ .

8. Formaldehyde in an aqueous solution is called _____ .

9. _____ is a highly reactive form of oxygen generated by high-voltage electric discharges.

10. A less irritating form of formaldehyde is _____ .

11. A compound that would only inhibit the growth of a fungus would be a fungi- _____ .

12. Steam _____ allows temperatures above boiling to be reached.

13. Steam under pressure is obtained in retorts, pressure cookers, and _____ .

14. Ultraviolet light causes bonds to form between adjacent thymines in _____ .

15. Pseudomonads are often resistant to antimicrobials because of the characteristics of their cell wall openings called _____ .

16. *Staphylococcus aureus, Pseudomonas aeruginosa*, and *Salmonella typhi* are tested under standard conditions for susceptibility to a test disinfectant in the _____ test.

17. The most effective wavelength of ultraviolet light for germicidal activity is about _____ nm.

18. Generally speaking, the group of organisms that is more resistant to osmotic pressure than bacteria is _____ .

19. Chemicals may denature proteins by reacting with their _____ bonds (or bridges).

20. A liquid chemical disinfectant that is sporicidal yet requires 3 to 10 hours of contact is _____ .

21. An antimicrobial found in many deodorant soaps that is mainly effective against gram-positive bacteria is _____ .

V. Critical Thinking

1. Discuss the use of heavy metals as antimicrobial agents. Give an example of an application of the use of heavy metals. Define oligodynamic action.

2. Explain how the type of microbe relates to efforts to control its growth.

3. What physical method of control would be most effective in each of the following situations?

    a. To eliminate endospore-forming pathogens.

    b. To sterilize milk for storage at room temperature.

    c. To sterilize vaccines.

    d. To sterilize microbiological media.

4. Complete the following table by placing an "X" in each of the appropriate boxes.

Chemical agent	Injures cell membrane	Denatures/ inactivates protein	Dissolves lipid	Inactivates enzymes
Ethylene oxide				
Phenolics				
Halogens				
Chlorhexidine				
Alcohols				

5. What chemical agent would be most effective in each of the following situations?

    a. A puncture wound acquired while gardening.

    b. For presurgical scrubbing.

    c. To sterilize packaged bandages.

    d. To prevent the growth of molds in liquid makeup.

6. Compare and contrast sterilization and sanitation.

7. What does "phenol coefficient" express? Does a phenol coefficient greater than or less than 1 indicate that an antiseptic or disinfectant is more effective than phenol under the same test conditions?

8. Discuss the advantages and disadvantages associated with each of the following physical methods of control.

    a. Osmotic pressure

    b. Desiccation

    c. Refrigeration

    d. Filtration

9. Discuss the advantages and disadvantages of UV light as a method to control microbial growth.

10. Define germicide, bacteriostasis, and degerming.

CHAPTER 8

Microbial Genetics

Learning Objectives

After completing this chapter, you should be able to:

- Define genetics, chromosome, gene, genetic code, genotype, and phenotype.
- Describe how DNA serves as genetic information.
- Describe DNA replication.
- Describe protein synthesis, including transcription, RNA processing, and translation.
- Explain the regulation of gene expression in bacteria by induction, repression, and catabolic repression.
- Relate gene regulation to bacterial survival.
- Define mutagen.
- Classify mutations by type, and describe how mutations are prevented or repaired.
- Outline methods of direct and indirect selection of mutants.
- Identify the purpose of and outline the procedure for the Ames test.
- Compare mechanisms of genetic recombination in bacteria.
- Define plasmid and transposon, and discuss their functions.
- Relate mechanisms for genetic change to microbial evolution.
- Discuss how genetic mutation and recombination provide material for natural selection to act on.

Chromosomes are cellular structures made up of genes that carry hereditary information. **Genetics** is the study of how genes carry information, how they are replicated and passed to other generations, and how they affect the characteristics of an organism.

STRUCTURE AND FUNCTION OF THE GENETIC MATERIAL

In Chapter 2 we saw that DNA is composed of repeating **nucleotides** containing the bases adenine (A), thymine (T), cytosine (C), or guanine (G); a deoxyribose sugar; and a phosphate group. Bases occur in specific complementary pairs, the hydrogen bonds from which connect strands of DNA: adenine with thymine, and cytosine with guanine. A **gene** is a segment of DNA that codes for a functional product. The information in DNA can be transcribed into RNA (*transcription*) and this information, in turn, translated into protein (*translation*).

Genotype and Phenotype

The **genotype** is an organism's genetic makeup, the information that codes for all the characteristics and potential properties of the organism. The genotype is its gene collection—its DNA. The **phenotype** refers to an organism's actual expressed properties, such as its ability to perform a chemical reaction. The phenotype is the collection of enzymatic or structural proteins.

DNA and Chromosomes

DNA in chromosomes is in the form of one long double helix. In procaryotes, DNA is not found within a nuclear membrane. The eucaryotic chromosome is complexed with histone proteins (a mixture called **chromatin**), which is not found in procaryotes.

DNA Replication

In DNA *replication* (Figure 8.2 in the text), the two helical strands unravel and separate from each other at a **replication fork** (Figure 8.5 in the text), where the synthesis of new strands begins. The complementary pairing of bases—for example, adenine with thymine—yields a complementary copy of the original DNA. Segments of new nucleotides are joined to form short strands of DNA by **DNA polymerase** enzymes. Short strands of DNA are then joined into continuous DNA by action of **DNA ligase** enzymes. Because each new double-stranded DNA molecule has one original strand and one new strand, the process is called **semiconservative replication.**

RNA and Protein Synthesis

Transcription

In **transcription,** a strand of **messenger RNA (mRNA)** is synthesized from the genetic information in DNA. (Adenine in the DNA dictates the location of uracil, which replaces thymine in mRNA.) If DNA has the base sequence ATGCAT, the mRNA will have UACGUA. The region where RNA polymerase (needed for synthesis) binds to DNA and transcription begins is known as the **promoter site.** The **terminator site** is where the RNA polymerase and newly formed mRNA are released from the DNA, signaling the endpoint for transcription of the gene. In eucaryotic organisms, transcription takes place in the nucleus. Eucaryotic genes are composed of **exons,** regions of DNA that are expressed in protein production, and **introns,** regions that are not expressed as protein. When the RNA leaves the nucleus and becomes messenger RNA, many introns are cut out by **ribozymes.** These RNA enzymes are nonprotein enzymes.

Translation

To form proteins from mRNA information **(translation),** one end of the mRNA becomes associated with a **ribosome.** Ribosomes consist of two subunits and contain protein and **ribosomal RNA (rRNA).** In the cytoplasm is a pool of 20 different amino acids. These are activated, in turn, by **transfer RNA (tRNA);** there is a different tRNA type for each amino acid type. The tRNA-amino acid attachment is made using an *amino acid activating enzyme* and energy from ATP. Also on each tRNA is an **anticodon,** a sequence of three bases (such as UAC) that matches a set of three bases (such as AUG) on mRNA called a **codon.** A group of codons is a *reading frame.* The tRNA–amino acid unit is brought into position where mRNA attaches to the ribosome. Proteins are synthesized in the $5' \rightarrow 3'$ direction (Figure 8.12 in the text). The ribosome moves along the mRNA strand, and amino acids are joined into a protein strand in which the amino acid sequence is dictated by the codon sequence in the mRNA. The amino acids are joined by **peptide bonds,** and tRNA is recycled. Several ribosomes may be attached at one time to a strand of mRNA. An mRNA strand with several attached ribosomes is called a **polyribosome** (or *polysome*).

The Genetic Code

Because there are 64 possible codons and only 20 amino acids, most amino acids are signaled by several codons (see Figure 8.12 in the text). This is called **degeneracy** of the code. There are 61 **sense codons** (each coding for an amino acid in the synthesized protein) and 3 **nonsense codons** (also called *stop* codons and coding for termination of synthesis of a protein). The start of protein synthesis is signaled by the **initiator codon.**

REGULATION OF GENE EXPRESSION IN BACTERIA

The cell conserves energy by making only those proteins needed at the time. If a gene produces a product at a fixed rate, it is *constitutive.*

Repression, Induction, and Attenuation

Repression and Induction

An **inducer** is a substance (substrate) whose presence results in the formation, or increase in the amount, of an enzyme. Such enzymes are called **inducible enzymes;** this genetically controlled response is termed **enzyme induction.** (Lactase production in response to lactose is an example.) Genetic regulation that decreases enzyme synthesis is **enzyme repression.** Repression occurs if cells are exposed to an overabundance of a particular end-product of a metabolic pathway.

Operon Model of Gene Expression

Protein synthesis in bacteria is controlled by a system called the **operon model.** For example, three enzymes are involved in uptake and utilization of the sugar lactose in *E. coli.* The genes for these enzymes, **structural genes,** are close together on the bacterial chromosome. There is also an **operator site** next to these structural genes, and more remotely located is a gene that codes for a **repressor** protein. The operator and promoter sites plus the structural genes are the **operon.**

When lactose is absent, the repressor protein prevents the operator from making lactose-utilizing enzymes. A small molecule, the **corepressor,** is required to bind repressor to operator gene. When lactose is present, some diffuses into the cells and binds with the repressor protein so it cannot bind to the operator site. The operator then induces the structural genes to produce enzymes to utilize lactose, an example of an **inducible enzyme** (Figure 8.13 in the text). Many genes are not regulated in this manner but are **constitutive,** and usually represent functions needed for major life processes. An example of constitutive enzymes are those for utilization of glucose. Cells prefer glucose to lactose, but if the level of glucose is too low, the cell responds with a series of events that cause formation of the lactose operon, enabling the cell to grow on lactose. This phenomenon is termed **catabolic repression** or the **glucose effect.**

In response to stress, bacteria can turn on or off sets of genes—called **global regulation.** Temperature increases and nutrient starvation are examples.

MUTATION: CHANGE IN THE GENETIC MATERIAL

A **mutation** is a change in the base sequence of DNA.

Types of Mutations

The most common mutation is a **base substitution,** or **point mutation,** in which a single base in DNA is replaced with a different one. Such a substitution is likely to result in the incorporation of an incorrect amino acid in the synthesized protein, a result known as a **missense mutation.** Such an error may create a stop codon, which stops protein synthesis before completion, resulting in a **nonsense mutation.** Deletion or addition of base pairs results in a **frameshift mutation.** In this mutation, there is a shift in the "translational reading frame" (the three-by-three grouping of nucleotides), and a long stretch of missense and an inactive protein product result. **Spontaneous mutations** occur without the known intervention of mutation-causing agents. Many chemicals and radiation bring about mutations; these are called **mutagens.**

Chemical Mutagens

Nitrous acid is a **base pair mutagen.** It causes adenine to pair with cytosine instead of thymine. Other mutagens are **base analogs,** which are structurally similar to bases and are incorporated into DNA by error. Examples are 2-aminopurine and 5-bromouracil, which are analogs of adenine and thymine. Some antiviral drugs are base analogs. Examples of **frameshift mutagens,** many of which are carcinogens, are benzpyrene (found in smoke and soot), aflatoxin (a mold toxin), and acridine dyes.

Radiation

Ionizing radiation, such as X rays and gamma rays, damage DNA and are mutagens. Ultraviolet light (an example of **nonionizing radiation**) is another mutagen that affects DNA. Certain photoreactivating enzymes can repair ultraviolet damage in a process stimulated by visible light. Dark excision repair uses enzymes to cut out distorted DNA and synthesize replacements.

Frequency of Mutation

The **mutation rate** is the probability of a gene mutation each time a cell divides. DNA replication is very faithful, and only about once in 1 billion base pair replications does an error occur. Mutagens increase the rate of such errors 10 to 1000 times.

Identifying Mutants

Mutants, which occur at low rates, can be identified more easily with bacteria because bacteria produce very large populations very quickly. **Positive (direct) selection** is illustrated by plating out bacteria on a medium containing penicillin. Survivors, which are resistant mutants, can be isolated. Nutritional mutants called **auxotrophs** are unable to synthesize a nutritional requirement such as an amino acid—an activity the parent type is capable of doing. Colonies growing on a master plate containing a complete medium can be transferred by **replica plating** (Figure 8.20 in the text). This is an example of **negative (indirect) selection.** A sterile velvet pad is pressed onto the master plate, and the colonies are transferred simultaneously to a **minimal medium,** which lacks essential nutrients such as the required amino acid. An auxotrophic mutant will fail to appear on the minimal medium.

Identifying Chemical Carcinogens

The **Ames test** (Figure 8.21 in the text) is based on the ability of a mutated cell to mutate again and to revert to its original form. An auxotroph of *Salmonella,* which has lost the ability to synthesize the amino acid histidine, is plated out on minimal medium without histidine. The test chemical (together with a rich source of activation enzymes found in rat liver extract) is placed on this plate also. Mutations of the *Salmonella* to the normal histidine-synthesizing form are indicated by colonies growing near the test chemical. High mutation rates are characteristic of the effect of carcinogens.

GENETIC TRANSFER AND RECOMBINATION

Genetic recombination is the rearrangement of genes to form new combinations. If two chromosomes break and are rejoined in such a way that some of the genes are reshuffled between the two chromosomes, the process is called **crossing over.** The original chromosomes each have been recombined so that they carry a portion of genes from the other chromosome. Recombination is more likely to be beneficial than mutation and can occur in several ways. In all cases, however, the **donor cell** gives a portion of its total DNA to a different **recipient cell.** The recipient, called the **recombinant,** has DNA from the donor added to its own DNA.

Transformation in Bacteria

In **transformation,** "naked" DNA in solution is transferred from one bacterial cell to another. The process occurs naturally among very few genera of bacteria, and it usually occurs when donor and recipient are closely related and in the log phase of growth. To take up the DNA fragment, the recipient cell must be **competent;** that is, its cell wall must be permeable or possess specific receptor sites.

Conjugation in Bacteria

Conjugation requires contact between living cells of opposite mating types. The donor cell, called an F$^+$ cell (corresponding to maleness), has extra DNA pieces called **F factors.** These are a type of plasmid (discussed below), a free genetic element in the cell. When F$^+$ and F$^-$ (corresponding to femaleness) cells are mixed, the F$^+$ cells attach by sex pili to F$^-$ cells. F factors are duplicated by the donor, and the new copy is transferred to the F$^-$ cell, which becomes an F$^+$ cell. The bacterial chromosome of the F$^+$ is not passed, and no recombinants are produced. However, some F$^+$ cells have the F factor integrated into their chromosome and are called **Hfr (high frequency of recombination) cells.** The chromosome of these cells may be transferred along with the integrated F factor if conjugation proceeds without interruption (a copy is retained by the donor). In this case, the recipient cell becomes an Hfr donor cell. Considerable amounts of the donor chromosome can be transferred even if interruption occurs. In this case, the recipient cell may acquire a sizable portion of the donor's genome; but unless conjugation is completed, the F factor will not be transferred.

Transduction in Bacteria

In **generalized transduction,** the **phage** (short for "bacteriophage," a bacterial virus) attaches to the bacterial cell wall and injects DNA into the bacterium. Normally this directs the synthesis of new viruses. Sometimes, however, bits of bacterial chromosome are accidentally incorporated into the viral DNA. When these viruses infect a new host cell, they can incorporate this genetic information from the previous bacterial host into the new bacterial host. In **specialized (restricted) transduction,** only certain bacterial genes are transferred.

Plasmids

Plasmids, circular pieces of DNA that replicate independently from the cell's chromosome, usually carry only genes that are not essential for growth of the cell. **Dissimilation plasmids** code for enzymes to utilize unusual sugars and hydrocarbons. Plasmids cause the synthesis of **bacteriocins,** toxic proteins that kill other bacteria. **Resistance factors (R-factors)** carry genes for antibiotic resistance and other antimicrobial factors. Sometimes R-factors contain two groups of genes: **resistance transfer factor,** coding for plasmid replication, and **r-determinant,** coding for enzymes that inactivate antimicrobials.

Transposons

Transposons (*transposable genetic elements*) are small segments of DNA that can move from one region of the chromosome to another (jumping genes). Such movement is uncommon, about the same as spontaneous mutation rates in bacteria. The simplest transposons are called **insertion sequences;** they code only for an enzyme (transposase) that cuts and ligates DNA in recognition sites. **Recognition sites** are short regions of DNA that the enzyme recognizes as recombination sites between transposon and chromosome. **Complex transposons** may carry other genes, such as those conferring antibiotic resistance. Transposons are a powerful mechanism for movement of genes from one chromosome to another, even between species.

Recombination in Eucaryotes

Eucaryotes differ from procaryotes in that genetic recombination is the result of a sexual reproductive process. Eucaryotic organisms have a characteristic number of chromosomes; humans have 46. This process involves the fusion of the haploid nuclei of two parental cells. The **haploid cells** (containing only one of each type of chromosome) are called **gametes** (male gametes are sperm; female gametes are ova). These produce a **zygote** (called a **diploid cell** because it has two sets of chromosomes), with half the chromosomes from one parent and half from the other. Before reproducing sexually, the eucaryote must reduce its chromosome numbers from diploid to haploid by **meiosis.** Crossing over, by which portions of the chromosome may be exchanged, may occur during this stage.

Self-Tests

In the matching section, there is only one answer to each question; however, the lettered options (a, b, c, etc.) may be used more than once or not at all.

I. Matching

1. Where the RNA polymerase and the newly formed mRNA are released.

2. Enzymes that assemble the nucleotides of DNA into chains.

3. Formation of protein from the genetic information contained in mRNA.

4. Formation of mRNA from the genetic information contained in DNA.

5. Enzymes that bind short strands of DNA together into longer strands.

6. Where transcription begins on mRNA.

7. Nonprotein enzymes.

a. DNA ligases

b. DNA polymerases

c. Transcription

d. Translation

e. Promoter site

f. Terminator site

g. Ribozymes

II. Matching

1. A sequence of three bases coding for the position of an amino acid in the assembly of a protein chain.

2. A cluster of related genes together with the operator and promoter sites on mRNA.

3. A sequence of three bases on tRNA that locates the codon on the mRNA at the ribosome.

4. A sequence of bases that does not code for an amino acid, but that terminates the protein chain.

a. Anticodon

b. Stop codon

c. Codon

d. Operon

e. Initiator codon

III. Matching

1. The actual template upon which the protein chain is assembled.

2. The product of transcription.

3. One of these is specific for each of the 20 amino acids.

4. The original genetic information in a bacterial cell.

a. mRNA

b. tRNA

c. rRNA

d. DNA

e. cDNA

IV. Matching

1. The probability of a gene mutation each time a cell divides.

2. Usually a result of the deletion or addition of a base pair.

3. A mutation caused by a chemical that is structurally similar to nucleotide components such as adenine or thymine.

4. A mutagen that would, for example, make the base adenine pair with cytosine instead of thymine.

a. Missense mutation

b. Frameshift mutation

c. Spontaneous mutation

d. Base pair type of mutagen

e. Base-analog type of mutagen

f. Base-substitution type of mutagen

g. Mutation rate

V. Matching

1. DNA transferred between cells in solution in the suspending medium.

2. Requires contact between living cells of opposite bacteria mating types.

3. Requires a sex pilus.

4. Hfr cells.

5. The method by which plasmids such as F factors are transferred between cells.

a. Conjunction in bacteria

b. Transformation in bacteria

c. Transduction in bacteria

VI. Matching

1. Contain genes coding for enzymes that catabolize unusual sugars or hydrocarbons, for example.

2. Contain genes for synthesis of toxic proteins lethal for other bacteria.

a. Dissimilation plasmids

b. Bacteriocinogenic plasmids

c. Conjugative plasmids

VII. Fill-in-the-Blanks

1. An example of nonionizing radiation is _____ .

2. A nutritional mutant is known as a(n) _____ .

3. Colonies growing on a master plate containing a complete medium can be transferred simultaneously to minimal medium by the _____ technique.

4. A segment of DNA that codes for a functional product is a _____ .

5. The *Salmonella* organism in the Ames test has lost the ability to synthesize the amino acid

    _____ .

6. The site at which the replicating DNA strands separate is called the _____ .

7. Several ribosomes may be attached at one time to an mRNA strand; this structure is called a

    _____ .

8. The fact that there is more than one possible codon for each amino acid is called

    _____ of the code.

9. The organism's entire genetic potential is the _____ .

10. A cell with a cell wall permeable to soluble DNA that has specific receptor sites for it is

    _____ .

11. Bacteria that have the F factor integrated into their chromosome and that tend to transfer F factor and chromosome together are called _____ cells.

12. The haploid cells in a eucaryote are called the _____ .

13. Reduction of chromosome numbers before a eucaryote reproduces sexually is known as

    _____ .

14. Enzymes that are always present in the cytoplasm are called _____ enzymes.

15. In the operon model, the regulator gene codes for a _____ protein.

16. The place on the mRNA at which the repressor binds to prevent transcription of structural genes into protein is known as the _____ site.

17. Some R factors have a set of genes called the r-determinant that codes for resistance, and another set of genes called the _____ that codes for replication and conjugation.

18. A bacterial virus is known, for short, as a _____ .

19. In making a new strand of DNA, where there is an adenine on the original strand there will be _____ on the new strand.

20. In making a strand of mRNA from DNA, where there is an adenine on the DNA there will be _____ on the RNA.

21. Small segments of DNA that can move from one region of the chromosome to another are called _____ .

22. Recognition sites are short regions of DNA that transposase recognizes as recombination sites between _____ and chromosome.

23. Noncoding (for protein) stretches of DNA in eucaryotes are called _____ .

VIII. Label the Art

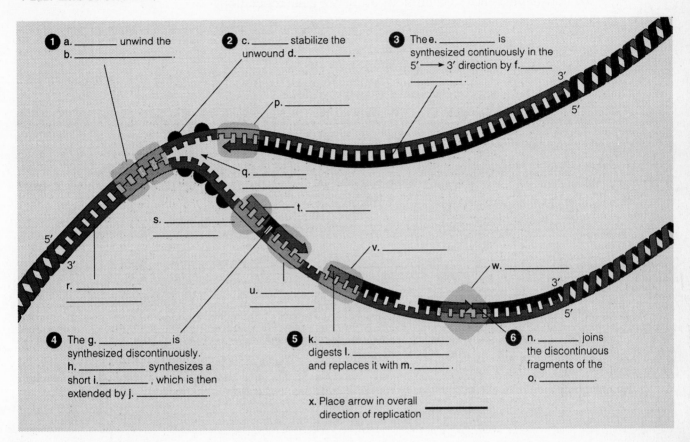

1. a. _____ unwind the
   b. _____ .

2. c. _____ stabilize the unwound d. _____ .

3. The e. _____ is synthesized continuously in the 5′ ⟶ 3′ direction by f._____ _____ .

   3′
   5′

p. _____

q. _____

t. _____

s. _____

5′
3′

r. _____
_____

v. _____

w. _____

3′
5′

4. The g. _____is synthesized discontinuously.
   h. _____ synthesizes a short i._____ , which is then extended by j. _____ .

5. k. _____
   digests l. _____
   and replaces it with m. _____ .

x. Place arrow in overall _____ direction of replication

6. n. _____ joins the discontinuous fragments of the o. _____ .

IX. Critical Thinking

1. What role do hydrogen bonds play in the structure of DNA molecules?

2. Discuss how synthesis of the leading strand and the lagging strand of DNA differs.

3. How do you account for the fact that *E. coli* can replicate faster than two replication forks can separate the strands of DNA in the circular chromosome?

4. Discuss the roles of mRNA and tRNA in the synthesis of protein.

5. Distinguish between base mutations (point mutations) and frameshift mutations. Which type of mutation is more likely to result in termination of translation?

6. How may environmental conditions influence gene regulation?

7. Distinguish between genotype and phenotype.

8. Discuss the mutations caused by ionizing and nonionizing radiation. By what mechanism can bacteria repair the damage caused by radiation?

9. What are "jumping genes," and what role do they play in the evolution of bacteria?

10. Explain the clinical significance of using antibiotics as a food supplement in animal feeds.

CHAPTER 9

Recombinant DNA and Biotechnology

Learning Objectives

After completing this chapter, you should be able to:

- Compare and contrast genetic engineering, recombinant DNA, and biotechnology.
- Identify the roles of a clone and a vector in genetic engineering.
- Define restriction enzymes, and outline how they are used to make recombinant DNA.
- Describe the use of plasmid and viral vectors.
- Describe five ways of getting DNA into a cell.
- Describe how a gene library is made.
- Differentiate between cDNA and synthetic DNA.
- Explain how each of the following is used to locate a clone: antibiotic-resistance genes, DNA probes, antibodies.
- List one advantage of engineering each of the following: *E. coli, Saccharomyces cerevisiae,* vaccinia virus, mammalian cells.
- List at least five applications of genetic engineering.
- Diagram the Southern blotting procedure and provide an example of its use.
- Diagram DNA fingerprinting and provide an example of its use.
- Outline the PCR and provide an example of its use.
- Outline genetic engineering with *Agrobacterium.*
- List the advantages of and problems associated with the use of genetic engineering techniques.
- Discuss some possible results of sequencing the human genome.

The Advent of Recombinant-DNA Technology

Recall from Chapter 8 that recombination, the reshuffling of genes between two DNA molecules, forming **recombinant DNA,** occurs naturally in microorganisms. It is also possible to artificially manipulate DNA to combine genes from two different sources, even from vertebrates to bacteria. Artificial gene manipulation is known as **genetic engineering,** and the term **biotechnology** usually means the industrial use of genetically engineered microorganisms.

Overview of Recombinant-DNA Procedures

The gene of interest is first inserted into **vector DNA.** This DNA molecule used as a carrier must be self-replicating, such as a plasmid or viral genome. This recombinant vector DNA must enter a cell where it can multiply, forming a **clone** of genetically identical cells. The gene itself may be the desired product, or it may be a protein product expressed by the gene.

Restriction Enzymes

A special class of DNA-cutting enzymes, **restriction enzymes,** are the technical basis of genetic engineering. The enzyme recognizes and cuts only one particular sequence of nucleotide bases. Many restriction enzymes make *staggered* cuts in the two DNA strands—cuts that are not directly opposite each other. The stretches of single-stranded DNA at the ends of the DNA fragments are called **sticky ends.** They stick to complementary stretches of single-stranded DNA by base pairing. If two fragments of DNA from different sources have been produced by the same restriction enzymes, the sets of sticky ends can be spliced (recombined) easily. **DNA ligase** then links the DNA pieces.

Cloning a Gene with a Plasmid Vector

Figure 9.1 diagrams the cloning of a DNA fragment by using a **plasmid** for a vector. The host cell can be induced to take up the plasmid vector by chemical treatment. Plasmids that can exist in several species are **shuttle vectors. Bacteriophages** can also be used as vectors.

Methods for Inserting Foreign DNA into Cells

The walls of cells can be enzymatically removed (forming a **protoplast**) and exchange DNA by **protoplast fusion.** Polyethylene glycol may be used to improve efficiency. Transfer of DNA in this manner can also be enhanced by using an electric field to form minute pores in the protoplast membranes—**electroporation.** Foreign DNA can be introduced into plant cells by coating microscopic particles of tungsten with DNA and firing it through the plant cell wall using a "gene gun." DNA can be introduced into a cell through a minute glass micropipette by **microinjection.**

Sources of DNA

Genes from a particular organism are isolated by cutting up the entire genome with restriction enzymes, splicing as many as possible into vectors, and then introducing the genes into bacterial cells. A collection of bacterial clones containing different DNA fragments is called a **gene library.** Another form of gene library is one in which the clones each contain **complementary DNA (cDNA),** which is DNA synthesized by using mRNA as a template.

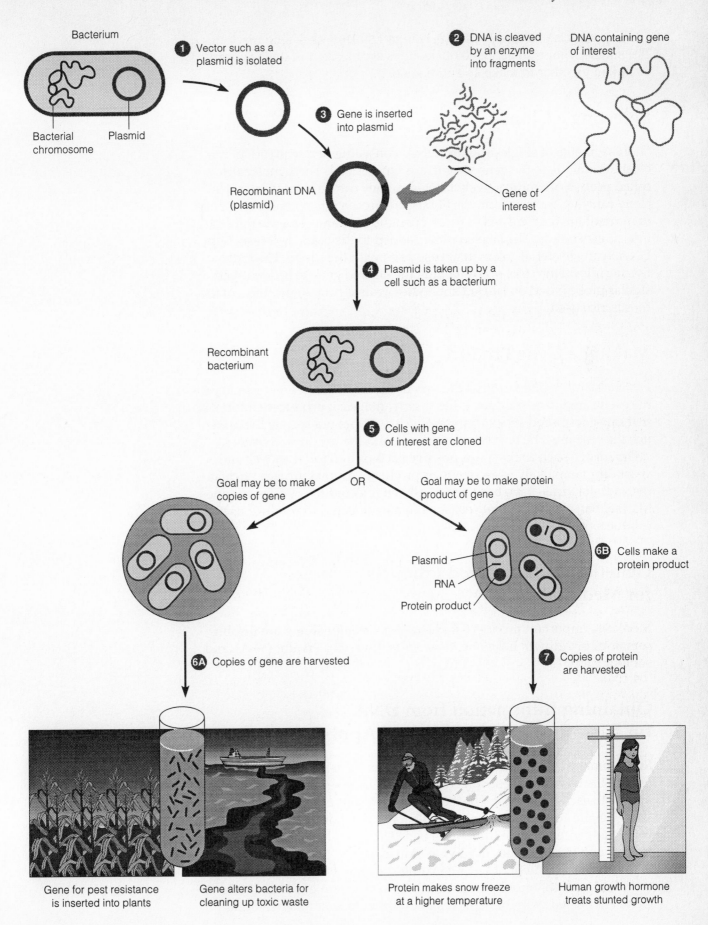

FIGURE 9.1 Typical genetic engineering procedures with examples of applications.

Genes can also be made with the help of DNA-synthesizing machines. The smaller chains of about 40 nucleotides synthesized in this way can be linked together to make an entire gene.

Selecting a Clone

Short segments of single-stranded DNA, consisting of a sequence of nucleotides unique to the gene sought, are synthesized. These molecules, called **probes,** are radioactively labeled so they can be located later. The clone-carrying bacteria from the library are grown into colonies on a plate of nutrient medium treated to break open the cells and separate the DNA into single strands. The labeled probe, added to the plate, will react with DNA in any bacterial colony that base pairs with the probe. The radioactive tag allows the colonies containing the desired gene to be identified. A similar probe based on labeled antibodies against protein products of the cells is also used.

Making a Gene Product

Escherichia coli is often used as the genetically engineered organism to produce a desired gene product. It has disadvantages; it produces endotoxins that cause fever and shock in animals. Also, it does not secrete the product; the cells must be harvested, ruptured, and the product recovered.

Organisms such as the gram-positive bacterium *Bacillus subtilis* and yeasts are more likely to secrete their products. Animal viruses, such as the vaccinia virus, have been genetically engineered to produce vaccines. Mammalian cells and plant cells are often engineered to produce useful products.

Genetically Engineered Products for Medical Therapy

Medically important products made by genetic engineering are insulin, somatotropin growth hormone, tissue-plasminogen activator (t-PA), and subunit vaccines.

Obtaining Information from DNA for Basic Research and Medical Applications

Recombinant-DNA technology can be used to make multiple copies of DNA—the basis of genetic fingerprinting. DNA, like a fingerprint, is unique to each individual. The technology is also the basis of DNA analysis of genetic abnormalities responsible for various diseases, and it contributes to advances in **gene therapy,** in which abnormal genes might be replaced with normal genes in a living individual.

To isolate a fragment of DNA containing a gene, the method of **Southern blotting** is commonly used. DNA fragments from each candidate clone are treated with the same restriction enzyme, and the resulting fragments separated by gel electrophoresis. The bands representing the fragments are then blotted onto a special filter, which is bathed with radioactively tagged probes of DNA complementary to the gene desired. The desired bands can be cut out of the gel and recovered by soaking in solvent. The genes may be studied by **DNA sequencing** to determine the sequence of nucleotides. This process is often highly automated. These procedures can be used for identification procedures called **DNA fingerprinting** of crime scene samples. A recent development in DNA analysis is the **polymerase chain reaction (PCR).** Starting with just one gene-sized piece of DNA, PCR can make billions of copies in only a few hours. The target piece of DNA is heated to separate the DNA strands, which serve as templates for DNA synthesis. **DNA polymerase** enzyme, which forms DNA by linking the nucleotides, is supplied with DNA's four nucleotides and short pieces of primer nucleic acid. Each newly synthesized piece of DNA serves in turn as a template for more new DNA.

Agricultural Applications of Recombinant-DNA Technology

The most elegant method of introducing recombinant DNA into a plant cell is by the **Ti plasmid.** A bacterium that infects plants normally carries this plasmid. The infection causes a tumorlike growth called a crown gall (Ti means tumor inducing). The plasmid also serves as a vehicle for insertion of genetically engineered DNA into a plant. Other applications are to make crop plants resistant to herbicides that then selectively kill weeds, and to improve the ability to fix nitrogen in certain symbiotic bacteria. A bacterium has been engineered to produce a toxin that kills certain plant pathogens that feed on the plant. A genetically engineered product, bovine growth hormone, increases milk production in dairy herds.

Safety Issues and the Ethics of Genetic Engineering

Laboratories engaged in recombinant-DNA research must meet rigorous safety standards to avoid accidentally releasing genetically engineered microbes. The microbes may also be engineered to contain suicide genes that prevent them from surviving outside the laboratory environment. Genetic screening for hereditary diseases and birth defects in the fetus introduces ethical questions not yet resolved.

The Future of Genetic Engineering

The **Human Genome Project,** currently under way, is aimed at mapping the entire human genome. The result will have a profound effect on biology and medical science.

Self-Tests

In the matching section, there is only one answer to each question; however, the lettered options (a, b, c, etc.) may be used more than once or not at all.

I. Matching

1. DNA-cutting enzymes that often form sticky ends.

2. A self-replicating DNA molecule used as a carrier to transmit a gene from one organism to another.

3. An enzyme that links short pieces of DNA into longer pieces.

4. An enzyme that links nucleotides to form DNA.

a. Restriction enzymes

b. DNA ligase

c. DNA polymerase

d. Vector

II. Fill-in-the-Blanks

1. The reshuffling of genes between two DNA molecules forms _____ DNA.

2. DNA synthesized by using mRNA as a template is called _____ DNA.

3. Sticky ends stick to each other by complementary stretches of single-stranded DNA by _____ pairing.

4. The two most common vectors used in genetic engineering are bacteriophage and _____ .

5. A collection of bacterial clones each containing a different DNA fragment is called a _____ .

6. Short segments of single-stranded DNA that recognize the DNA sequence of a gene and labeled so that they can be used to locate the gene are called _____ .

7. To isolate a fragment of DNA containing a gene, DNA fragments of clones are separated by gel electrophoresis. This is an early step in the _____ technique for DNA analysis.

8. The procedure by which billions of copies of a sequence of DNA can be made in a few hours is called the _____ reaction.

9. The most elegant way of introducing recombinant DNA into a plant cell is by the _____ plasmid.

10. Polyethylene glycol is used to improve the efficiency of DNA exchange between cells by _____ fusion.

III. Label the Art

I.

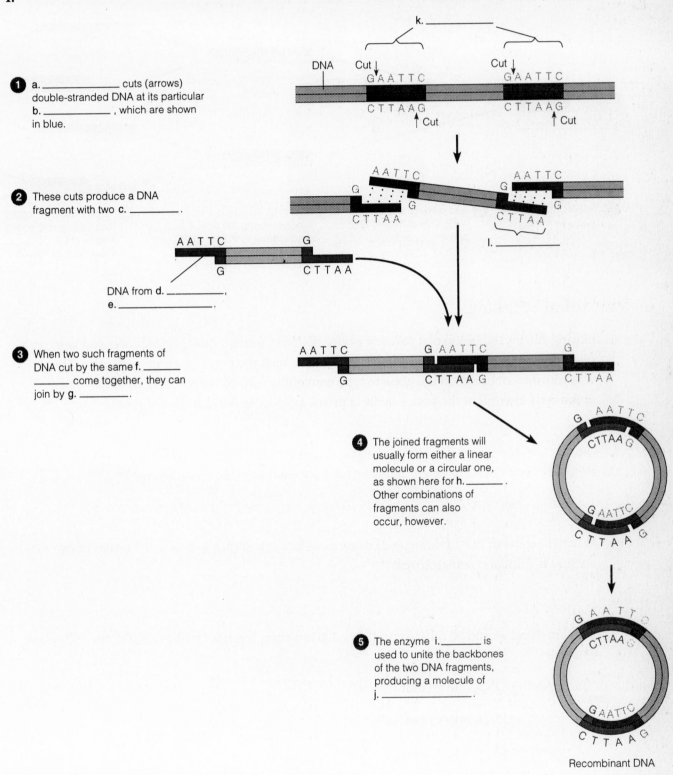

1 a. _____ cuts (arrows) double-stranded DNA at its particular b. _____ , which are shown in blue.

DNA   Cut ↓
G A A T T C
C T T A A G
↑ Cut

Cut ↓
G A A T T C
C T T A A G
↑ Cut

k. _____

2 These cuts produce a DNA fragment with two c. _____ .

A A T T C
G
G
C T T A A

A A T T C
G
G
C T T A A

l. _____

A A T T C
G   C T T A A

DNA from d. _____ ,
e. _____ .

3 When two such fragments of DNA cut by the same f. _____ come together, they can join by g. _____ .

A A T T C   G A A T T C   G
G   C T T A A G   C T T A A

4 The joined fragments will usually form either a linear molecule or a circular one, as shown here for h. _____ . Other combinations of fragments can also occur, however.

G A A T T C
C T T A A G
G A A T T C
C T T A A G

5 The enzyme i. _____ is used to unite the backbones of the two DNA fragments, producing a molecule of j. _____ .

G A A T T C
C T T A A G
G A A T T C
C T T A A G

Recombinant DNA

(continued)

II.   POLYMERASE CHAIN REACTION

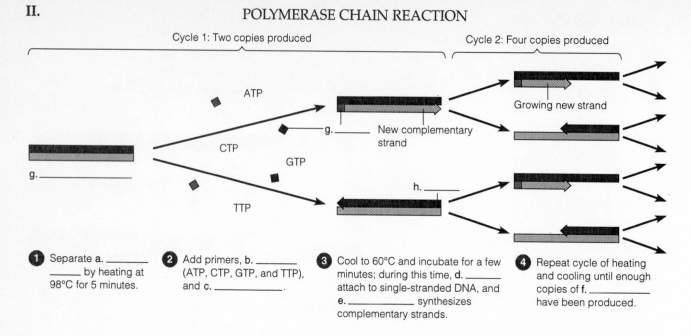

① Separate a. _____ _____ by heating at 98°C for 5 minutes.

② Add primers, b. _____ (ATP, CTP, GTP, and TTP), and c. _____ .

③ Cool to 60°C and incubate for a few minutes; during this time, d. _____ attach to single-stranded DNA, and e. _____ synthesizes complementary strands.

④ Repeat cycle of heating and cooling until enough copies of f. _____ have been produced.

IV. Critical Thinking

1. During the investigation of a robbery, police discover a small quantity of blood and skin on a piece of broken glass from the window through which the perpetrator gained entry. There was insufficient blood on the glass to type conventionally. No usable fingerprints were obtained. How might the police tie their prime suspect to the crime?

2. What are restriction enzymes, and how are they used to make recombinant DNA?

3. What are some of the advantages of using genetic engineering to produce human hormones such as insulin and somatotropin?

4. What method would be most appropriate for inserting foreign DNA in each of the following examples?

   a. Inserting DNA into an animal cell.

   b. Inserting DNA into a plant cell.

   c. Inserting DNA into a yeast.

5. What is cDNA? How is cDNA related to the expression of eucaryotic genes by bacterial cells?

6. Discuss how R plasmids are used to identify genetically engineered bacterial clones.

7. You are asked to develop a protocol for the industrial production of a protein from a genetically engineered bacterium. A bacterium of what Gram designation would most likely be the best choice? Why?

8. What are subunit vaccines? Are they safer than avirulent vaccines? Why?

9. Discuss three applications of genetic engineering.

10. What are "suicide genes," and what is their role in genetic engineering?

CHAPTER 10

Classification of Microorganisms

Learning Objectives

After completing this chapter, you should be able to:

- Define taxonomy.
- Discuss the limitations of a two-kingdom classification system.
- List the characteristics of the Kingdom Procaryotae that differentiate it from other kingdoms.
- List the major characteristics used to differentiate among kingdoms in the five-kingdom system and in the proposed three-kingdom system.
- Explain why scientific names are used.
- List the major taxa.
- Define bacterial species.
- Compare and contrast classification and identification.
- Explain the purpose of *Bergey's Manual*.
- Describe how staining and biochemical tests are used to identify bacteria.
- Explain how serological tests and phage typing can be used to identify an unknown bacterium.
- Describe how a new bacterium can be classified by the following molecular methods: amino acid sequencing, protein profiles, base composition, DNA fingerprinting, PCR, and hybridization.
- Differentiate between Western blotting and Southern blotting.
- Provide the rationale for the use of numerical taxonomy.

The science of classification, especially of living forms, is called **taxonomy**.

The Study of Phylogenetic Relationships

Organisms are arranged into taxonomic categories called taxa, which reflect degrees of relatedness among organisms. The hierarchy of taxa shows evolutionary, or **phylogenetic** (from a common ancestor), relationships.

A Five-Kingdom System of Classification

In 1969, H. R. Whittaker proposed the **five-kingdom system** of biological classification. The main division in the system is made between procaryotic and eucaryotic cells. Included with the procaryotic organisms (called **Procaryotae,** or **Monera**) are the bacteria and cyanobacteria. In the classification of eucaryotic cells, the simpler organisms, mostly unicellular, are grouped as **Protista** and include funguslike water molds, slime molds, protozoa, and the more primitive algae. Branching from the Protista are three kingdoms of more complex organisms distinguished primarily by their nutritional mode. **Fungi** are eucaryotic organisms that include yeasts, molds, and mushrooms. These absorb dissolved organic matter. **Plantae** are multicellular algae, mosses, ferns, conifers, and flowering plants. These obtain energy by photosynthesis. **Animalia** are multicellular animals, including sponges, worms, insects, and vertebrate animals. These obtain nutrition by ingestion of organic matter, as by mouth.

A Three-Kingdom System

In 1978 Carl Woese proposed a three-kingdom system that is based on modern techniques in molecular biology and biochemistry. Primarily, the distinction is that there are two types of procaryotic cells, the **eubacteria** and the **archaeobacteria.** The third kingdom consists of all the eucaryotes.

Archaeobacteria differ from eubacteria in that their cell walls never contain peptidoglycan. They live in extreme environments and carry out unusual metabolic processes; examples are the anaerobic methanogens, extreme halophiles requiring high concentrations of salt, and thermo-acidophiles found in hot, acidic environments.

Classification of Organisms

Scientific Nomenclature

The system of scientific nomenclature used today was developed in the eighteenth century by Carolus Linnaeus. All organisms have two names: the **genus name** and the **specific epithet.** Both names are underlined or italicized. This system of two names for an organism is called **binomial nomenclature.**

The Taxonomic Hierarchy

For eucaryotic organisms, a **species** is a group of closely related organisms that breed among themselves and not with other species. A **genus** (plural, *genera*) consists of different species related by descent. Related genera make up a **family.** A group of similar families constitutes an **order,** and a group of similar orders makes up a **class.** Related classes in turn constitute a **division.** (In zoology, **phylum** is a term comparable to division.) All phyla or divisions that are related to each other make up a **kingdom.**

A Phylogenetic Hierarchy

There is little fossil record to determine evolutionary relationships among bacteria. However, DNA and RNA hybridization and RNA sequencing studies help in such understanding.

Classification of Bacteria

The term *species,* as applied to higher organisms, does not apply well to bacteria. Bacterial morphological traits are limited, and sexuality involves unilateral transfer of genetic material rather than breeding, which is the fusion of egg and sperm. A **bacterial species** is, accordingly, defined as a population of cells with similar characteristics. Sometimes a pure culture of the same species contains distinguishable groups within the species called **strains.** Presumably these are derived from a single cell and are identified by numbers, letters, or names following the species name.

Bergey's Manual of Systematic Bacteriology is the standard reference for bacterial taxonomic classification. In *Bergey's Manual,* bacteria are grouped according to shared properties such as morphology, staining characteristics, nutrition, and metabolism.

Classification of Viruses

Viruses (Chapter 13) are not classified under any of the kingdoms in which bacteria are grouped. However, a **viral species** is a population of viruses with similar characteristics (morphology, genes, and enzymes) occupying a particular ecological niche.

Criteria for Classification and Identification of Microorganisms

Morphological Characteristics

Literally hundreds of bacterial species appear in the shape of small rods or small cocci. Nonetheless, these characteristics are still of some use in classifying bacteria. Sometimes the presence of endospores or flagella can be helpful. One of the first steps in identification is doing a Gram stain. Other stains, such as the acid-fast stain, are sometimes useful for several bacterial groups.

Biochemical Tests

Enzymatic activity is widely used to differentiate among bacteria. Tests that determine the ability to ferment an assortment of selected carbohydrates are particularly common. Bacteria also may be subjected to other biochemical tests. The use of selective or differential media (Chapter 6) has been discussed for use in bacterial isolation or identification.

Serology

Microorganisms are antigenic; that is, they are capable of stimulating antibody formation when injected into animals. The immune system of an animal injected with a bacterial species will produce antibodies highly specific for that species of bacteria. Solutions of such antibodies, called **antisera** (singular, **antiserum**), are commercially available for identification of many medically important bacteria. In a demonstration of such methods, an unknown bacterium is placed in a drop of salt solution on a microscope slide and mixed with a drop of known antiserum. In such a **slide agglutination test,** the bacteria will agglutinate, or clump together, when mixed with antibodies produced against that strain or species. **Enzyme-linked immunosorbent assays (ELISAs)** are widely used; they are fast and can be computer-read. The **Western blot** test identifies antigens in a patient's serum.

Phage Typing

Phages (Chapter 8), like antibodies, are highly specific for species or strains within a species. By means of stocks of different phages, this characteristic can be used to identify bacteria or even to trace epidemics. **Phage typing** is commonly used in outbreaks of illness caused by staphylococci.

Amino Acid Sequencing

The amino acid sequence of a protein is a direct reflection of the base sequence of the encoding gene. Thus, by comparing the amino acid sequence of proteins from two different organisms, it is possible to determine the degree of their relatedness. The more similar the proteins, the more closely related are the organisms.

Protein Analysis

Protein profiles are obtained by using **polyacrylamide gel electrophoresis (PAGE).** Disintegrated cell proteins are distributed by an electric current (electrophoresis) in a gel. The rate of movement, which is visible by staining, "fingerprints" the cells' proteins and can show relationships between bacterial species.

Base Composition of Nucleic Acids

A technique that may be able to suggest evolutionary relationships among bacteria is the determination of the base composition of the DNA, the **G + C ratio.** The number of guanine (G) and cytosine (C) base pairs can be determined as a percentage. By inference, this will give the percentage of adenine (A) and thymine (T) as well. If there is a difference of more than 10% in the percentage of G + C pairs, the organisms are probably unrelated. On the other hand, two organisms with the same percentage of G + C need not be closely related; other supporting data are needed.

Restriction enzymes (Chapter 9) can cut bacterial DNA into certain base sequences. The restriction fragments produced can be separated by electrophoresis and the DNA patterns of different organisms compared—**DNA fingerprints.** The amount of DNA can be increased to improve sensitivity, especially if the microbe cannot be cultivated, by the **polymerase chain reaction** (Chapter 9).

Nucleic Acid Hybridization

If a double-stranded molecule of DNA is heated, the complementary strands will separate as the hydrogen bonds connecting the bases break.

However, because the strands are complementary, they will reunite if allowed to incubate together. If the separated strands of two different organisms are incubated together, the closer the relationship, the greater the amount of DNA that will join together by complementary base-pairing. This technique is known as **nucleic acid hybridization** and assumes that if two species are similar or related, a major portion of their nucleic acid sequences also will be similar. **DNA probes,** based on these techniques, are being developed for rapid identification of bacteria.

Flow Cytometry

Some bacteria can be identified without culturing. A moving fluid containing the bacteria is forced through a small opening (Figure 18.14 in the text). A laser beam scattered by the bacteria gives information about the bacteria, which may also be fluorescently labeled.

Numerical Taxonomy

In **numerical taxonomy,** which is used to help discover relationships among organisms, many microbial characteristics are listed, and the presence or absence of each characteristic is scored for each organism. (A weakness of the method is that all characteristics are given equal weight.) When different bacteria are matched against each other (with the aid of a computer), the greater the number of characteristics shared, the closer the taxonomic relationship is assumed to be. A match of 90% or more usually indicates a single taxonomic unit or species.

Self-Tests

In the matching section, there is only one answer to each question; however, the lettered options (a, b, c, etc.) may be used more than once or not at all.

I. Matching

1. Bacteria.

2. Elephants.

3. Multicellular algae.

4. Mushrooms.

5. Yeasts.

6. Cyanobacteria.

a. Procaryotae

b. Protista

c. Animalia

d. Plantae

e. Fungi

II. Matching

1. A distinguishable group within a species of bacteria.

2. Made up of all phyla or divisions that are related to each other.

3. In bacteria, all cells with similar characteristics in common.

4. In eucaryotes, closely related organisms that are capable of breeding among themselves.

5. A group of similar orders in eucaryotic cells.

a. Strain

b. Species

c. Genus

d. Kingdom

e. Class

f. Order

III. Matching

1. A serological test for bacterial identification.

2. Makes use of bacterial viruses to identify bacteria.

3. The accumulation of a microbe's characteristics, each with equal weight. The more characteristics shared, the closer the taxonomic relationship.

4. Determines the taxonomic relationship of bacteria by allowing complementary strands of DNA from different organisms to reassemble as a complementary pair.

5. Taxonomic relationship of bacteria found by base pairs percentage in DNA.

6. Used in analysis of bacterial protein.

a. Numerical taxonomy

b. Phage typing

c. Nucleic acid hybridization

d. Amino acid sequencing

e. G + C ratio

f. Slide agglutination test

g. Polyacrylamide gel electrophoresis

IV. Fill-in-the-Blanks

1. The science of classification, especially of living forms, is called _____ .

2. The system we use for naming biological organisms with two names is called _____ .

3. The taxonomic categories into which organisms are arranged and that reflect degrees of relatedness among them are called _____ .

4. The taxonomic classification scheme for bacteria may be found in the book called _____ *Manual.*

5. _____ typing is a technique commonly used to trace the epidemiology of outbreaks of illness caused by staphylococci.

6. The hierarchy of taxa shows evolutionary, or _____ (from a common ancestor), relationships.

7. Solutions of antibodies against specific bacteria are called _____ .

8. A technique that may be able to suggest evolutionary relationships among bacteria is the determination of the base composition of the DNA, the _____ ratio.

9. The plural of genus is _____ .

10. The cell walls of archaeobacteria differ from those of eubacteria by a lack of _____ .

V. Label the Art

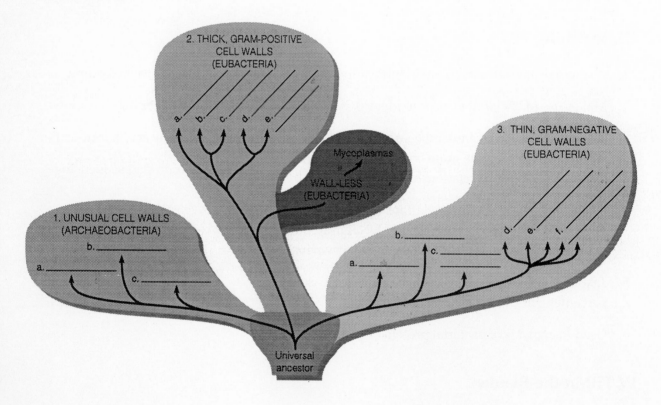

VI. Critical Thinking

1. List and discuss three applications of taxonomy.

2. What is the endosymbiont theory? What evidence exists to support the endosymbiont theory?

3. Distinguish among the five kingdoms of life on the basis of nutritional patterns.

4. How do archaeobacteria differ from eubacteria?

5. Distinguish between a eucaryotic species, a bacterial species, and a strain.

6. Discuss the differences in organization of bacteria between Bergey's determinative and systematic manuals.

7. Define the following: slide agglutination test, enzyme-linked immunosorbent assay, and Western blot.

8. What are three applications of DNA fingerprinting?

9. How are DNA probes used to identify bacteria?

10. Why aren't viruses included in the five kingdoms of life? Discuss the two hypotheses that attempt to explain the origin of viruses.

CHAPTER 11

Bacteria

Learning Objectives

After completing this chapter, you should be able to:

- List at least six characteristics used to classify and identify bacteria according to *Bergey's Manual of Systematic Bacteriology.*
- List two major characteristics of each section of *Bergey's Manual* described in this chapter.
- Identify the sections in *Bergey's Manual* that contain species of medical importance.

Bacterial Groups

Spirochetes

Spirochetes have a helical shape and are motile by means of **axial filaments.** One end of each axial filament is attached near a pole of the cell. Axial filaments are wound around the body of the cell in the space between an outer sheath and the body of the cell, and the cell moves by stretching and relaxing the axial filament. *Treponema pallidum,* the cause of syphilis; the genus *Borrelia,* which causes relapsing fever and Lyme disease; and *Leptospira* species, which cause leptospirosis—are all spirochetes.

Aerobic/Microaerophilic, Motile, Helical/Vibrioid, Gram-Negative Bacteria

Some helical bacteria are not included with the spirochetes because they lack an axial filament. They are motile instead by means of flagella, which may occur singly or in tufts at either or both poles of the cell. Many of these bacteria are harmless aquatic organisms such as *Spirillum volutans,* which are probably adapted to the low concentrations of organic matter in their habitat. *Azospirillum* bacteria fix nitrogen in close association with the roots of grasses. Among pathogens in this group, *Campylobacter fetus* causes abortion in domestic animals, *Campylobacter jejuni* causes foodborne enteritis, and *Helicobacter pylori* causes stomach ulcers. **Vibrioid** is the term applied to helical bacteria that do not have a complete turn, such as *Bdellovibrio* spp., which preys on other bacteria, reproducing within them.

Gram-Negative Aerobic Rods and Cocci

The genus *Pseudomonas* is characterized by a polar flagellum. Many species excrete extracellular, water-soluble pigments. *Pseudomonas aeruginosa* produces a blue-green pigment and can infect the urinary tract, burns, and wounds. Pseudomonads are common in soil and can often grow at refrigerator temperatures. They are able to decompose chemicals such as pesticides in soil, and their ability to grow in antiseptic solutions and whirlpool baths as well as their resistance to antibiotics make them troublesome. Many pseudomonads substitute nitrate for oxygen during anaerobic respiration, depleting nitrate fertilizers of nitrogen, which escapes in the form of nitrogen gas.

Legionella is the cause of legionellosis. *Azotobacter* and *Azomonas* are nitrogen-fixing soil organisms. *Rhizobium* and *Bradyrhizobium* form a nitrogen-fixing symbiotic relationship with legumes. *Acetobacter* and *Gluconobacter* convert ethanol into vinegar. *Zoogloea* forms flocculant slimy masses essential to the operation of activated-sludge sewage treatment systems. *Neisseria gonorrhoeae* causes gonorrhea, and *Neisseria meningitidis* causes meningococcal meningitis. *Moraxella lucanata* causes conjunctivitis. *Brucella* causes brucellosis and has the unusual ability to survive phagocytosis. *Bordetella pertussis* causes whooping cough. *Francisella tularensis* causes tularemia. *Agrobacterium tumefaciens* is a plant pathogen causing crown gall; a plasmid from this organism is used in the genetic engineering of plants.

Facultatively Anaerobic Gram-Negative Rods

The three families in this group are discussed below.

Enterobacteriaceae (Enterics).　The **enterics** are clinically important, and there are many biochemical tests used (such as the IMViC tests) to distinguish between them. Many enterics have fimbriae for attachment to mucous membranes and sex pili for conjugation. Many also produce proteins called **bacteriocins,** which kill related species of bacteria.

Escherichia coli is one of the most common inhabitants of the human intestinal tract and a familiar laboratory bacterium. It is used as an indicator organism for fecal pollution. It can cause urinary tract infections. Occasionally enterotoxins produced by *E. coli* cause traveler's diarrhea and even very serious foodborne disease.

Almost all members of the genus *Salmonella* are pathogenic. Typhoid fever is caused by *Salmonella typhi*. Taxonomically, the salmonellae are divided into hundreds of **serovars** (serotypes) by serological means; serovars are further differentiated by biochemical, physiological properties into **biovars** (biotypes). No individual species are recognized in spite of their specieslike name. Many are known by antigenic formulas, according to the Kauffmann–White scheme. For example, *Salmonella typhimurium* is represented as $\underline{1}$, 4 (5), 12, i, 1, 2.

Species of *Shigella* cause bacillary dysentery or shigellosis; *Klebsiella pneumoniae* causes a form of pneumonia; and *Serratia marcescens* produces a red pigment and causes infections of the urinary and respiratory tracts in hospital patients. Members of the genus *Proteus* are actively motile and may cause infections of urinary tracts and wounds, as well as infant diarrhea. *Yersinia pestis* causes bubonic plague. *Erwinia* species are primarily plant pathogens, producing enzymes that dissolve the pectin between individual plant cells (soft-rots). *Enterobacter cloacae* and *Enterobacter aerogenes* cause urinary tract infections.

Vibrionaceae.   Morphologically, bacteria in the genus *Vibrio* are characterized as slightly curved rods. Important pathogens are *Vibrio cholerae,* the cause of cholera, and *Vibrio parahaemolyticus,* the cause of a form of gastroenteritis transmitted mostly by shellfish.

Pasteurellaceae.   The organisms of the genus *Pasteurella* are mostly pathogens of domestic animals. *Pasteurella multocida* can be transmitted to humans by dog and cat bites. *Hemophilus influenzae* is a common cause of meningitis, otitis media, epiglottitis, and a number of other important diseases. *Hemophilus* bacteria are cultured on media enriched by hemoglobin or culture media containing X and V factors. *Gardnerella vaginalis* causes vaginitis.

Anaerobic, Gram-Negative, Straight, Curved, and Helical Rods

Members of the genus *Bacteroides* live in the human intestinal tract. Some are found in the oral cavity, genital tract, and upper respiratory tract. They do not form endospores. *Fusobacterium* are long and slender with pointed ends. In humans they can cause dental abscesses.

Dissimilatory Sulfate-Reducing or Sulfur-Reducing Bacteria

These organisms are not medically important. They are obligately anaerobic bacteria that use oxidized forms of sulfur as electron acceptors, which produces hydrogen sulfide. They are found in anaerobic sediments and human and animal intestinal tracts. *Desulfovibrio* is the best known genus.

Anaerobic Gram-Negative Cocci

Bacteria of the genus *Veillonella* are part of the normal mouth microbiota and are components of dental plaque.

Rickettsias and Chlamydias

Both of these are obligate, intracellular parasites, as are viruses, but in morphological and biochemical aspects they resemble, and are classified as, bacteria. They are usually pathogenic.

  Rickettsias are gram-negative. They are often transmitted to humans by insects or ticks. *Coxiella burnetii* causes Q fever. Other rickettsial diseases are epidemic typhus caused by *Rickettsia prowazekii,* endemic murine typhus caused by *Rickettsia typhi,* and Rocky Mountain spotted fever caused by *Rickettsia rickettsii.* They are usually cultivated in the yolk sac of chicken embryos.

Chlamydia are gram-negative, coccoid bacteria. They do not require arthropods (such as insects) for transmission. The infectious form, the **elementary body,** attaches to a host cell and is phagocytized and housed in a cell vacuole. Within the host cell the elementary body becomes a larger, less infective **reticulate body** that divides successively. Eventually these condense into infectious elementary bodies that are released to infect surrounding host cells. *Chlamydia trachomatis* causes trachoma, as well as the venereal diseases nongonococcal urethritis and lymphogranuloma venereum. *Chlamydia psittaci* causes psittacosis (ornithosis). Chlamydia are cultivated in laboratory animals, in cell cultures, or in the yolk sac of chicken embryos. *Chlamydia pneumoniae* causes a mild form of pneumonia.

Mycoplasmas

Mycoplasmas are bacteria that do not form cell walls. *Mycoplasma pneumoniae* causes primary atypical pneumonia. *Spiroplasma* are serious plant pathogens. *Ureaplasma* may cause urinary tract infections. *Thermoplasma,* which are also grouped with archaeobacteria, are sometimes found in hot water systems. To grow mycoplasmas on artificial media, sterols must be provided.

Gram-Positive Cocci

Staphylococcus bacteria occur in grapelike clusters. The most important species is *Staphylococcus aureus.* Staphylococci are able to survive and grow at high osmotic pressures. *S. aureus* produces a yellow pigment and many toxins. Among the toxins is an *enterotoxin* that causes food poisoning. Rapid development of antibiotic resistance is another troublesome characteristic.

   Streptococcus bacteria cause a great variety of diseases. They do not use oxygen, but are mostly aerotolerant. **Alpha-hemolytic** species form a greenish zone around a colony on blood agar. **Beta-hemolytic** species form a clear zone of hemolysis. **Nonhemolytic** (gamma-hemolytic) streptococci are not hemolytic. The streptococci produce substances that destroy phagocytic cells, enzymes that digest the host's connective tissue and enzymes that lyse fibrin.

Endospore-Forming Gram-Positive Rods and Cocci

Bacillus species are aerobic or facultative anaerobes. *Bacillus anthracis* causes anthrax. *Bacillus thuringiensis* is an insect pathogen. *Clostridium* species are mostly obligate anaerobes. *Clostridium tetani* causes tetanus, *Clostridium botulinum* causes botulism, and *Clostridium perfringens* causes gas gangrene. Cocci of the genus *Sporosarcina* are soil bacteria.

Regular Nonsporing Gram-Positive Rods

The genus *Lactobacillus* represents a group of aerotolerant bacteria that produce lactic acid from simple carbohydrates and grow in acidic environments. They are found in the human vagina and oral cavity. Industrially, they produce products such as sauerkraut and yogurt. The pathogen

Listeria monocytogenes survives within phagocytic cells, can grow at refrigerator temperatures, and can cause serious damage to a fetus.

Irregular Nonsporing Gram-Positive Rods

Corynebacterium diphtheriae causes diphtheria. (Coryne is from the Greek for club, which reflects the irregular morphology of the genus.) *Propionibacterium acnes* is an anaerobic skin organism responsible for acne. *Actinomyces* spp. are filamentous, anaerobic bacteria that can fragment into coryneform cells. *Actinomyces israelii* causes actinomycosis.

Mycobacteria

The prefix "myco" suggests "fungus," so named because of the occasional filamentous growth of this rod-shaped genus. Most species are acid-fast. *Mycobacterium tuberculosis*, the cause of tuberculosis, and *Mycobacterium leprae,* the cause of leprosy, are important pathogens in this genus. A number of nonpathogenic soil organisms are also found among the mycobacteria.

Nocardioforms

Members of the genus *Nocardia* are filamentous bacteria that resemble the Actinomyces morphologically, but are aerobic. They are common in soil. *Nocardia asteroides* is a cause of pulmonary nocardiosis, a chronic lung infection, and mycetoma, a localized destructive infection of the feet or hands.

Gliding, Sheathed, and Budding and/or Appendaged Bacteria

Appendaged Bacteria.   This group of bacteria is linked taxonomically by the presence of **prosthecae,** protrusions such as stalks and buds. *Caulobacter* are the best studied, having stalks that anchor them to surfaces. This increases their nutrient uptake in the low-nutrient environments where they are found. *Caulobacter* reproduction results in one stalked cell and one flagellated swarmer cell, which eventually becomes another stalked cell.

Gliding Nonfruiting Bacteria.   **Gliding bacteria** are motile by gliding over surfaces. *Cytophaga* are important cellulose-degraders in the soil. *Beggiatoa* morphologically resembles cyanobacteria but is not photosynthetic. It uses hydrogen sulfide as an energy source.

Gliding Fruiting Bacteria.   The **myxobacteria** have a remarkable life cycle. Large numbers of vegetative cells converge on a single point, where they aggregate and differentiate into a stalked body that carries resting cells called *myxospores.* These myxospores eventually germinate and renew the cycle.

Budding Bacteria. **Budding bacteria** form buds; that is, the parent cell retains its identity, and the bud increases in size until it separates as a new cell. The best known are members of the genus *Hyphomicrobium.*

Sheathed Bacteria. **Sheathed bacteria,** such as *Sphaerotilus natans,* form a filamentous sheath that surrounds the bacteria. They are found in fresh water and sewage.

Chemoautotrophic Bacteria

These organisms are capable of using inorganic chemicals as energy sources and carbon dioxide as the only source of carbon. *Nitrobacter* oxidizes ammonium (NH_4^+) to nitrite (NO_2^-). *Nitrosomonas* oxidizes nitrite to nitrate (NO_3^-). Nitrate is mobile and the form of nitrogen best available to plants. For their source of energy, *Thiobacillus* and other sulfur-oxidizing bacteria oxidize reduced forms of sulfur, such as hydrogen sulfide (H_2S) or elemental sulfur (S^0), into sulfates (SO_4^{2-}).

Archaeobacteria

This interesting group of organisms includes the extreme **halophiles** such as *Halobacterium* or *Halococcus* that actually require high concentrations of sodium chloride to grow. *Sulfolobus* also thrives in extreme environments such as acidic, sulfur-rich hot springs. The **methane-producing** bacteria that derive energy by combining hydrogen (H_2) with CO_2 to form methane (CH_4) are important in anaerobic sewage treatment.

Phototrophic Bacteria

There are three groups of bacteria that are **phototrophic;** that is, they use light as an energy source.

The **purple** or **green sulfur bacteria** are generally anaerobes that inhabit deep aquatic sediments. They are **anoxygenic;** their photosynthesis does not produce oxygen. Instead of splitting water (H_2O) to release oxygen, they split hydrogen sulfide (H_2S) to form elemental sulfur (S).

Other phototrophs, the **purple nonsulfur** and **green nonsulfur bacteria,** use organic compounds such as acids and carbohydrates for the photosynthetic reduction of carbon dioxide, rather than using the hydrogen in water or in hydrogen sulfide.

The photosynthetic **cyanobacteria** are aerobes that carry out oxygen-producing photosynthesis much like higher plants. Many species of cyanobacteria are capable of fixing nitrogen from the atmosphere. They do so using enzymes carried in structures called **heterocysts. Gas vacuoles** in many aquatic species help maintain buoyancy for best access to light. There are unicellular, colonial, and filamentous forms of cyanobacteria.

Actinomycetes

Actinomycetes have a moldlike growth habit with long, branched filaments. The diameter of the filaments is of bacterial dimensions. Most reproduce by forming asexual spores. One genus, *Frankia,* causes formation of nitrogen-fixing nodules on alder tree roots. *Streptomyces* form conidiospores at the ends of filaments. Strict aerobes, these bacteria are important degraders of proteins, starch, and cellulose in soil. They produce **geosmin,** a gas that gives soil its odor. Most commercial antibiotics are produced by *Streptomyces*.

A Giant Procaryote

In 1991 a bacterium, assigned the name *Epulopiscium,* was discovered. Half a millimeter in length, it is visible to the unaided eye.

Self-Tests

In the matching section, there is only one answer to each question; however, the lettered options (a, b, c, etc.) may be used more than once or not at all.

I. Matching

1. The cause of syphilis.

2. Do not form cell walls.

3. An important genus in this group is *Pseudomonas*.

4. *Staphylococcus aureus.*

5. *Streptomyces* spp.

6. Motile by means of axial filament.

a. Spirochetes

b. Gram-negative aerobic rods and cocci

c. Gram-negative anaerobic bacteria

d. Gram-positive cocci

e. Actinomycetes

f. Mycoplasma

II. Matching

1. Obligate intracellular parasites.

2. *Clostridium* spp.

3. *Bacteroides* spp.

4. The cause of gonorrhea.

5. *Lactobacillus* spp.

6. *Veillonella* spp.

7. *Escherichia coli.*

8. *Sporosarcinae* spp.

a. Helical/vibrioid, gram-negative bacteria

b. Facultatively anaerobic, gram-negative rods

c. Gram-negative, aerobic rods and cocci

d. Anaerobic, gram-negative, straight, curved, and helical rods

e. Anaerobic, gram-negative cocci

f. Endospore-forming, gram-positive rods and cocci

g. Regular, nonsporing, gram-positive rods

h. Rickettsias and chlamydias

III. Matching

1. *Vibrio* spp.

2. *Neisseria* spp.

3. *Streptococcus* spp.

4. *Streptomyces* spp.

5. *Chlamydia* spp.

6. *Hemophilus* spp.

7. *Gluconobacter* spp.

a. X and V factors required

b. Cholera

c. Group A β-hemolytic

d. Botulism, tetanus, gas gangrene

e. Form conidiospores on filaments

f. Form infectious elementary bodies

g. Gonorrhea

h. Ethanol into vinegar

IV. Matching

1. Cause of whooping cough.

2. Produces a food-poisoning enterotoxin.

3. Rickettsial organism.

4. Bubonic plague.

5. Water soluble; blue-green pigment.

6. Essential for the operation of an activated-sludge sewage system.

7. Gliding bacterium.

8. A filamentous bacterial pathogen.

a. *Pseudomonas aeruginosa*

b. *Bordetella pertussis*

c. *Escherichia coli*

d. *Yersinia pestis*

e. *Staphylococcus aureus*

f. *Coxiella burnetti*

g. *Zoogloea* spp.

h. *Nocardia asteroides*

i. *Cytophaga* spp.

V. Matching

1. A genus of gliding bacteria that is an important cellulose degrader.

2. A sheathed bacterium found in water and sewage.

3. A chemoautotrophic bacterium that participates in nitrification in soil.

4. Photosynthetic bacteria that may fix nitrogen.

5. Photosynthetic, anoxygenic bacteria; use reduced sulfur compounds instead of water and produce sulfur granules rather than oxygen.

a. *Nitrosomonas*

b. Cyanobacteria

c. *Sphaerotilus natans*

d. Purple sulfur, or green sulfur, bacteria

e. *Cytophaga*

f. *Beggiatoa*

VI. Fill-in-the-Blanks

1. Spiral and curved bacteria, as grouped in *Bergey's Manual*, are motile by means of

   _____ .

2. The genus *Salmonella* consists almost entirely of pathogens and is taxonomically divided into hundreds of _____ , rather than species.

3. *Serratia marcescens* colonies produce a _____ colored pigment.

4. A number of _____ species are plant pathogens, causing plant soft-rot diseases.

5. The cyanobacteria are often able to fix atmospheric nitrogen by means of enzymes contained in structures called _____ .

6. Anaerobic, gram-negative, long, slender rods with pointed ends are in the genus
   _____ .

7. Spherical bacteria of the genus _____ occur in grapelike clusters.

8. _____ hemolytic types of bacteria form a narrow, greenish zone of hemolysis
   on blood agar plates.

9. _____ hemolytic types of bacteria form a clear zone of hemolysis on blood
   agar plates.

10. The rod-shaped members of the genus _____ form endospores and are
    obligate anaerobes.

11. *Actinomyces* resemble _____ but are obligate anaerobes rather than aerobes.

12. After an elementary body of a *Chlamydia* penetrates a host cell, it becomes a(n)
    _____ , which then divides successively.

13. The odor of soil is due to a gas, _____ , produced by *Streptomyces* bacteria.

14. The rickettsial organism that causes Q fever is named _____ .

15. Members of the genus _____ produce most of our commercial antibiotics.

16. Some appendaged aquatic organisms such as those in the genus _____ have
    stalks by which they anchor themselves to surfaces.

17. The disease leptospirosis is caused by spirochetes of the genus _____ .

18. A genus of bacteria resistant to most antibiotics and capable of using unusual substrates such
    as pesticides as nutrients is _____ .

19. Gram-positive cocci that are catalase-negative are members of the genus
    _____ .

20. The term _____ is applied to bacteria shaped as slightly curved rods.

VII. Critical Thinking

1. Why do many ulcer patients respond favorably to treatment with antibiotics?

2. Between July 3rd and August 17th of one year, 103 cases of dysentery occurred in a Southern California community. Fever was seen in approximately 50% of the cases. Most of the patients recovered without medication in just a few days, although a few persistent cases required antibiotics. Gram-negative, facultatively anaerobic, nonmotile bacilli were isolated from the patients' stools. The bacilli fermented glucose without gas and failed to ferment lactose. Eventually, the outbreak was linked to cantaloupes grown in Mexico. What genus of bacterium caused the outbreak of dysentery? How were the cantaloupes contaminated?

3. Two days after consuming raw oysters, a young woman developed abdominal pain, severe cramping, explosive watery diarrhea, and a low-grade fever. The patient recovered in a few days without antibiotics. What is the genus and species of the most likely culprit?

4. Discuss the difficulties in using a taxonomic rather than a systematic approach when classifying bacteria. What characteristics of *Gardnerella vaginalis* make it especially difficult to classify taxonomically?

5. What genus of bacterium is most likely to grow well on the following media or media containing the following ingredients?

    a. Media containing blood, V factor, and nicotinamide adenine dinucleotide.

    b. Buffered charcoal-yeast extract agar.

    c. Media containing sterols.

6. What is unusual about the metabolism of *Lactobacillus*? How does this organism's metabolism help it to create an ecological niche for itself?

7. How do chemoautotrophic bacteria contribute to agriculture?

8. What genus of bacterium is most likely to be found living in each of the following situations?

    a. In a salt extractor in the southwestern United States.

    b. In nodules on the roots of a soybean plant.

    c. In anaerobic, sulfur-rich mudflats.

9. Compare and contrast the following microorganisms:

    a. Spirochetes and facultatively anaerobic gram-negative rods.

    b. The rickettsias and chlamydias.

    c. Mycoplasmas and mycobacteria.

10. Discuss the system of organization used in *Bergey's Manual of Systematic Bacteriology.* How are bacteria divided into divisions? Into sections?

CHAPTER 12

Fungi, Algae, Protozoa, and Multicellular Parasites

Learning Objectives

After completing this chapter, you should be able to:

- List the defining characteristics of fungi.
- List the defining characteristics of the four phyla of fungi described in this chapter.
- Differentiate between asexual and sexual reproduction, and describe each of these processes in fungi.
- List the defining characteristics of algae.
- List the outstanding characteristics of the six phyla of algae discussed in this chapter.
- List the distinguishing characteristics of lichens, and describe their nutritional needs.
- Describe the roles of the fungus and the alga in a lichen.
- Compare and contrast cellular slime molds and plasmodial slime molds.
- List the defining characteristics of protozoa.
- Describe the outstanding characteristics of the four phyla of protozoa discussed in this chapter, and give an example of each.
- Differentiate between an intermediate host and a definitive host.
- List the distinguishing characteristics of parasitic helminths.
- Provide a rationale for the elaborate life cycles of parasitic worms.
- List the characteristics of the three groups of parasitic helminths, and give an example of each.
- Describe a parasitic infection in which humans serve as a definitive host, as an intermediate host, and as both.
- Define arthropod vector.
- Differentiate between a tick and a mosquito, and name a disease transmitted by each.

FUNGI

The study of fungi is called **mycology. Molds** are multicellular, filamentous organisms that include mildews, rusts, and smuts. **Fleshy fungi** include the mushrooms and puffballs. **Yeasts** are unicellular fungi.

128

Characteristics of Fungi

Vegetative Structures

Fungal filaments are called **hyphae** and form the body (or **thallus**) of molds or fleshy fungi. Crosswalls in the hyphae are **septa; coenocytic hyphae** have no septa. A mass of hyphae is a **mycelium;** the **vegetative mycelium** is concerned with obtaining nutrients, and the **reproductive** or **aerial mycelium** is involved in reproduction. Fragments of hyphae can elongate to form new hyphae.

Yeasts usually reproduce by **budding,** in which a protuberance on the cell enlarges and eventually separates as another cell. If daughter cells do not detach immediately, they form a short chain, a **pseudohypha.** Many yeasts are facultative anaerobes and ferment carbohydrates to produce CO_2 and ethanol. Fungi that, under different conditions, grow in either yeastlike or moldlike forms are known as **dimorphic fungi.** Features differentiating fungi from bacteria are summarized in Table 12.2 in the text.

Life Cycle

Asexual spores arise from one organism only. **Sexual spores** result from the fusion of nuclei from two opposite mating strains.

Asexual Spores. **Arthrospores** are formed by the fragmentation of a septate hypha into single cells. **Chlamydospores** form from rounding and enlargement of segments within a hypha. **Sporangiospores** are produced within a sac **(sporangium)** at the end of an aerial hypha called a **sporangiophore. Conidiospores** are produced in a chain at the end of a **conidiophore** and are not enclosed. **Blastospores** are buds.

Sexual Spores. A **zygospore** results from the fusion of the nuclei of two cells morphologically similar to each other. **Ascospores** result from the fusion of the nuclei of two cells morphologically either similar or dissimilar. The latter are produced in a sac **(ascus). Basidiospores** are formed externally on a base pedestal called a **basidium.** Sexual spores are the main basis for grouping the fungi taxonomically.

Nutritional Adaptations

1. Fungi usually prefer an acidic pH, too acid for most bacteria.
2. Molds are almost all aerobic. Yeasts are facultative anaerobes.
3. Fungi are relatively resistant to osmotic pressures; for example, they grow well at high sugar and salt concentrations.
4. Fungi are capable of growing on substances with a low moisture content, where bacteria are unable to grow.
5. Fungi require less nitrogen for equivalent weight of growth than do bacteria.
6. Fungi are capable of using complex carbohydrates such as lignins.

Medically Important Phyla of Fungi

Deuteromycota

The **Deuteromycota** also are known as the **Fungi Imperfecti,** meaning no sexual spores have yet been demonstrated. Asexual reproduction may involve chlamydospores, arthrospores, conidiospores, or budding.

Zygomycota

The **Zygomycota,** or conjugation fungi, form sexual zygospores and asexual sporangiospores. Their hyphae are coenocytic.

Ascomycota

Members of the **Ascomycota,** or sac fungi, form septate hyphae. Their sexual spores are ascospores produced in an ascus.

Basidiomycota

The **Basidiomycota,** or club fungi, have septate hyphae. They form sexual basidiospores, and some form asexual conidiospores. Table 12.4 in the text summarizes the characteristics of some of the fungi.

Fungal Diseases

A fungal infection is a **mycosis.** A **systemic mycosis** occurs deep within the patient in various tissues and organs. Infections usually are transmitted by inhalation and often begin in the lungs. Fungal infections just beneath the skin are **subcutaneous mycoses.** They usually result from puncture wounds and often form disfiguring subcutaneous abscesses.

Dermatophytes are fungi that infect the epidermis, hair, and nails, causing **cutaneous mycoses.** They secrete *keratinase,* which degrades a protein (keratin) found in hair, skin, and nails.

Superficial mycoses are localized along hair shafts and the superficial epidermal cells. **Mucormycosis** is an opportunistic mycosis caused by *Rhizopus* and *Mucor,* primarily in patients with ketoacidosis resulting from diabetes, leukemia, or treatment with immunosuppressive drugs. **Aspergillosis** is caused by *Aspergillus* and usually occurs in individuals with lung disease or cancer. **Candidiasis** is usually caused by an overgrowth of *Candida albicans.* **Thrush** is a mucocutaneous candidiasis, usually occurring in the mouths and throats of newborns; vulvovaginal candidiasis may occur during pregnancy.

Economic Effects of Fungi

All the listed nutritional adaptations are important in understanding the reasons for fungal spoilage of dry cereals; acidic foods with high sugar content, such as jelly; growth on painted walls or on shower curtains; and the importance of fungi as plant pathogens. Examples of plant fungal diseases are Dutch elm disease and the chestnut blight; the latter essentially eliminated chestnut trees in the United States.

ALGAE

Characteristics of Algae

Algae are photosynthetic autotrophs; that is, they use light to convert atmospheric carbon dioxide into carbohydrates for energy. Oxygen is a by-product of photosynthesis. The body of a multicellular alga is a **thallus,** which may collectively function as **holdfasts** to anchor them. They often have stemlike **stipes** and leaflike **blades** (Figure 12.10b in the text). Algae are buoyed by gas-filled **bladders.**

Asexual reproduction in multicellular or filamentous algae is accomplished by fragmentation of the thallus. Following mitosis, unicellular algae divide by simple fission. Some algae reproduce sexually.

Selected Phyla of Algae

Brown algae, or **kelp,** may be 50 meters in length. *Algin,* a product used as a food thickener and in the production of other goods, is extracted from their cell walls. **Red algae** are branched and multicellular and live at greater depths than do other algae. **Agar** and **carrageenan** are gelatinous thickeners derived from red algae. **Green algae** are usually classified as microscopic plants.

Dinoflagellates are unicellular algae, referred to as **planktonic** because they are free-floating. Some produce neurotoxins, which may be ingested by shellfish and eventually poison humans (paralytic shellfish poisoning). **Euglenoids** are unicellular, flagellated algae that are facultative chemoheterotrophs. They resemble protozoa. **Diatoms** are unicellular or filamentous algae with walls of pectin or silica that fit together as do the two halves of a Petri plate.

Roles of Algae in Nature

Algae are important to the aquatic food chain because they convert carbon dioxide into consumable organic molecules. A large increase in numbers of planktonic algae is called a **bloom.** Blooms of dinoflagellates cause *red tides* in the ocean. When large numbers of algae die and decompose, oxygen dissolved in water is depleted. Some algae are symbionts of animals, providing carbohydrates for a large clam, for example. Petroleum may be partly derived from ancient algae.

LICHENS

A **lichen** is a combination of a green alga (or cyanobacterium) and a fungus. This is a *mutualistic* relationship, from which each member benefits. **Crustose** lichens grow flush on a surface; **foliose** lichens are leaflike in shape; **fruticose** lichens have fingerlike projections. The photosynthetic alga provides carbohydrates, and the fungus provides protection from desiccation and a means of attachment.

SLIME MOLDS

Slime molds have both fungal and animal characteristics and are classified as protists. **Cellular slime molds** resemble amoebas at one stage. When conditions are unfavorable for growth, large numbers of amoeboid cells aggregate to form a single structure, a **slug.** Cyclic AMP produced by some amoebas is the attractant toward which they migrate to form the slug. The slug moves toward light and eventually forms a stalked structure with a spore cap at the top. Under favorable conditions this spore cap differentiates into single amoebalike spores, repeating the cycle.

  Plasmodial (acellular) slime molds are a mass of protoplasm called a **plasmodium.** The entire plasmodium moves like a giant amoeba and engulfs organic debris and bacteria. **Cytoplasmic streaming,** which apparently distributes oxygen and nutrients, can be observed in these slime molds. When conditions are unfavorable for growth, the plasmodium separates, and stalked sporangia with spores are formed. When conditions improve, the spores germinate and the plasmodium is again formed.

PROTOZOA

Characteristics of Protozoa

Protozoa are one-celled eucaryotic organisms that mostly feed on bacteria and small particulate nutrients. They are classified largely on the basis of their means of motility.

Life Cycle

Protozoa reproduce asexually by fission, budding, or **schizogony.** The last involves multiple fission of the nucleus prior to cell division. Daughter cells eventually form around each nucleus. Sexual reproduction such as **conjugation** has been observed in some ciliates. Two cells fuse, and a haploid nucleus from each migrates to the other cell. Both are now fertilized and produce daughter cells. Other protozoa produce **gametes** or **gametocytes**—haploid sex cells that fuse to form a zygote.

Encystment

Some protozoa produce a protective, hardened capsule called a **cyst.**

Nutrition

Most protozoa are aerobic heterotrophs, but most intestinal protozoa are capable of anaerobic metabolism. A few, like Euglena, contain chlorophyll and are photoautotrophs. Some protozoa absorb food through the cell membrane, but many have a coating, the **pellicle,** that does not allow them to do so. Ciliates wave cilia (other than those used for motility) to bring food to an opening called the **cytostome.** Amoebas engulf food and phagocytize it. Digestion in all protozoa takes place in **vacuoles,** and waste is eliminated through an **anal pore.**

Medically Important Phyla of Protozoa

Amoeboflagellates

The Sarcomastigophora are protozoa that are motile by pseudopods or flagella.

Amoebas. The **amoebas** move by extending lobelike projections of the cytoplasm called **pseudopods.** An important disease is amoebic dysentery.

Flagellates. Members of the **flagellates** move by means of sinuous movements of flagella. Some have an undulating membrane, which is perhaps a modified flagellum. The vegetative form of these protozoa is called a **trophozoite.** Representative diseases caused by this protozoan group are giardiasis, Chagas' disease, African sleeping sickness, and trichomoniasis.

Ciliates

Cilia, the organs of propulsion for the phylum Ciliata, are shorter than flagella and are arranged over the cell, moving in synchrony to propel it through the medium. Dysentery caused by *Balantidium coli* is the only ciliate-caused disease.

Apicomplexans

Apicomplexans are nonmotile in their mature forms and are obligate intracellular parasites. An example is the genus *Plasmodium,* which causes malaria. *Plasmodium* has a complex life cycle involving an animal host and a mosquito. The sexual cycle takes place in the mosquito, resulting in **sporozoites** that enter a host by means of mosquito bites. There they enter the liver cells, undergo **schizogony,** and leave the liver cells as **merozoites.** These merozoites infect red blood cells, again undergo schizogony, and are released as more merozoites when the red blood cells rupture. This release is periodic and causes the fever and chills of malaria. Some merozoites develop into sexual forms called **gametocytes.** These are picked up by feeding *Anopheles* mosquitoes, in which a new sexual cycle begins.

Toxoplasma gondii is another pathogenic sporozoa, causing the disease toxoplasmosis. It reproduces sexually only in cats. The feces of infected cats contain oocysts which, if ingested, infect humans or animals. This disease is dangerous to pregnant women, as it can cause congenital infections of the fetus. Table 12.6 in the text summarizes some of the parasitic protozoa.

Cruptosporidium is an apicomplexan causing a diarrheal illness.

Microsporans

Microsporans are obligate intracellular parasites lacking mitochondria and microtubules. Protozoa of this type may cause chronic diarrhea or inflammation of the conjunctiva of the eye.

HELMINTHS

Helminths are multicellular eucaryotic animals. Many are parasitic.

Life Cycle

Adult helminths may be **dioecious;** that is, there are male and female individuals. Some, however, are **hermaphroditic;** one animal has both male and female reproductive organs. The **definitive host** harbors the adult, sexually mature helminth. **Intermediate hosts** may be necessary for larval or developmental stages.

Platyhelminths

The **platyhelminths,** or flatworms, are dorsoventrally flattened. The digestive system is incomplete; there is only one opening for entry of food and exit of waste. Trematodes and cestodes are members of this group.

Trematodes

Trematodes, or **flukes,** have flat, leaf-shaped bodies with a ventral sucker and an oral sucker to hold them in place and suck fluids from the host. They also may obtain food by absorption through the outer covering, the cuticle. A typical life cycle is that of the lung fluke, *Paragonimus westermani*. It begins with the excretion of an egg into water. A **miracidial larva** develops from the egg and enters a snail. There it produces **rediae,** each of which develops into a **cercaria,** which, in turn, penetrates a crayfish. There it encysts as a **metacercaria.** Humans are infected by ingesting undercooked crayfish. Cercariae of diseases such as schistosomiasis are not ingested but burrow through the skin of the human host.

Cestodes

Cestodes, or **tapeworms,** are intestinal parasites. The head or **scolex** has suckers and usually attachment hooks. The worm consists of segments called **proglottids,** which contain both male and female organs. Mature proglottids contain fertilized eggs that are shed in the feces. The egg-laden proglottids of the beef tapeworm, for example, are ingested by grazing animals. The eggs hatch in the intestine, releasing larvae that migrate to muscles, where the larval form is encysted as **cysticerci.** These may be ingested by humans, thus initiating an infestation. In some dog and cat tapeworms, humans are only intermediate hosts. The eggs are picked up from contamination of the hands by dog feces or a dog's tongue. The cyst forming in these infections of the host tissue, usually the lung or liver, is called a **hydatid cyst** in the case of the tapeworm *Echinococcus granulosus.*

Nematodes

The **nematodes,** or **roundworms,** are cylindrical and tapered at the ends. They have a complete digestive system of mouth, intestine, and anus. Human parasites are found in the nematode group.

   Parasitic nematodes (many are free-living in soil and water) do not have the succession of larval stages found in flatworms. In some cases, the eggs are infective for humans. The pinworm *Enterobius vermicularis* lives near the human anus. *Ascaris lumbricoides* lives in the intestines on semidigested food and excretes eggs in the feces. After the ingested eggs hatch in the intestine, the larvae mature in the lungs before migrating to the intestines. Hookworms are an example in which larvae are the infective form. Hookworm larvae in the soil penetrate the human skin on contact, eventually reaching the intestines. Eggs shed in feces become larvae in the soil and continue the cycle. In trichinosis, the encysted larvae are ingested in meats such as pork. In the human host, the cysts mature into adults in the intestine, and eventually eggs from the adults become larvae and encyst in human muscles. A summary of representative parasitic helminths is presented in Table 12.7 in the text.

ARTHROPODS AS VECTORS

Arthropods are jointed-legged animals and include the **Arachnida,** which have eight legs (spiders, mites, ticks); **Crustacea** (crabs, crayfish); **Chilopoda** (centipedes); **Diplopoda** (millipedes); and **Insecta,** which have six legs (bees, flies). Ticks and mites (Arachnida) and lice, fleas, and mosquitoes (Insecta) are particularly important in medicine as vectors of disease; that is, they carry and transmit disease-causing microorganisms.

Self-Tests

In the matching section, there is only one answer to each question; however, the lettered options (a, b, c, etc.) may be used more than once or not at all.

I. Matching

1. The sac produced at the end of an aerial hypha called a sporangiophore.

2. Formed by the fragmentation of a hypha.

3. Buds.

4. Produced in a chain; not enclosed; asexual.

5. Formed on a base pedestal called a basidium.

6. A sexual spore resulting from the fusion of nuclei of two cells, morphologically either similar or dissimilar; produced in a sac.

7. A result of fusion of the nuclei of two cells morphologically similar to each other.

a. Arthrospores

b. Blastospores

c. Conidiophores

d. Conidiospores

e. Chlamydospores

f. Sporangiospores

g. Sporangium

h. Basidiospores

i. Zygospores

j. Ascospores

II. Matching

1. Tapeworm.

2. Roundworm.

3. Scolex.

4. Proglottids.

5. Hydatid cysts.

6. Fluke.

a. Trematode

b. Nematode

c. Cestode

III. Matching

1. A resistant capsule formed by a protozoa.

2. Organelles of movement by amoebas.

3. An organism (e.g., a mosquito) that transmits a disease-causing organism.

4. A term describing helminths with both male and female reproductive organs.

5. An outer covering on some helminths.

6. An outer covering on some protozoa.

a. Dioecious

b. Cuticle

c. Hermaphroditic

d. Pellicle

e. Vector

f. Pseudopods

g. Ergot

h. Sclerotia

i. Cyst

j. Schizogony

IV. Matching

1. Nonmotile in their mature forms.

2. Toxoplasma gondii.

3. *Balantidium coli.*

4. The cause of malaria.

a. Ciliates

b. Flagellates

c. Amoebas

d. Apicomplexans

V. Matching

1. Walls of pectin or silica fit together much like the two halves of a Petri plate.

2. May be 50 meters in length; algin is extracted from them.

3. Agar is a product of this type of organism.

4. Crustose, foliose, fruticose.

5. Cyclic AMP is involved in aggregation of individual cells.

6. Composed of a mass called a plasmodium.

7. Some species are responsible for red tides in the ocean.

8. A flagellated alga sometimes classified as a form of protozoa.

a. Dinoflagellates

b. Cellular slime molds

c. Kelp

d. Red algae

e. Plasmodial (acellular) slime molds

f. Diatoms

g. Euglenoids

h. Lichens

VI. Fill-in-the-Blanks

1. In the life cycle of the protozoa causing malaria, the organism grows in the red blood cell, which eventually ruptures and releases the form called _____ . Then, in the mosquito the male and female gametocytes of the protozoan mate and produce _____ , which the mosquito transmits to a new human host.

2. The _____ , or conjugation fungi, form sexual zygospores and asexual conidiospores.

3. A fungal infection is called a _____ .

4. The scientific name of the club fungi is _____ .

5. _____ are fungi that infect the epidermis, hair, and nails.

6. Fungi that are sometimes yeastlike and other times filamentous are called _____ .

7. _____ is the name of the opportunistic disease caused by an overgrowth of *Candida albicans*.

8. The _____ are also known as the Fungi Imperfecti, meaning no sexual spores have yet been demonstrated.

9. The body of a multicellular alga is called a _____ .

10. Insects have _____ legs.

11. _____ , a larval form of the disease organism causing schistosomiasis, are not ingested but burrow through the skin of the human host.

12. The _____ host harbors the adult, sexually mature helminth. The _____ hosts may be necessary for larval or developmental stages.

13. An animal with both male and female reproductive organs is called _____ .

14. The vegetative form of the flagellated protozoa is called a _____ .

15. Fungal filaments are called _____ , and a mass of hyphae is called _____ .

16. _____ are unicellular fungi.

17. Hyphae with no septa are termed _____ hyphae.

18. The sac in which an ascospore is formed is an _____ .

19. Fungal infections just beneath the skin, usually resulting from a puncture wound, are called _____ mycoses.

20. The common name for candidiasis of the mouth and throat is _____ .

21. Digestion in all protozoa takes place in _____ , and wastes are eliminated through an anal pore.

22. One form of division by protozoa involves repeated fission of nuclei prior to cell division. This is called _____ .

23. A form of sexual reproduction called _____ is found in some ciliated protozoa and involves two cells fusing together and exchanging haploid nuclei.

24. In the life cycle of the lung fluke, the rediae in the snail host develop into a _____ , which in turn penetrates a crayfish and forms a metacercaria.

25. The stemlike structures of a multicellular alga are called _____ .

26. A lichen represents a _____ type of association between an alga and a fungus.

27. The structures that buoy an alga in water are called _____ .

VII. Label the Art

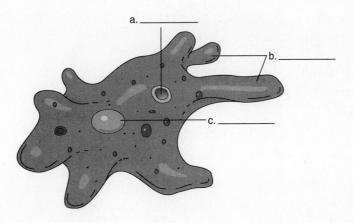

1. *Amoeba proteus*

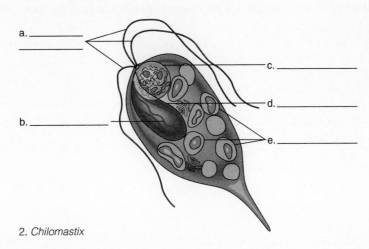

2. *Chilomastix*

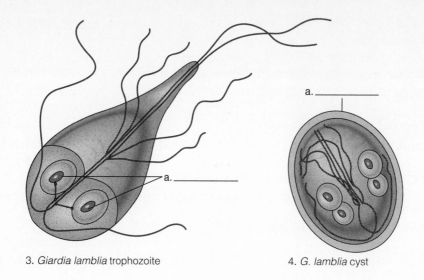

3. *Giardia lamblia* trophozoite
4. *G. lamblia* cyst

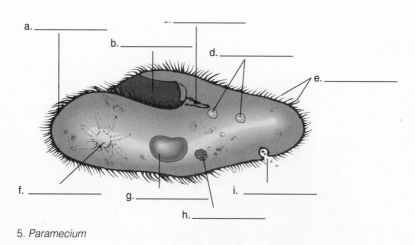

5. *Paramecium*

VIII. Critical Thinking

1. How are fungi classified into phyla? Why? List the four phyla of fungi discussed in the text. For each phylum give an example of an asexual spore formed by members of this phylum and a representative genus.

2. Why has it been difficult to determine the taxonomic position of *Pneumocystis carinii*? Where does current research suggest it should be classified? Why?

3. By what two criteria are mycoses classified? Complete the following table:

Category of mycosis	How transmitted	Level of tissue
Systemic		
Subcutaneous		
Cutaneous		
Opportunistic		
Superficial		

4. List and discuss at least three ways in which algae are economically important.

5. What type of symbiotic relationship is displayed by lichens? What two organisms make up a lichen? What does each partner contribute to the relationship?

6. Discuss the role of the cyst in the life cycle of protozoa.

7. A patient is admitted to the hospital with dysentery and abdominal pain and cramps. Examination of a stool sample revealed protozoan trophozoites containing human RBCs and cysts with four nuclei. What is the genus and species name of the protozoa? How is this protozoa transmitted?

8. Outline the life cycle of *Plasmodium*.

9. Discuss two ways to break the chain of infection of humans with *Paragonimus westermanni*.

10. Give an example of each of the following:

    a. A nematode with an egg that is infectious for humans.

    b. A nematode with a larvae infective for humans.

    c. A cestode for which humans serve as an intermediate host.

CHAPTER 13

Viruses

Learning Objectives

After completing this chapter, you should be able to:

- Differentiate between a virus and a bacterium.
- Describe the chemical composition and physical structure of both an enveloped and a nonenveloped virus.
- Define viral species.
- Give an example of a family, genus, and common name for a virus.
- Describe how bacteriophages are cultured.
- Describe how animal viruses are cultured.
- List three techniques used to identify viruses.
- Describe the lytic cycle of T-even bacteriophages.
- Describe the lysogenic cycle of T-even bacteriophages.
- Compare and contrast the multiplication cycle of DNA- and RNA-containing animal viruses.
- Define oncogene and transformed cell.
- Explain four methods of activating oncogenes.
- Discuss the relationship of DNA- and RNA-containing viruses to cancer.
- Provide an example of a latent viral infection.
- Differentiate between slow viral infections and latent viral infections.
- Discuss three possible explanations for infectious diseases with no apparent cause.
- Differentiate among virus, viroid, and prion.
- Name a virus that causes a plant disease.

In the early days of microbiology, the term **filterable agents** or **filterable virus** (the word virus derives from the word poison) was used to designate an infectious agent that passed through filters that retained bacteria. Later, the term **virus** alone came into use. At the time, no one was sure of the particulate nature of these submicroscopic agents.

General Characteristics of Viruses

Viruses are **obligatory intracellular parasites** that require a living host cell in order to multiply. The term **host range** refers to the spectrum of host cells the virus can infect. Most viruses are much smaller than bacteria, although some larger ones approach the size of very small bacteria. The size ranges from about 20 to 14,000 nm (see Figure 13.1 in the text).

Viral Structure

A **virion** is a fully developed complete viral particle.

Nucleic Acid

Viral nucleic acid may be either DNA or RNA in double-stranded or single-stranded forms. It may be linear, or even in several separate molecules.

Capsid and Envelope

The protein coat is the **capsid**; it is made up of protein subunits, the **capsomeres.** The capsid may be covered by an **envelope** of some combination of lipids, proteins, and carbohydrates. Envelopes may be covered with **spikes** projecting from the surface. Some viruses use these spikes to adhere to red blood cells, causing a clumping called **hemagglutination.** Viruses not covered by an envelope are known as **nonenveloped viruses.**

General Morphology

Viruses may be classified into several morphological types. **Helical viruses** resemble long rods, their capsids a hollow cylinder with a helical structure. Examples are the tobacco mosaic virus or bacteriophage M 13. **Polyhedral viruses** usually have a capsid in the shape of an *icosahedron* (a polyhedron of 20 regular triangular faces). Examples are the adenovirus and poliovirus. **Enveloped viruses** have their capsid covered by an envelope. They are roughly spherical but pleomorphic (variable in shape). A helical virus (in this case the helical capsid is folded, not extended in rod-like form) such as the influenza virus is referred to as **enveloped helical.** A polyhedral virus such as herpes simplex, with a capsule, is an **enveloped polyhedral virus. Complex viruses,** such as the poxviruses, do not contain identifiable capsids. They may have several coats around the nucleic acid or, like many bacteriophages, have a polyhedral head and a helical tail.

Taxonomy of Viruses

In this text we group viruses according to host range; that is, animal, bacterial, or plant viruses. Current classification systems are based on type of nucleic acid, morphology, presence or absence of an envelope, and so on (see Table 13.1). A **viral species** is a group of viruses sharing the same genetic information and ecological niche.

TABLE 13.1 Classification of Animal Viruses by Nucleic Acid Replication

CHARACTERISTICS	VIRAL FAMILY	VIRAL GENUS (WITH REPRESENTATIVE SPECIES) AND UNCLASSIFIED MEMBERS*	DIMENSIONS OF VIRION (DIAMETER IN NM)	CLINICAL OR SPECIAL FEATURES
Single-stranded DNA, nonenveloped	Parvoviridae	*Dependovirus*	18–25	Depend on coinfection with adenoviruses; cause fetal death, gastroenteritis.
Double-stranded DNA, nonenveloped	Adenoviridae	*Mastadenovirus* (adenovirus)	70–90	Medium-sized viruses that cause various respiratory infections in humans; some cause tumors in animals.
	Papovaviridae	*Papillomavirus* (human wart virus) *Polyomavirus*	40–57	Small viruses that induce tumors; the human wart virus (papilloma) and certain viruses that produce cancer in animals (polyoma and simian) belong to this family. Refer to Chapters 21 and 26.
Double-stranded DNA, enveloped	Poxviridae	*Orthopoxvirus* (vaccinia and smallpox viruses) *Molluscipoxvirus*	200–350	Very large, complex, brick-shaped viruses that cause diseases such as smallpox (variola), molluscum contagiosum (wartlike skin lesion), cowpox, and vaccinia; vaccinia virus gives immunity to smallpox. Refer to Chapter 21.
	Herpesviridae	*Simplexvirus* (herpes simplex viruses 1 and 2) *Varicellavirus* (varicella-zoster virus) *Cytomegalovirus* *Lymphocryptovirus* (Epstein-Barr virus) Human herpes virus 6	150–200	Medium-sized viruses that cause various human diseases, such as fever blisters, chickenpox, shingles, and infectious mononucleosis; implicated in a type of human cancer called Burkitt's lymphoma. Refer to Chapters 21, 23, and 26.
	Hepadnaviridae	*Hepadnavirus* (hepatitis B virus)	42	After protein synthesis, hepatitis B virus uses reverse transcriptase to produce its DNA from mRNA; causes hepatitis B and liver tumors. Refer to Chapter 25.
Single-stranded RNA, nonenveloped + strand	Picornaviridae	*Enterovirus* *Rhinovirus* (common cold virus) Hepatitis A virus	28–30	At least 70 human enteroviruses are known, including the polio-, coxsackie-, and echoviruses; more than 100 rhinoviruses exist and are the most common cause of colds. Refer to Chapters 22, 23, 24, and 25.
Single-stranded RNA, enveloped + strand	Togaviridae	*Alphavirus* *Rubivirus* (rubella virus)	60–70	Included are many viruses transmitted by arthropods (*Alphavirus*); diseases include eastern equine encephalitis (EEE) and western equine encephalitis (WEE). Rubella virus is transmitted by the respiratory route. Refer to Chapters 21, 22, 23, and 26.
	Flaviviridae	*Flavivirus* *Pestivirus* Hepatitis C virus	40–50	Can replicate in arthropods that transmit them; diseases include yellow fever, dengue, St. Louis encephalitis, and Japanese encephalitis. The unclassified hepatitis C virus is most likely in this family. Refer to Chapters 22 and 23.
	Coronaviridae	*Coronavirus*	80–160	Associated with upper respiratory tract infections and the common cold. Refer to Chapter 24.

*Unclassified viruses have not been assigned to genera; therefore, only their common names are listed.

(continued)

TABLE 13.1   (continued)

CHARACTERISTICS	VIRAL FAMILY	VIRAL GENUS (WITH REPRESENTATIVE SPECIES) AND UNCLASSIFIED MEMBERS*	DIMENSIONS OF VIRION (DIAMETER IN NM)	CLINICAL OR SPECIAL FEATURES
− strand, one strand of RNA	Rhabdoviridae	*Vesiculovirus* (vesicular stomatitis virus) *Lyssavirus* (rabies virus)	70–180	Bullet-shaped viruses with a spiked envelope; cause rabies and numerous animal diseases. Refer to Chapter 22.
	Filoviridae	*Filovirus*	80–14,000	Enveloped, helical viruses; Ebola and Marburg viruses are filoviruses.
	Paramyxoviridae	*Paramyxovirus* *Morbillivirus* (measles virus)	150–300	Paramyxoviruses cause parainfluenza, mumps, and Newcastle disease in chickens. Refer to Chapters 21 and 25.
− strand, multiple strands of RNA	Orthomyxoviridae	*Influenzavirus* (influenza viruses A and B) Influenza C virus	80–200	Envelope spikes can agglutinate red blood cells. Refer to Chapter 24.
	Bunyaviridae	*Bunyavirus* (California encephalitis virus) *Hantavirus*	90–120	Hantaviruses cause hemorrhagic fevers such as Korean hemorrhagic fever and *Hantavirus* pulmonary syndrome; associated with rodents. Refer to Chapter 24.
	Arenaviridae	*Arenavirus*	50–300	Helical capsids contain RNA-containing granules; cause lymphocytic choriomeningitis and hemorrhagic fevers. Refer to Chapter 24.
Produce DNA	Retroviridae	Oncoviruses *Lentivirus* (HIV)	100–120	Includes all RNA tumor viruses and double-stranded RNA viruses. Oncoviruses cause leukemia and tumors in animals; the *Lentivirus* HIV causes AIDS. Refer to Chapter 19.
Double-stranded RNA, nonenveloped	Reoviridae	*Reovirus* Colorado tick fever virus	60–80	Involved in mild respiratory infections and infantile gastroenteritis; an unclassified species causes Colorado tick fever. Refer to Chapter 23.

*Unclassified viruses have not been assigned to genera; therefore, only their common names are listed.

Isolation, Cultivation, and Identification of Viruses

Growth of Bacteriophages in the Laboratory

Bacteriophages require a specific host bacterium for growth. The growth medium for the host may be liquid or solid, but solid media are used for detection and counting of viruses by the **plaque method.** For this method, a melted agar suspension of host cells and bacteriophage (**phage,** for short) are poured in a thin layer over an agar surface on a Petri plate. The bacteria develop into a turbid lawn except where they are destroyed by proliferating phage, forming a circular clearing called a **plaque.** Counts of phage are made in terms of **plaque-forming units.**

Growth of Animal Viruses in the Laboratory

Viruses can be cultured in suitable *living animals,* and some can be grown only in this way. Because signs of disease in the animal are often significant, this method can be used in diagnosis. **Embryonated eggs** can be inoculated by a hole drilled in the shell. Growth may be detected by death of the embryo or formation of pocks or lesions on the membranes. Most recently, **cell culture** has been the method of choice for viral cultivation. Animal (or plant) cells may be separated and grown as homogeneous collections of cells, not unlike bacterial cultures. In containers, the cells tend to adhere to surfaces and form a **monolayer** of cells. Cell infection by a virus causes observable death or damage known as **cytopathic effects (CPE),** which can be used, much as plaques are, for counting or detecting viruses.

Primary cell lines are derived directly from tissue and tend to die after a few generations, but a few specialized human cell lines may be cultivated for 100 generations or so. **Diploid cell lines,** developed from embryonic human cells, are used to culture rabies virus for human diploid cell vaccines. They can be maintained for about 100 generations. **Continuous cell lines,** often cancer cells such as the HeLa cells, can be maintained an indefinite number of generations.

Identification of viruses is difficult. Immunological methods using specific antibodies are most commonly used. Viruses can be seen only with an electron microscope.

Viral Multiplication

The virus has only a few genes. Most enzymes encoded in viral nucleic acid are not part of the virion but are synthesized and function only within the host cell. The viral enzymes mostly replicate viral nucleic acid and seldom the viral proteins. Thus, the virus must invade a host cell and take over its metabolic machinery.

Multiplication of Bacteriophages

The most familiar example of a viral life cycle is that of the T-even phages (T_2, T_4, T_6). The tail of the phage is **adsorbed** or **attached** to the host cell. This is a highly specific reaction depending on a complementary receptor site. The phage forms a hole in the cell wall using *phage lysozyme* and drives the tail core through the cell wall **(penetration)**; it then injects the DNA of the virus into the cytoplasm. The head (capsid) remains outside. The viral DNA causes transcription of RNA from viral DNA and thus commandeers the metabolic machinery of the host cell for its own biosynthesis. For several minutes following infection, complete phages cannot be found; this is called the **eclipse period.** During the **maturation period** that follows, the phage DNA and capsid, formed separately, are assembled into virions. When complete, the host cell **lyses** and **releases** these virions. The time required from phage adsorption to release is the **burst time** (about 20–40 minutes), and the number released is the burst size (about 50–200). The stages of phage multiplication can be demonstrated with the **one-step growth experiment** (see Figure 13.15 in the text).

Lysogeny.   Sometimes the lytic cycle just described does not occur. The phage DNA becomes incorporated as a **prophage** into the host's DNA, a state called **lysogeny.** Such phages are **lysogenic** or **temperate** phages (such as bacteriophage lambda (λ)) and their bacterial host, **lysogenic cells.** The lysogenized cell may exhibit new properties such as toxin production (examples are scarlet fever, diphtheria, and botulism). The prophage is reproduced along with the bacterial chromosome but can be induced to complete the lytic cycle. In **specialized transduction,** a lysogenic phage incorporates small amounts of host DNA along with its own DNA and can confer this DNA to a newly infected cell.

Multiplication of Animal Viruses

Multiplication of animal viruses follows the general pattern just described, but with important differences. Animal viruses have no tail, so adsorption is by spikes, small fibers, and so on. Penetration occurs by fusion of the envelope with the host plasma membrane or by the nonenveloped virus somehow entering the cytoplasm. Penetration by **endocytosis** requires the cell's plasma membrane to fold inward as vesicles. The host enfolds the virus into this vesicle, bringing it into the cell. An alternative method of penetration is **fusion.** The viral envelope fuses with the plasma membrane and releases the capsid into the host cell's cytoplasm. Once inside the host cell, the viral nucleic acid separates from the protein coat, a process called **uncoating.**

Biosynthesis of DNA-Containing Viruses.   Multiplication of DNA viruses may occur entirely in the cytoplasm (poxviruses). Or, the DNA may be formed in the nucleus and the protein in the cytoplasm, final assembly taking place in the nucleus.

Adenoviruses. Cause of some common colds; named after adenoids.

Poxviruses. Infections such as smallpox. Pox are pus-filled sacs on skin.

Herpesviruses. Cause of cold sores, chickenpox (varicella-virus), and infectious mononucleosis (Epstein-Barr virus).

Papovaviruses. Named for *pa*pillomas (warts), *po*lyomas (tumors), and *va*cuolation (cytoplasmic vacuoles produced by some of these viruses).

Hepadnaviruses. Named because they cause *hepa*titis and contain *DNA*; unusual because they synthesize DNA by copying RNA with reverse transcriptase, which will be discussed soon with retroviruses.

Biosynthesis of RNA-Containing Viruses. Multiplication of RNA viruses is essentially similar to that of DNA viruses but takes place in the cytoplasm.

Picornavirus (from *pico*, meaning small, and *RNA*). These are single-stranded RNA viruses such as poliovirus. The single strand (**+** or **sense strand**) acts as mRNA. It serves as a means to make *RNA-dependent RNA polymerase,* which catalyzes the synthesis of another strand of RNA (**–** or **antisense strand**). The – strand serves as a template for + strands that in turn serve as a means to produce viral RNA or viral protein.

Togavirus (from *toga*, or covering). Also containing a single + strand of RNA, togaviruses differ from picornaviruses in that two types of mRNA are transcribed from the – strand. One codes for capsid proteins and the other for envelope proteins.

Rhabdovirus (from *rhabdo-*, or rod). These are usually bullet-shaped (such as the rabies virus) and contain a single – strand of RNA. Because rhabdoviruses already contain RNA-dependent RNA polymerase, they do not have to synthesize this enzyme. The RNA polymerase produces a + strand, which serves as mRNA and a template for synthesis of viral RNA.

Reovirus (from first letters of respiratory, enteric, and orphan).
These contain double-stranded RNA. One of the capsid proteins of these viruses serves as RNA-dependent RNA polymerase. After the capsid enters a host cell, mRNA is produced inside the capsid and released into the cytoplasm, where it is used to synthesize more viral proteins. One of these proteins acts as RNA-dependent RNA polymerase to produce – strands of RNA. These – strands and the mRNA + strands form the double-stranded RNA in reoviruses.

Retrovirus (from *reverse transcriptase*). Some retroviruses cause cancers, and one type has been implicated as the cause of **acquired immune deficiency syndrome (AIDS).** These viruses carry a polymerase **(reverse transcriptase)** that uses the RNA of the virus to make a complementary strand of DNA. This DNA becomes integrated into the DNA of a host cell **(provirus),** and transcription into mRNA may then take place normally.

Maturation and Release. For enveloped cells, the envelope develops around the capsid from the plasma membrane by a process called **budding,** which occurs as the nucleic acid enclosed in the capsid pushes out through the plasma membrane. Budding does not necessarily kill the host cell. Nonenveloped viruses released by host cells usually cause **lysis** and death of the host cell.

Viruses and Cancer

When cells multiply in an uncontrolled way, the excess tissue is called a tumor, which is **malignant** if cancerous and **benign** if not. **Leukemias** are not solid tumors but an excess production of white cells. Chicken **sarcoma** (cancer of connective tissue) and **adenocarcinoma** (cancer of glandular tissue) can be transmitted by viruses.

Transformation of Normal Cells into Tumor Cells

It is believed that cancer-causing changes in cellular DNA are directed by parts of the genome called **oncogenes.** Under normal conditions oncogenes probably code for necessary cellular proteins **(protooncogenes),** but mutations can cause oncogenes to bring about cancerous transformations of cells. The most common oncogene products are **protein kinases,** which remove a phosphate group from ATP and add it to certain amino acids in a protein. Oncogenes can be activated by chemicals, oncogenic viruses, and radiation. A tumor cell formed by activation of an oncogene is said to be **transformed** and is distinctly different from normal cells. Viruses that activate oncogenes are called **oncogenic viruses,** or **oncoviruses.** Sometimes the provirus remains latent much like lysogeny and replicates only with the host cell, or it may become transcribed and produce new, infective viruses. Finally, it may convert the host cell into a tumor cell. Transformed cells lose **contact inhibition;** that is, they do not stop reproduction when in contact with neighbor cells. They also contain **tumor-specific transplantation antigens (TSTA)** on the surface, or **T antigens** in the nucleus.

Activation of Oncogenes

Cancer appears to be a multistep process in which at least two oncogenes must be activated. Four mechanisms for activation have been proposed.

1. A **single mutation** alters the amino-acid sequence of the gene product, possibly affecting regulation of the oncogene. A possible result is uncontrolled production of a growth factor and uncontrolled cellular growth.
2. **Transduction** of oncogenes by a virus could remove the oncogenes from normal cellular controls. The viral regulatory proteins could cause abnormal formation of oncogene products.
3. **Translocation** of an oncogene could place it at a position on the chromosome where normal controls are not active.
4. **Gene amplification** could cause the gene to be replicated several times, causing the cell to produce unusually large amounts of oncogenic products.

DNA Oncogenic Viruses

Adenoviruses, herpesviruses, poxviruses, and papovaviruses groups all contain **oncogenic viruses.** Among the papovaviruses are the papilloma viruses that cause warts. Herpesviruses, including the Epstein-Barr (EB) virus, may cause Burkitt's lymphoma or nasopharyngeal carcinoma. The hepadnavirus causing hepatitis B also has a role in liver cancer.

RNA Oncogenic Viruses

Only the retroviruses seem to be oncogenic among the RNA types. Their oncogenic activity seems related to the production of reverse transcriptase. The DNA synthesized from viral RNA becomes integrated into the host cell's DNA (provirus). In some cases, this may convert the host cell into a tumor cell.

Latent Viral Infections

Sometimes the virus remains latent in the nerve cells of the host for long periods without causing disease. Stress or some other cause may trigger their reappearance. This is the case with the herpes simplex virus, which causes cold sores, and the chickenpox virus, which causes shingles.

Slow Viral Infections

The term *slow viral infection* refers to a disease process that occurs gradually over a long period. It refers to the slow progress of the disease. An example may be *subacute sclerosing panencephalitis,* in which the measles virus continues to reproduce slowly, causing this rare encephalitis.

Prions

A number of diseases may be caused by **prions,** which have characteristics unique to biology. The prototype of these diseases is *scrapie,* a neurological disease of sheep. The prion appears to be pure protein and to lack nucleic acids. Among hypotheses to explain their reproduction is that the prion protein is a gene found in normal host DNA, and that an abnormal form of the protein causes the disease condition. Another possibility is that the prion may contain an undetectably small amount of nucleic acid. Some think that the infectious agent may be a bacterium that passes through filters that would normally serve to separate bacteria from viruses. Diseases caused by these agents, other than scrapie in sheep, include Creutzfeldt–Jakob disease, kuru, and Gerstmann–Straüssler syndrome.

Plant Viruses and Viroids

Some plant diseases are caused by **viroids.** These are very short pieces of nonenveloped RNA with no protein coat.

Self-Tests

In the matching section, there is only one answer to each question; however, the lettered options (a, b, c, etc.) may be used more than once or not at all.

I. Matching

1. A complete, assembled virus.

2. The subunits making up the protein outer coating of most viruses.

3. The protein outer coating of most viruses.

4. A term derived from the word for poison.

5. A combination of lipids, proteins, and carbohydrates covering the protein coating of a virus.

a. Virion

b. Capsid

c. Capsomere

d. Envelope

e. Virus

II. Matching

1. Describes the morphology of the capsid of many viruses.

2. A method by which a virus enters an animal host cell.

3. A cell line derived from tissue that normally reproduces for relatively few generations.

4. The HeLa cell line would be placed in this group.

5. A clearing in a "lawn" of susceptible bacterial cells.

6. The number of bacteriophages produced by one bacterial host cell.

7. Presumed agent causing diseases such as sheep scrapie.

8. A bacterial virus.

9. A short strand of RNA virus without a capsid.

a. Burst size

b. Burst time

c. Primary cell line

d. Continuous cell line

e. Plaque

f. Cytopathic effect

g. Icosahedral

h. Endocytosis

i. Phage

j. Viroid

k. Diploid cell line

l. Prion

III. Matching

1. Describes a method by which an enveloped virus leaves the host cell while acquiring the envelope.

2. Describes growth characteristics of normal cell cultures in glass or plastic containers.

3. A term meaning cancer-causing.

4. Observable changes in a virus-infected cell.

5. The time during which the capsids and DNA of a phage, already formed, are now assembled into complete viruses.

a. Replicative form

b. Maturation period

c. Budding

d. Oncogenic

e. Cytopathic effect

f. Endocytosis

g. Monolayer

h. Eclipse period

IV. Matching

1. Cancer of connective tissue.

2. The clumping of red blood cells due to adherence to spikes on viruses.

3. Equivalent to mRNA in a single-stranded RNA virus.

4. RNA to DNA.

   a.   Sarcoma

   b.   + or sense strand

   c.   Reverse transcription

   d.   Interferon

   e.   Hemagglutination

V. Fill-in-the-Blanks

1. The virus, once inside the host cell, separates the viral nucleic acid from the capsid; this is called _____ .

2. Another term for a lysogenic phage is _____ phage.

3. _____ are not solid tumors but an excessive production of white blood cells.

4. Many viruses can be grown in _____ eggs.

5. The herpes simplex virus remains _____ in nerve cells of the host for long periods without causing disease.

6. Counts of phage are made in terms of _____ units.

7. An oncogene might become active when placed on the chromosome in a position where normal controls are not active; this is termed _____ .

8. The term _____ refers to the spectrum of host cells the virus can infect.

9. When cells multiply in an uncontrolled way, the excess tissue is called a _____ .

10. Oncogenic viruses are those that _____ cells into tumor cells.

11. The type of virus implicated as a cause of AIDS is a _____ .

12. The acronym TSTA stands for tumor-specific _____ antigens.

13. For several minutes following infection by a phage, no complete phages can be found in the host cell; this is called the _____ period.

14. The _____ of the phage is adsorbed to the host cell.

15. The phage forms a hole in the cell wall using phage _____ and drives the tail core through the cell wall.

16. Sometimes the lytic cycle does not occur upon phage infection of a host bacterium. The phage DNA becomes incorporated as a _____ into the host's DNA.

17. When the phage DNA is incorporated into the host's DNA, this state is called _____ .

18. Transformed cells lose _____ ; that is, they do not stop reproduction when in contact with neighbor cells.

19. The hepadnavirus has genetic material called _____ NA.

20. Picornaviruses have genetic material called _____ NA.

21. Tumors are malignant when cancerous and _____ when not cancerous.

9. During 1993, several deaths caused by a virus occurred in the southwestern United States. Eventually, other cases surfaced in other parts of the country. What method was used to isolate the viral agent? What genus of virus caused the outbreak?

10. Briefly outline the lytic cycle of the T-even bacteriophages.

CHAPTER 14

Principles of Disease and Epidemiology

Learning Objectives

After completing this chapter, you should be able to:

- Define pathogen, etiology, infection, host, and disease.
- Define normal microbiota.
- Compare and contrast normal, transient, and opportunistic microbes.
- Compare commensalism, mutualism, and parasitism, and give an example of each.
- List Koch's postulates.
- Differentiate between a communicable and a noncommunicable disease.
- Categorize diseases according to incidence and according to severity.
- Define reservoir of infection; contrast human, animal, and nonliving reservoirs, and give one example of each.
- Explain four methods of disease transmission.
- Define nosocomial infections and explain their importance.
- Provide an example of a compromised host.
- Identify four predisposing factors for disease.
- Put the following terms in proper sequence in terms of the pattern of disease: period of decline, period of convalescence, period of illness, prodromal period, period of incubation.
- Define epidemiology, and describe three types of epidemiologic investigation.

Pathology, Infection, and Disease

We are susceptible to **pathogens** (disease-causing microorganisms). **Pathology** is the science that deals with the study of disease. It involves the **etiology** (cause) of the disease, the manner in which a disease develops **(pathogenesis),** the structural and functional changes brought about by the disease, and the final effects on the body. **Infection** means invasion or colonization of the body (the host, in this case) by potentially pathogenic microorganisms. **Disease** itself is any change from a state of health, an abnormal state in which the body is not properly adjusted or capable of performing its normal functions.

Normal Microbiota

Microorganisms that establish permanent residence without producing disease are known as **normal flora** or **normal microbiota.** Other microorganisms that may be present for a time and then disappear are called **transient microbiota.** Normal microbiota can benefit the host by preventing the overgrowth of harmful microorganisms, a process called **microbial antagonism.** An example is the production of **bacteriocins** by *E. coli* cells in the large intestine, which inhibits pathogens such as *Salmonella* and *Shigella.* The relationship between the normal microbiota of a healthy person and that person is called **symbiosis.** If, in the symbiosis, one of the organisms is benefited and the other unaffected, the relationship is known as **commensalism.** If both organisms are benefited it is called **mutualism,** and if one organism is benefited at the expense of the other, it is **parasitism.** Under certain conditions these relationships can change, and members of the normal microbiota can become **opportunistic pathogens.**

Etiology of Infectious Disease

Not all diseases are caused by microorganisms. We have **inherited (genetic)** diseases (like hemophilia) and **degenerative diseases** (osteoarthritis). Here we will discuss **infectious diseases**—those caused by microorganisms.

Koch's Postulates

Koch's postulates must be fulfilled in order to demonstrate that a specific microorganism is the cause of a specific disease.

1. The same pathogen must be present in every case of the disease.
2. The pathogen must be isolated from the diseased host and grown in pure culture.
3. The pathogen from the pure culture must cause the disease when inoculated into a healthy, susceptible laboratory animal.
4. The pathogen must again be isolated from the inoculated animal and must be shown to be the same pathogen as the original organism.

A few diseases, such as syphilis, are caused by organisms that cannot be isolated and virulent strains grown on laboratory media. However, evidence based on the first postulate over many years' experience has been adequate to show that a certain organism is associated with the disease. Some diseases have such clearly defined symptoms—diphtheria or tetanus, for example—that associations between specific microorganisms and the disease are obvious. Other diseases—such as pneumonia, peritonitis, and meningitis—may be caused by a variety of microorganisms, and the specific etiology is not easy to determine from the symptoms. Furthermore, some pathogens may infect a number of different organs or tissues and cause very different diseases or disease symptoms. A good example is *Streptococcus pyogenes,* which can cause sore throats, scarlet fever, erysipelas, and puerperal fever.

Classifying Infectious Diseases

Symptoms are changes in body function felt by the patient, such as pain and malaise (a vague feeling of body discomfort), that are *subjective* and not apparent to an observer. The patient also may exhibit **signs,** *objective* changes that the physician can observe and measure. A specific group of symptoms or signs accompanying a particular disease is called a **syndrome.**

Communicable diseases are spread directly or indirectly from one host to another; typhoid fever and tuberculosis are examples. **Noncommunicable diseases** are caused by microorganisms such as *Clostridium tetani,* which only produces tetanus when it is introduced into the body by contamination of wounds. A **contagious disease** is easily spread from one person to another.

The **incidence** of a disease is the fraction of a population that contracts it during a particular length of time. The **prevalence** of a disease is the fraction of the population that has the disease at a given time. If a disease occurs only occasionally, it is called **sporadic;** when it is constantly present, as is the common cold, it is termed **endemic.** If many people in a given area acquire a certain disease in a short period of time, it is referred to as an **epidemic** disease, such as influenza. A worldwide epidemic is referred to as a **pandemic** disease.

An **acute disease** is one that develops rapidly but lasts only a short time—influenza, for example. A **chronic disease** develops more slowly and the body reactions are often less severe, but it is continuous or recurrent for long periods of time. Tuberculosis, syphilis, and leprosy are such diseases. Diseases intermediate between acute and chronic are described as **subacute.** A **latent disease** is one in which the pathogen is inactive for a time but then becomes active to produce the symptoms. Shingles is an example.

Extent of Host Involvement

A **local infection** is one in which the invading microorganisms are limited to a relatively small area of the body, as, for example, in boils or abscesses. In a **systemic,** or **generalized,** infection, microorganisms or their products are spread throughout the body by the blood or lymphatic system. A **focal infection** is one in which a local infection, such as infected teeth, tonsils, or sinuses, enters the blood or lymph and spreads to other parts of the body. The presence of bacteria in the blood is known as **bacteremia,** and the rapid multiplication of bacteria in the blood is called **septicema.** **Toxemia** is the presence of toxins in the blood, and **viremia** is the presence of viruses in the blood. A **primary infection** is an acute infection that causes the initial illness. A **secondary infection** is one caused by an opportunist only after the primary infection has weakened the body's defenses. An **inapparent,** or **subclinical, infection** is one that does not cause any noticeable illness.

Spread of Infection

Reservoirs

A continual source of the pathogen, such as an animal or inanimate object, is a **reservoir of infection.**

Human Reservoirs. Many people harbor pathogens and transmit them to others, directly or indirectly. These people may be diseased and obvious transmitters, but others, called **carriers,** do not exhibit symptoms.

Animal Reservoirs. Diseases that occur primarily in wild and domestic animals but that can be transmitted to humans are called **zoonoses.** Transmission may be by direct contact with infected animals; contamination of food and water; insect vectors; contact with contaminated hides, fur, or feathers; or consumption of infected animal products.

Nonliving Reservoirs. Examples of nonliving reservoirs for infectious diseases are soil, which harbors the agent of botulism, and water contaminated by human or animal feces, which transmits gastrointestinal pathogens.

Transmission of Disease

Contact Transmission. Infections may be spread more or less directly from one host to another by *direct contact,* such as kissing, handshaking, bites, or sexual intercourse. *Droplet infection,* in which agents of disease are spread very short distances while contained in droplets of saliva or mucus from coughing or sneezing, also is considered a form of contact transmission. *Indirect contact* transmission involves a nonliving object, such as a drinking cup or towel, called a **fomite.**

Vehicle Transmission. Inanimate reservoirs such as food, water, or blood may transmit diseases to large numbers of individuals. Diseases spread by agents of infection traveling on droplets or dust for a distance of more than a meter are considered to be by *airborne transmission.* Spores produced by fungi also can be transmitted by the airborne route.

Vectors. Arthropods are the most important group of disease **vectors**—animals that carry pathogens from one host to another. In **mechanical transmission,** insects, such as flies, carry pathogens on their bodies to food that is later swallowed by the host. In **biological transmission,** the arthropod may pass the pathogen in a bite, or it may pass the pathogen in its feces, which later enters the wound caused by the arthropod's bite.

Nosocomial (Hospital-Acquired) Infections

Hospital patients often are **compromised hosts** with lowered resistance and increased susceptibility to infection. The three most likely conditions to compromise a host are: (1) broken skin or mucous membrane, (2) a suppressed immune system, and (3) impaired defense cell activity. Many nosocomial infections are caused by opportunistic pathogens. At one time nosocomial diseases were caused mostly by gram-positive bacteria, such as streptococci or staphylococci. Although antibiotic-resistant strains of these organisms are still significant causes of nosocomial diseases, today most are caused by gram-negative bacteria, such as *Pseudomonas aeruginosa* or *Escherichia coli.*

Patterns of Disease

Predisposing factors, such as gender, genetic background, climate, age, and nutrition, can greatly affect the occurrence of disease in individuals.

Development of Disease

The development of disease follows a certain sequence of steps. The **period of incubation** is the time between actual infection and the first appearance of signs or symptoms. The **prodromal period** follows the incubation period in some diseases and is characterized by mild symptoms of the disease. During the **period of illness,** the overt symptoms of the disease are apparent. During the **period of decline,** the signs and symptoms subside. The patient regains his or her prediseased state during the **period of convalescence.**

Epidemiology

The science that deals with the transmission of diseases in the human population, and where and when they occur, is called **epidemiology.** An epidemiologist determines not only the etiology of a disease, but also data such as geographical distribution, and gender, age, and so on, of persons affected. First, an epidemiologist collects data such as the place(s) where a disease occurred and time(s) when it occurred. Persons affected by the disease would provide information on age, sex, personal habits, and so on. This process is known as **descriptive epidemiology.**

These data are then studied **(analytical epidemiology)** to look for common factors among the affected persons that might have preceded the disease outbreak.

Experimental epidemiology tests a hypothesis, such as assumed effectiveness of a drug. Randomly selected groups receive either one of two substances, the selected drug or a placebo (a substance that has no effect). The groups are compared to discern the effectiveness of the drug versus the placebo.

The Centers for Disease Control and Prevention (CDC), a branch of the U.S. Public Health Service, located in Atlanta, Georgia, are a central source of epidemiological information in the United States. The CDC issues the *Morbidity and Mortality Weekly Report* (*MMWR*), which contains data on morbidity (relative incidence of a disease) and mortality (deaths from a disease). **Notifiable diseases** are those for which physicians must report cases to the Public Health Service.

Self-Tests

In the matching section, there is only one answer to each question; however, the lettered options (a, b, c, etc.) may be used more than once or not at all.

I. Matching

1. Invasion or colonization of the body by potentially pathogenic microorganisms.

2. The cause of a disease.

3. A change from a state of health, in which the body is not properly adjusted or capable of performing its normal functions.

4. The manner in which a disease develops.

a. Infection

b. Pathogenesis

c. Disease

d. Etiology

II. Matching

1. One organism is benefited at the expense of another.

2. The general relationship between the normal microbiota and the host.

3. One of the organisms is benefited and the other unaffected.

4. A symbiosis that benefits both organisms.

a. Symbiosis

b. Opportunistic

c. Commensalism

d. Mutualism

e. Parasitism

III. Matching

1. First mild symptoms appear.

2. The individual regains strength, and the body returns to its prediseased state.

3. The time between infection and the first appearance of signs and symptoms.

a. Prodromal period

b. Period of convalescence

c. Period of incubation

d. Period of illness

IV. Matching

1. Easily spread from one person to another person.

2. Tetanus is an example.

a. Contagious disease

b. Communicable disease

c. Noncommunicable disease

V. Matching

1. An inanimate object that may transmit disease.

2. A group of symptoms associated with a disease.

3. Identification of a disease.

4. Objective changes caused by a disease that the physician may observe.

5. An arthropod, for example, that carries malaria.

6. The rapid growth of a bacterium in the blood or lymph.

a. Septicemia

b. Bacteremia

c. Syndrome

d. Diagnosis

e. Signs

f. Vector

g. Fomite

VI. Fill-in-the-Blanks

1. Diseases acquired in a hospital are called _____ diseases.

2. People who transmit diseases, but who do not exhibit any symptoms of illness, are called _____ .

3. In _____ transmission of disease, an insect such as a fly carries the pathogen on its body to human food.

4. Diseases that occur in animals and can be transmitted to humans are called _____ .

5. The _____ of a disease is the fraction of the population that contracts it during a particular period of time.

6. _____ disease is one that develops rapidly but lasts for only a short time.

7. A disease that occurs only occasionally is called _____ .

8. The simple presence of bacteria in the blood is known as _____ .

9. A _____ infection is one caused by an opportunist after the primary infection has weakened the body's defenses.

10. _____ are changes in body function felt by the patient and subjective in nature, such as pain.

11. The science that deals with transmission of diseases in the human population, and when and where they occur, is called _____ .

12. The acronym CDC stands for _____ .

13. An abscess is an example of a _____ type of infection.

14. An infection in which the microorganisms or their products are spread throughout the body in the blood or lymphatic system is known as a _____ infection.

15. An inapparent, or _____ , disease is one that does not cause any noticeable illness.

16. The _____ of a disease is the fraction of the population having the disease at a given time.

17. Diseases between acute and chronic are described as being _____ .

18. A worldwide epidemic is referred to as a _____ disease.

19. A pathogen is found in all cases of a certain disease and grown in pure culture; then it is inoculated into a laboratory animal. What is the next step in Koch's postulates?

_____

_____

20. _____ is the science that deals with the study of disease.

VII. Label the Art

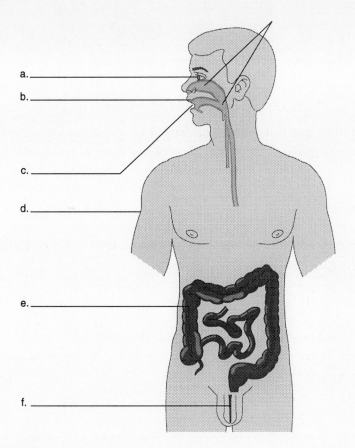

a. _____

b. _____

c. _____

d. _____

e. _____

f. _____

VIII. Critical Thinking

1. What type of symbiotic relationship exists between normal microbiota and the host? Give two examples of contributions made by normal microbiota to the human host.

2. Discuss how an infant is colonized with normal microbiota.

3. What is microbial antagonism, and how does it contribute to a healthy host? List and briefly discuss three examples of microbial antagonism.

4. List at least four procedures that may transmit nosocomial infections via fomites.

5. Explain the role of the infection-control nurse.

6. List and briefly describe the five steps in the development of disease.

7. Using at least two examples, explain microbial synergism.

8. Under what circumstances is it difficult to use Koch's postulates to determine the etiologic agent of an infectious disease? How has this problem been overcome?

9. What are emerging infectious diseases? Discuss the factors that contribute to their development.

10. Compare and contrast mechanical and biological transmission.

CHAPTER 15

Microbial Mechanisms of Pathogenicity

Learning Objectives

After completing this chapter, you should be able to:

- Define portal of entry, pathogenicity, virulence, and LD_{50}.
- Identify a microbe that enters the human body via (1) the mucous membranes, (2) the skin, and (3) the parenteral route.
- Using examples, explain how microbes adhere to host cells.
- Explain how capsules, cell wall components, and enzymes contribute to pathogenicity.
- Compare the effects of hemolysins, leukocidins, coagulase, kinases, hyaluronidase, and collagenase.
- Contrast the nature and effects of exotoxins and endotoxins.
- Outline the mechanisms of action of diphtheria toxin, botulinum toxin, tetanus toxin, cholera toxin, and lipid A.
- Identify the importance of the *Limulus* amoebocyte lysate assay.
- Using examples, describe the roles of plasmids and lysogeny in pathogenicity.
- List nine cytopathic effects of viral infections.
- Discuss the causes of symptoms in fungal, protozoan, helminthic, and algal diseases.

Pathogenicity is the ability to cause disease in a host. First, however, the pathogen must enter the host's body. **Virulence** is the degree of pathogenicity.

Entry of a Microorganism into the Host

The avenue by which a microbe gains access to the body is a **portal of entry**.

Mucous Membranes

To gain access to the body, pathogens can penetrate mucous membranes lining the conjunctiva of the eye and the respiratory, gastrointestinal, and genitourinary tracts. The respiratory tract is the easiest, most frequently used route of entry for infectious microorganisms. Diseases contracted by this route are the common cold, pneumonia, tuberculosis, influenza, measles, and smallpox. Microorganisms contracted from food, water, or fingers enter the body by the gastrointestinal tract. Although many are destroyed by stomach acids and intestinal enzymes, diseases such as poliomyelitis, hepatitis A, typhoid fever, amoebic dysentery, bacillary dysentery, and cholera are transmitted in this manner. The organism causing syphilis can penetrate intact mucous membranes.

Skin

With a few exceptions, such as the hookworm, microorganisms cannot penetrate unbroken skin. Some fungi, however, grow on the keratin of the skin, and other microorganisms gain access by penetrating into openings such as hair follicles and sweat ducts.

Parenteral Route

When the skin and mucous membranes are punctured or injured (traumatized), microorganisms can gain access to body tissues. This route is referred to as the **parenteral route.**

Preferred Portal of Entry

Whether or not disease results after entry of microorganisms depends on many factors. The organism must enter by a preferred route. *Salmonella typhi,* for example, must enter the gastrointestinal tract to cause typhoid, rather than being rubbed onto the skin. *Clostridium tetani* must penetrate the skin to cause tetanus, and *Corynebacterium diphtheriae* must enter the respiratory tract to cause diphtheria.

Numbers of Invading Microbes

A measure of virulence is the LD_{50} **(lethal dose),** which is the dose of pathogen that will kill half of the test animals. If the pathogen causes only a nonfatal disease, the test is referred to as ID_{50} **(infectious dose).** The fewer microorganisms required, the higher the virulence.

Adherence

For most pathogens, **adherence** is necessary for pathogenicity. The attachment between pathogen and host makes use of surface **ligands (adhesins)** and complementary surface receptors on the host cells. Ligands are mostly glycoproteins or lipoproteins, which are frequently associated with structures such as fimbriae.

How Pathogens Penetrate Host Defenses

Capsules

For many pathogens, such as *Streptococcus pneumoniae* and *Klebsiella pneumoniae,* **capsules** confer a resistance to phagocytosis.

Components of the Cell Wall

Cell surfaces sometimes contribute to invasiveness. *Streptococcus pyogenes,* for example, contains a protein called M protein on the fibrils and cell surface that helps it resist phagocytosis and improve adherence. Waxes in the cell wall of *Mycobacterium tuberculosis* resist digestion by macrophages.

Enzymes

Extracellular **enzymes** (exoenzymes) have the ability to break open cells, dissolve material between cells, and form or dissolve blood clots. These may contribute to invasiveness. **Leukocidins** are substances produced by some bacteria that destroy certain phagocytic cells. **Hemolysins** are enzymes that cause lysis of red blood cells. Staphylococci, streptococci, and *Clostridium perfringens,* the agent of gas gangrene, are important pathogens producing hemolysins. **Coagulases** produced by some members of the genus *Staphylococcus* are enzymes that coagulate blood. The fibrin clot formed may protect the bacterium from phagocytosis. **Bacterial kinases** are enzymes that break down fibrin and dissolve clots formed by the body to isolate infections. **Streptokinase (fibrinolysin),** produced by streptococci, and **staphylokinase,** produced by staphylococci, are the best known. **Hyaluronidase** is an enzyme secreted by certain bacteria that digests hyaluronic acid. This compound is a mucopolysaccharide that holds together certain body cells, much like mortar holds together bricks, especially in connective tissue. Both the organisms that produce gas gangrene and a number of streptococci produce this enzyme, which may promote spread of the infections to adjoining tissues. **Collagenase,** produced by several species of *Clostridium,* breaks down the collagen framework of the muscle tissue. **Necrotizing factors,** which cause the death of body cells, **hypothermic factors** affecting body temperature, **lecithinase** that destroys plasma membrane, **protease** that breaks down proteins, especially in muscle tissue, and **siderophores,** which scavenge iron from host tissue, also may be important to virulence.

How Bacterial Pathogens Damage Host Cells

Toxins

Toxins are poisonous substances produced by certain microorganisms. The capacity to produce them is called **toxigenicity.** The presence of toxins in the blood and lymph is called **toxemia.**

Exotoxins.   **Exotoxins** are proteins released from bacteria into the surrounding medium. Most bacteria involved are gram-positive, and the genes are probably all carried on bacterial plasmids or phages. Examples are **botulinum toxin,** which acts to prevent transmission of nerve impulses; **tetanus toxin** (tetanospasmin), which blocks the relaxation pathway and results in muscle spasms; **diphtheria toxin,** which inhibits protein synthesis in eucaryotic cells; **erythrogenic toxin,** which is produced by *Streptococcus pyogenes* and causes the red skin rash of scarlet fever; **staphylococcal enterotoxin,** which causes diarrhea and vomiting; and *Vibrio* **enterotoxin,** which causes diarrhea by altering the host's water and electrolyte balance.

Endotoxins.   **Endotoxins** are not secreted by the cell like the exotoxins. They are not proteins, but are lipopolysaccharide (LPS) components of the cell walls of most gram-negative bacteria. **Lipid A,** a portion of the LPS, is the endotoxin. They are released upon the death of the cell. They cause fever, weakness, aches, and sometimes shock. Fever related to endotoxins is triggered by **interleukin-1 (IL-1),** produced by macrophages after ingestion of bacteria. **Septic shock** is caused by endotoxins, which when ingested by a phagocyte cause it to secrete **tumor necrosis factor (cachectin).** There are no effective antitoxins, and antibodies are not very effective. Diseases associated with endotoxins are typhoid fever, bacillary dysentery, and epidemic meningitis. The *Limulus* amoebocyte lysate assay can detect even minute amounts of endotoxin.

Plasmids, Lysogeny, and Pathogenicity

Several diseases are caused only when the pathogen has lysogenized or carries a particular plasmid.

Pathogenic Properties of Nonbacterial Microorganisms

Viruses

Some viruses cause **cytocidal effects** (changes resulting in cell death), whereas other viruses are noncytocidal. Some **cytopathic effects,** or observable changes, are:

1. Cellular macromolecular synthesis may be stopped.
2. Enzymes are released from lysosomes and cause autolysis of the cell.
3. **Inclusion bodies** are formed, viral parts being assembled into complete virions.
4. Host cells fuse to produce **polykaryocytes,** or "giant" cells.
5. Antigenic changes are induced on the surface of infected cells, thus making destruction of the cell by the host's immune system more likely.
6. Chromosomal changes are induced in the host cell. Cancer-causing oncogenes may also be introduced.

7. Host cells are *transform*ed into spindle-shaped cells that do not have *contact inhibition* and grow without regulation.
8. Some virus-infected cells produce **interferon,** which helps protect neighboring cells from viral infection.

Fungi, Protozoa, Helminths, and Algae

Some protozoa and helminths grow on host tissues, causing cellular damage to the host. Their waste products may contribute to symptoms of disease in the host. Some algae produce neurotoxins.

Some toxins are associated with **fungi**—for example, **ergot,** which causes **ergotism.** Ergot is contained in sclerotia (resistant mycelia) of the plant pathogen *Claviceps purpurea.* Ergotism can result in hallucinations, or it may constrict blood capillaries and cause gangrene in the extremities. **Aflatoxin** is produced by the mold *Aspergillus flavus* and can cause cancer of the liver in animals. Toxic mushrooms such as *Amanita phalloides* contain the dangerous neurotoxins **phalloidin** and **amanitin.**

Some algae produce neurotoxins called **saxitoxin.** For example, people become ill after ingesting mollusks that feed on the algae.

Self-Tests

In the matching section, there is only one answer to each question; however, the lettered options (a, b, c, etc.) may be used more than once or not at all.

I. Matching

1. Produced by some members of the genus *Staphylococcus;* forms a fibrin clot around the bacterium.

2. A substance produced by some bacteria that destroys certain phagocytic cells.

3. Enzymes that cause lysis of red blood cells.

4. Enzymes that break down fibrin and dissolve clots.

5. The fibrinolysins produced by the streptococci.

6. May cause hallucinations or gangrene.

7. Tumor necrosis factor.

8. Giant cells caused by viral infection.

a. Leukocidins

b. Collagenase

c. Kinases

d. Hyaluronidase

e. Coagulase

f. Hemolysins

g. Cachectin

h. Aflatoxin

i. Amanitin

j. Polykaryocytes

k. Ergot

II. Matching

1. A protein secreted by a bacterium.   a. Exotoxin

2. The tetanus toxin is a good example.   b. Endotoxin

3. A lipopolysaccharide component of the cell wall of
   many gram-negative bacteria.

4. Released upon lysis of the cell.

5. Detected by *Limulus* amoebocyte assay.

III. Matching

1. The capacity to form toxins.   a. Pathogenicity

2. Provide immunity to exotoxins.   b. Toxemia

3. The presence of a toxin in the bloodstream.   c. Antitoxin

                                                                   d. Toxigenicity

IV. Matching

1. A way to measure virulence.   a. Traumatized

2. The degree of pathogenicity.   b. Virulence

3. The ability of a pathogen to cause disease in a host.   c. Pathogenicity

                                                                   d. LD_{50}

V. Fill-in-the-Blanks

1. The route followed when the skin or mucous membrane is punctured and pathogens gain
   access to body tissues is called the _____ route.

2. The term LD_{50} refers to the dose of pathogen that will kill half of the test
   _____ .

3. Capsules aid in invasiveness because they confer a resistance to _____ on
   many pathogens.

4. _____ , produced by several species of *Clostridium*, breaks down the collagen framework of the muscle tissue.

5. Hyaluronidase is an enzyme secreted by certain bacteria that digests _____ acid.

6. Diseases associated with _____ -toxins are typhoid fever, bacillary dysentery, and epidemic meningitis. (Supply the prefix.)

7. Some viruses cause _____ effects, which result in the cell's death.

8. One effect of the viral infection of an animal cell may be the release of enzymes from _____ and autolysis of the cell.

9. One effect of the viral infection of an animal cell may be to transform cells to spindle-shaped cells that have no _____ .

10. Many bacteria and viruses can penetrate the intact mucous membranes, such as the organism that causes _____ .

11. Contributing to invasiveness by *Streptococcus pyogenes* is a cell wall protein called _____ protein.

12. Also affecting virulence by bacteria are _____ factors that decrease body temperature.

13. With few exceptions, microorganisms cannot penetrate the unbroken skin; however, some fungi grow on _____ of the skin.

14. The avenue by which a microorganism gains access to the body is called a _____ .

15. If the pathogen causes a nonfatal disease, the equivalent to the LD_{50} test is referred to as _____ .

16. The toxin ergot is contained in _____ (resistant mycelia) of a plant pathogen.

17. A product released by macrophages after ingestion of bacterial endotoxins that triggers the appearance of fever is called _____ .

18. A product produced by a virus-infected cell that inhibits viruses from infecting neighboring cells is _____ .

VI. Label the Art

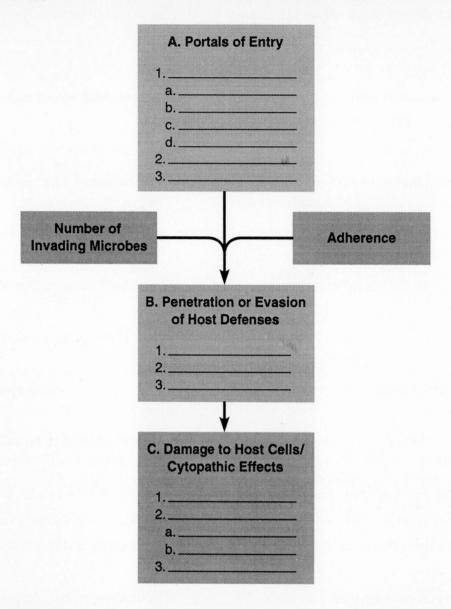

A. Portals of Entry

1. _____
   a. _____
   b. _____
   c. _____
   d. _____
2. _____
3. _____

Number of Invading Microbes

Adherence

B. Penetration or Evasion of Host Defenses

1. _____
2. _____
3. _____

C. Damage to Host Cells/ Cytopathic Effects

1. _____
2. _____
   a. _____
   b. _____
3. _____

VII. Critical Thinking

1. Using examples, discuss how adhesins contribute to the specificity of pathogenic microorganisms.

2. Explain the relationship between the waxy cell wall of *Mycobacterium tuberculosis* and virulence.

3. A 14-year-old male patient with a sore throat is examined by his physician. A throat swab cultured on blood agar produces beta-hemolytic streptococci. Antibodies against M protein are detected in the blood.

    a. What is the etiologic agent?

    b. What is the probable port of entry?

    c. What is the probable route of transmission?

4. Discuss two toxins associated with *Clostridium* infections. What is the route of entry for each? Which of these toxins is produced only by phage-infected *Clostridium* cells?

5. Compare and contrast hemolysins and coagulases.

6. List and briefly discuss seven ways that pathogenic species of *Streptococcus* can avoid the host's defensive mechanisms.

7. A patient begins a course of antibiotics on Tuesday morning. By Tuesday evening her symptoms are worse and include fever, chills, body ache, and weakness. The fever responds to aspirin and by Wednesday the symptoms have subsided.

    a. What is the Gram designation of the infecting organism?

    b. What specific property of the pathogen caused the symptoms?

    c. Why did the fever respond to aspirin?

    d. Why did the symptoms improve by the second day of antibiotic therapy?

8. Discuss the sequence of events that lead to shock in some gram-negative infections.

9. How does the architecture of HIV help it to avoid the action of antibodies? Does this structure serve any other purpose?

10. Compare and contrast the effects of ergot and aflatoxin.

CHAPTER 16

Nonspecific Defenses of the Host

Learning Objectives

After completing this chapter, you should be able to:

- Define resistance and susceptibility.
- Define nonspecific resistance.
- Describe the role of the skin and mucous membranes in nonspecific resistance, and differentiate between mechanical and chemical factors.
- Define phagocytosis, and include the stages of adherence and ingestion.
- Classify phagocytic cells and describe the roles of granulocytes and monocytes.
- Describe the stages of inflammation and their relation to nonspecific resistance.
- Describe phagocyte migration and phagocytosis.
- Discuss the role of fever in nonspecific resistance.
- Discuss the function of complement.
- Discuss the role of interferon.

Pathogenic microorganisms are endowed with special properties that, given the right opportunity, enable them to cause disease. Our bodies, however, have defenses against these microorganisms. In general, our ability to ward off disease through our defenses is called **resistance,** and our vulnerability or lack of resistance is known as **susceptibility. Nonspecific resistance** refers to defenses that tend to protect us from any kind of pathogen. **Specific resistance,** or immunity based on antibody production, is a defense against a particular microorganism.

Skin and Mucous Membranes

Mechanical Factors

The intact skin consists of the **dermis,** an inner, thicker portion composed of connective tissue, and the **epidermis,** an outer, thinner portion consisting of several layers of epithelial cells arranged in continuous sheets. The top layer of epidermal cells contains the protein **keratin.** Intact, these barriers are seldom penetrated by microorganisms. **Mucous membranes** line the body cavities that open to the exterior; these include the digestive, respiratory, urinary, and reproductive tracts. Mucous membranes consist of an **epithelial layer** and an underlying connective tissue layer. Cells in mucous membranes secrete **mucus,** which prevents the cavities from drying out. Some pathogens such as *Treponema pallidum* (syphilis) grow in mucus and are able to penetrate the membrane, which is generally less protective than the skin. Another mechanical factor involved in protection is the **lacrimal apparatus,** which manufactures and drains away tears. This apparatus has a cleansing effect on the eye surface. **Saliva,** produced by salivary glands, washes microorganisms from the surfaces of the teeth and mucous membrane of the mouth. Mucus tends to trap microorganisms that enter the respiratory and digestive tracts. Cells of the mucous membrane of the lower respiratory tract contain cilia, microscopic hairlike projections that move synchronously **(ciliary escalator)** and propel inhaled microorganisms trapped in mucus out of the respiratory system. The **flow of urine** tends to remove microorganisms from the urinary system, and **vaginal secretions** tend to move microorganisms out of the female body.

Chemical Factors

Sebaceous oil glands of the skin produce **sebum,** which forms a protective film over the skin surface. The unsaturated fatty acids of sebum inhibit the growth of certain pathogens. The low pH (3 to 5) of the skin is partly due to these secretions. Body odor is the result of the decomposition of both sloughed-off skin cells and sebum secretions by commensal bacteria. **Sweat glands** produce **perspiration,** which flushes microorganisms from the skin surface. Perspiration contains **lysozyme,** an enzyme that breaks down the cell walls of gram-positive bacteria. The enzyme is also found in saliva, tears, nasal secretions, and tissue fluids. **Gastric juice** is a mixture of hydrochloric acid, enzymes, and mucus, with an acidity of pH 1.2 to 3. These stomach secretions destroy many ingested microorganisms and toxins. Important exceptions are the toxins of botulism and *Staphylococcus aureus.* The bacterium *Helicobacter pylori* neutralizes stomach acid and can grow in the stomach, where it causes gastritis and ulcers. Resident microbiota usually compete more successfully for available nutrients and may produce metabolic end-products that inhibit colonization by pathogens. Blood contains iron-binding proteins **(transferrins)** that inhibit bacterial growth by reducing available iron.

Phagocytosis

Operating within the body is the host's second line of defense, of which **phagocytosis** is an important element. The word phagocytosis derives from the Greek words for "eat" and "cell" and refers to the ingestion of a microorganism or any particulate matter by a cell. **Phagocytes** are blood cells or derivatives of blood cells that perform this function (Figure 16.1).

Formed Elements in Blood

Blood fluid is called **plasma,** and cells and cell fragments of the blood are the **formed elements.** Most important for the present discussion are the **leukocytes,** or white blood cells. A **differential white blood cell count** detects leukocyte number changes. Leukocytes are subdivided into the following three categories:

1. **Granulocytes** have granules in the cytoplasm and are differentiated into **neutrophils,** which stain red and blue with a mixture of acidic and basic dyes; **basophils,** which stain blue with basic methylene blue; and **eosinophils,** which stain red with the acidic dye eosin. Neutrophils also are known as **polymorphonuclear leukocytes (PMNs),** or polymorphs.

Neutrophils (which account for 60% to 70% of leukocytes) are important in phagocytosis; basophils (0.5% to 1% of leukocytes) form heparin and histamines, a factor in inflammation and allergic responses; and eosinophils perform some phagocytosis but mainly attach to the outer surface of helminthic parasites and discharge lethal peroxide ions. Eosinophils increase in numbers during parasitic infections and allergic reactions.

An increase in the number of white blood cells during infection is called **leukocytosis.** Diseases causing this increase include meningitis, mononucleosis, appendicitis, pneumonia, and gonorrhea. Other diseases, such as typhoid fever, cause decreased leukocyte counts called **leukopenia.**

2. **Lymphocytes** (important to specific immunity) are not phagocytic. They occur in lymphoid tissue.

3. **Monocytes** also lack granules and are phagocytic only after maturing into **macrophages.**

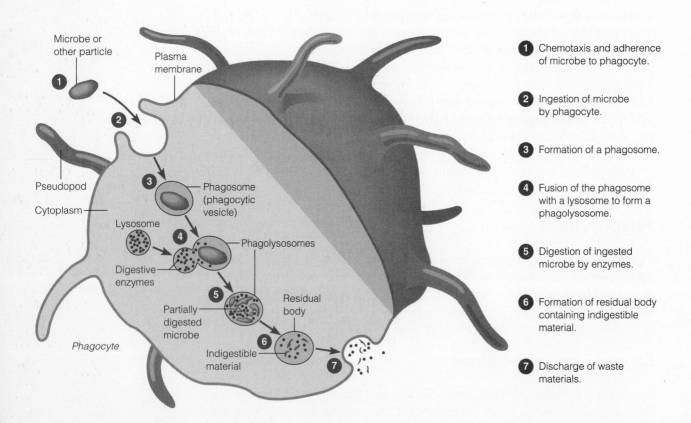

FIGURE 16.1 **The mechanism of phagocytosis in a phagocyte.**

Granulocytes and monocytes migrate to sites of infection. Granulocytes are mostly neutrophils that wander in the blood and can pass through capillary walls to reach trauma sites. Macrophages are highly phagocytic cells called **wandering macrophages** because of their ability to migrate. **Fixed macrophages** enter tissues and organs and remain there. The fixed macrophages are referred to as the **mononuclear phagocytic (reticuloendothelial) system.** Granulocytes predominate early in infections, as indicated by differential counts; as infections subside, monocytes predominate. Table 16.1 shows the classification of phagocytes.

TABLE 16.1 Formed Elements in Blood

TYPE OF CELL	ILLUSTRATION	NUMBERS PER CUBIC MILLIMETER (mm^3)	FUNCTION
Erythrocytes (red blood cells)		4.8 to 5.4 million	Transport of O_2 and CO_2
Leukocytes (white blood cells)		5000 to 9000	
A. Granulocytes (stained) 1. Neutrophils (PMNs) (60% to 70% of leukocytes)			Phagocytosis
2. Basophils (0.5% to 1%)			Production of heparin and histamine
3. Eosinophils (2% to 4%)			Production of toxic proteins against certain parasites; some phagocytosis
B. Monocytes (3% to 8%)			Phagocytosis (when they mature into macrophages)
C. Lymphocytes (20% to 25%)			Antibody production (B lymphocytes); cell-mediated immunity (T lymphocytes)*
Platelets		250,000 to 400,000	Blood clotting

*Discussed in Chapter 17.

Mechanism of Phagocytosis

Chemotaxis and Adherence. **Adherence,** or attachment between the cell membrane of the phagocyte and the organism, is facilitated by **chemotaxis,** which is the attraction of microorganisms to chemicals. Capsules help organisms avoid phagocytosis. **Opsonization**—coating a microorganism with plasma proteins such as antibodies and complement—promotes phagocytosis.

Ingestion and Digestion. Following adherence, projections of the cell membrane of the phagocyte **(pseudopods)** engulf the microorganism and then fold inward, forming a sac around it called a **phagosome** or **phagocytic vesicle.** This process is called **ingestion.** Within the cell, enzyme-containing phagosomes and lysosomes of the cell fuse to form a larger structure, the **phagolysosome,** within which the bacteria are usually quickly killed (Figure 16.1). Enzymes in lysosomes include lysozyme, various hydrolytic enzymes, and myeloperoxidases. Indigestible material in the phagolysosome is called the **residual body** and moves to the cell boundary, where it is discharged. Some microorganisms, such as the agents of tularemia, brucellosis, and tuberculosis, may remain dormant in phagocytes or even grow actively.

Inflammation

Inflammation is a host response to tissue damage, characterized by redness, pain, heat, swelling, and perhaps loss of function. It is generally beneficial, serving to destroy the injurious agent, to confine or wall it off, and to repair or replace damaged tissue.

Vasodilation and Increased Permeability of Blood Vessels

Vasodilation is the first stage of inflammation; it involves an increase in blood vessel diameters in the injured area. Increased **permeability** allows defense substances in the blood to pass through the walls of the blood vessels. Vasodilation is also responsible for the redness, heat, **edema** (swelling), and pain of inflammation. **Histamines** are released by injury and increase permeability as well. **Kinins** cause vasodilation and also increase permeability and attract phagocytes. **Prostaglandins** (substances released by damaged cells) and **leukotrienes** (substances related to prostaglandins) cause vasodilation. **Clotting elements** in the blood help prevent the spread of the infection; a focus of infection walled off in this fashion is an **abscess.**

Phagocyte Migration and Phagocytosis

Within an hour of the beginning of inflammation, phagocytes appear at the site. Blood flow decreases as the phagocytes stick to the inner lining of blood vessels **(margination)** and then pass through the vessel wall to the damaged area. This migration (cell walking) is called **diapedesis.** Monocytes appear later in the inflammatory response and mature into macrophages. Macrophages ingest dead tissue and invading microorganisms. Collections of dead phagocytic cells and fluids are called **pus.**

Tissue Repair

The final stage of inflammation is **tissue repair,** which involves production of new cells by the **stroma** (supporting connective tissue) and the **parenchyma** (functioning part of tissue). Scar tissue results from stroma-type repair.

Fever

The hypothalamus controls body temperature (normally 37°C or 98.6°F). The setting can be altered by ingestion of gram-negative bacteria by phagocytes. The release of endotoxins causes release of interleukin-1 (endogenous pyrogens). A **chill** (cold skin and shivering) is a sign of rising temperature. **Crisis** refers to a very rapid fall in temperature. Fevers are often beneficial as aids to body tissue repair and inhibitors of microbial growth.

Antimicrobial Substances

The Complement System

Complement consists of a group of at least 20 different proteins found in serum. It participates in lysis of foreign cells, inflammation, and phagocytosis. The system can be activated by an immune reaction in the classical pathway. In the alternative pathway, complement interacts directly with a bacterium; no antibodies are involved.

The complement proteins act in an ordered sequence, or **cascade;** one protein activates another. Protein C3, as shown in Figure 16.2, plays a central role in both the classical and the alternative pathways.

In the **classical pathway,** activity is initiated when antibody molecules bind to the antigen—a bacterial cell, for example. In the **alternative pathway,** which does not involve antibodies, complement proteins, and proteins called factors B, D, and P, combine with certain microbial polysaccharides. Especially affected are the lipopolysaccharide cell wall portions (endotoxins) of gram-negative enteric bacteria.

Consequences of Complement Activation.   Cytolysis: Complement protein then binds to two adjacent antibodies and initiates a sequence known as the **membrane attack complex.** Circular lesions called **transmembrane channels** are formed and cause the eventual lysis of the cell, to which the antibodies are attached. **Inflammation** also can develop from complement. Other complement proteins combine with mast cells and trigger the release of histamine, which increases blood vessel permeability. One protein chemotactically attracts phagocytes to the site. **Opsonization,** or immune adherence, promotes attachment of a phagocyte to the microbe. Complement involvement results from interaction with special receptors on the phagocytes.

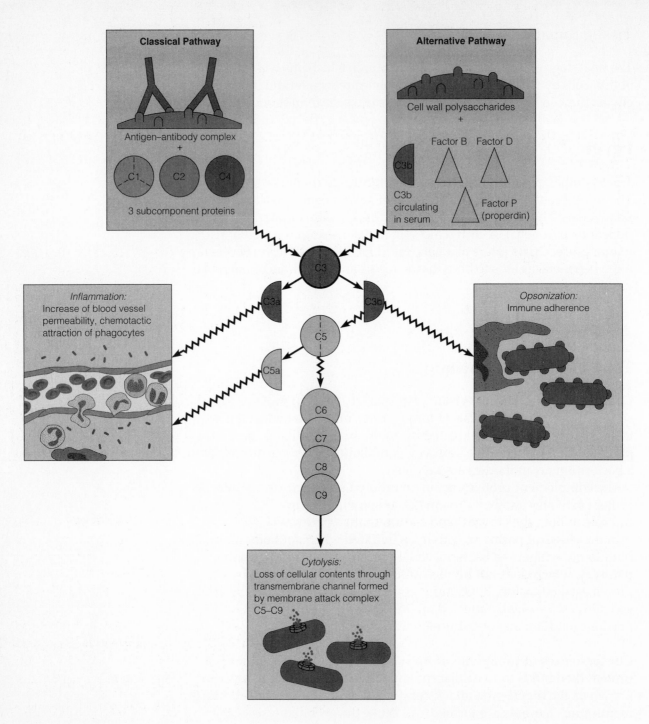

FIGURE 16.2   Complement activation via classical and alternative pathways. Note the cleavage of C3 into C3a and C3b. These fragments induce three kinds of consequences destructive to microorganisms: cytolysis, inflammation, and opsonization.

Interferons (IFNs)

Interferons of three types (alpha, beta, and gamma) are proteins produced by virus-infected animal cells. They tend to interfere with viral multiplication by inducing the uninfected cell to manufacture mRNA for synthesis of **antiviral proteins.** Interferons are most important in protection against acute virus-caused infections, such as influenza. Interferons can be produced by biotechnological methods and have been tested to determine their antiviral and anticancer effects. The three types have different effects.

Self-Tests

In the matching section, there is only one answer to each question; however, the lettered options (a, b, c, etc.) may be used more than once or not at all.

I. Matching

1. Produces tears.

2. The outer layer of the skin.

3. An oily substance forming a protective film over the skin surface.

4. Secreted by cells in mucous membrane; prevents the cavities from drying.

5. The inner portion of the skin, composed of connective tissue.

a. Dermis

b. Epidermis

c. Mucus

d. Mucous membrane

e. Lacrimal apparatus

f. Sebum

II. Matching

1. The blood fluid.

2. Cells and cell fragments of the blood.

3. Immunity based on antibodies.

4. Movement by a microorganism toward an attractant chemical.

5. An increase in the diameter of blood vessels.

6. A collection of dead phagocytic cells and fluids.

7. Vulnerability to a pathogen.

a. Serum

b. Plasma

c. Formed elements

d. Susceptibility

e. Specific resistance

f. Chemotaxis

g. Vasodilation

h. Pus

i. Opsonization

III. Matching

1. Neutrophils.

2. Monocytes.

3. Lymphocytes.

a. No granules in cellular cytoplasm; important to specific immunity

b. Granulocytes

c. Mature into macrophages

IV. Matching

1. An increase in the number of white blood cells.

2. Projections of the cell membrane of a phagocyte.

3. A larger structure formed when lysosome and phagosome fuse.

4. A decrease in the number of white blood cells.

a. Pseudopods

b. Phagolysosome

c. Phagosome

d. Lysozyme

e. Lysosome

f. Leukocytosis

g. Leukopenia

V. Fill-in-the-Blanks

1. Some cells of the mucous membrane of the lower respiratory tract contain _____ , which are microscopic hairlike projections.

2. _____ glands produce perspiration.

3. _____ are also known as polymorphonuclear leukocytes.

4. The most numerous of the granulocytes are the _____ .

5. Complement acts in a sequence called a _____ .

6. Percentages of the various types of white blood cells are determined by doing a _____ .

7. In the membrane attack complex associated with complement, circular lesions called _____ channels are formed.

8. The various macrophages of the body are referred to as the _____ system.

9. Phagocytes may stick to the inner lining of blood vessels before passing through the vessel wall to the damaged area; this action is called _____ .

10. The _____ controls body temperature.

11. _____ is a group of at least 20 proteins found in blood serum.

12. Granulocytes that stain with basic methylene blue dyes are known as _____ .

13. Lymphocytes and monocytes do not have _____ in their cytoplasm.

14. The attachment between the cell membrane of the phagocyte and the organism is called _____ .

15. The coating of a microorganism with plasma proteins such as complement is called _____ and promotes phagocytosis.

16. The host response to tissue damage characterized by redness, pain, heat, and swelling is called _____ .

17. The migration of phagocytes through the vessel wall to the damaged area is called _____ .

18. Compounds produced by the body that cause vasodilation include _____ .

19. Scar tissue results from _____ -type repair.

20. The _____ pathway of the complement system does not involve antibodies.

21. Eosinophils stain red with the acidic dye _____ .

22. _____ attach to the outer surface of helminthic parasites and discharge lethal peroxide ions.

23. C-proteins react with mast cells and attached antibodies and cause the release of _____ , which increase blood vessel permeability.

24. A protein in blood that inhibits microbial growth by reducing the amount of available iron is _____ .

VI. Label the Art

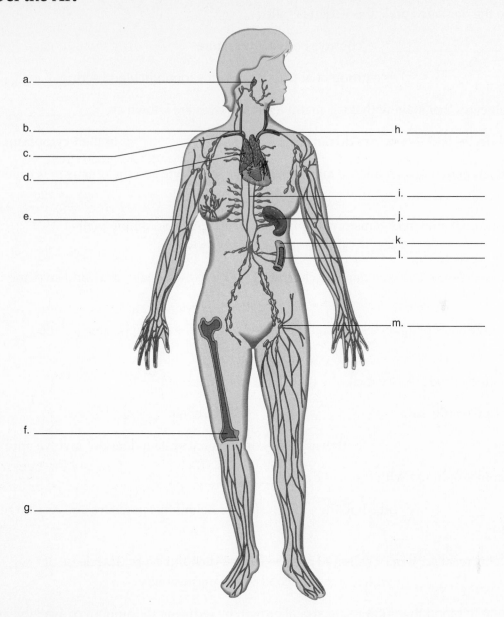

a. _____

b. _____

c. _____

d. _____

e. _____

f. _____

g. _____

h. _____

i. _____

j. _____

k. _____

l. _____

m. _____

VII. Critical Thinking

1. List and briefly discuss the nonspecific defense mechanisms that help prevent infection of the respiratory tract.

2. Discuss the role of sebum as a chemical factor contributing to the nonspecific defenses of the body. How do normal microbiota avoid the effects of sebum?

3. How do some bacteria overcome the strongly acidic environment of the stomach to cause disease?

4. Define phagocytosis. How do some bacteria avoid or survive the action of phagocytes?

5. Compare and contrast the roles of wandering macrophages and fixed macrophages.

6. Define opsonization. How does opsonization help enhance phagocytosis?

7. Discuss the role of histamines and kinins in vasodilation.

8. What is diapedesis? What happens to phagocytes after the majority of invading microorganisms and damaged tissue have been eliminated?

9. What roles do stroma and parenchyma play in tissue repair? What is the result when only parenchyma cells are active during tissue repair? When only stroma repair cells are active?

10. How does C3 contribute to the destruction of invading microorganisms?

CHAPTER 17

Specific Defenses of the Host: The Immune Response

Learning Objectives

After completing this chapter, you should be able to:

- Define specific defenses, innate resistance, and immunity.
- Contrast the four types of acquired immunity.
- Differentiate between humoral and cell-mediated immunity.
- Explain what an antigen is and what a hapten is.
- Explain what an antibody is, and describe the structural and chemical characteristics of antibodies.
- Describe at least one function for each of the five classes of antibodies.
- Describe the clonal selection theory.
- Explain how an antibody reacts with an antigen; identify the consequences of the reaction.
- Distinguish between a primary and a secondary response.
- Describe the genetic basis and molecular mechanisms that account for antibody diversity.
- Outline the procedure for making monoclonal antibodies.
- Describe at least one function for each of the four types of T cells, memory cells, and natural killer cells.
- Identify at least one function of each of the following in cell-mediated immunity: cytokine, APC, MHC.
- Compare and contrast T-dependent antigens and T-independent antigens.
- Describe the role of antibody and natural killer cells in antibody-dependent cell-mediated cytotoxicity.
- Compare and contrast humoral and cell-mediated immunity.

In addition to the nonspecific host defenses discussed in the previous chapter, humans have **innate resistance** to some illnesses. We do not get certain animal diseases; some racial groups, probably because of historical exposure, are comparatively more susceptible to some diseases. In this chapter, we will discuss the specific defenses of the host—that is, the immune system.

Types of Acquired Immunity

Acquired immunity refers to immunity developed in response to **antigens,** which are organisms or substances that provoke the production of special proteins called **antibodies** or certain **specialized lymphocytes,** in particular **B cells** and **T cells.**

Naturally Acquired Immunity

Naturally acquired active immunity is obtained by natural exposure to antigens—for example, disease organisms. **Subclinical infections** (no evident symptoms) also can confer immunity. **Naturally acquired passive immunity** involves transfer of antibodies formed by the mother to her infant. This may be done by **transplacental transfer** and renders the infant immune to most of the diseases to which the mother was immune. Breast milk, especially the first secretions called **colostrum,** provides some immunity. These passive immunities last only a few weeks or months.

Artificially Acquired Immunity

Artificially acquired active immunity results from vaccination (immunization) in which **vaccines** composed of inactivated bacterial toxins **(toxoids),** killed microorganisms, or living but attenuated (weakened) microorganisms are injected. **Artificially acquired passive immunity** involves injection of antibodies formed in the **serum** (the fluid remaining when blood has clotted and blood cells and other matter have been removed) of other people or animals. Most antibodies remain in the serum; the term **antiserum** means blood-derived fluids containing antibodies. **Serology** is the study of antibodies and antigens. The antibody-rich serum component is **gamma globulin** or **immune serum globulin.**

The Duality of the Immune System

The **humoral immune system** involves antibodies that are dissolved in various body fluids or secretions (humors). This system responds when **B cells** (specialized lymphocytes) are exposed to antigens. Humoral immunity defends mostly against bacteria, bacterial toxins, and viruses circulating in body fluids.

The **cell-mediated immune system** involves specialized lymphocytes called **T cells.** T cells do not secrete antibodies but have antigen receptors attached to their surfaces. The cell-mediated immune system is most effective against bacteria and viruses located within phagocytic or other host cells, and against fungi, protozoa, and helminths. It also responds to what it perceives as foreign, such as transplanted tissue and cancer cells.

Antigens and Antibodies

Most **antigens** (sometimes called **immunogens**) are proteins, lipoproteins, glycoproteins, or large polysaccharides. These may be part of microorganisms or antigens such as pollen, egg white, blood cells, or transplanted tissues or organs. Antibodies usually recognize and interact with **antigenic determinants,** or **epitopes,** on the antigen, rather than an entire antigen. A bacterium or virus may have numerous antigenic determinants. **Haptens** are low-molecular-weight antigens that are not antigenic unless first attached to a carrier molecule. Once antibody against the hapten has formed, the hapten will react independently of its carrier. Penicillin is a good example of a hapten.

The Nature of Antibodies

Antibodies belong to proteins called **immunoglobulins (Ig).** They are highly specific and react with only one type of antigenic determinant. Each antibody has at least two antigen-binding sites. The **valence** is the number of such sites on the antibody.

Antibody Structure.   Figure 17.1a shows a typical monomer-type antibody. It has four protein chains—two identical **light (L) chains** and two identical **heavy (H) chains.** At the ends of the arms of Y-shaped molecules are **variable (V) regions,** with a structure that accounts for the ability of different antibodies to recognize and bind with different antigens (Figure 17.1b)—the **antigen-binding sites.**

    The stem of the antibody monomer and lower parts of the Y arms are called **constant (C) regions.** The stem of the Y-shaped monomer is the **Fc (stem) region.** The antibody molecule can attach to a host cell by the Fc region. Complement can bind to the Fc region.

Immunoglobulin Classes. (Refer to Figure 17.2.)   **IgG** antibodies account for 80% of all antibodies in serum. They are monomers and readily cross blood vessel walls to enter tissue fluids. Maternal IgG crosses

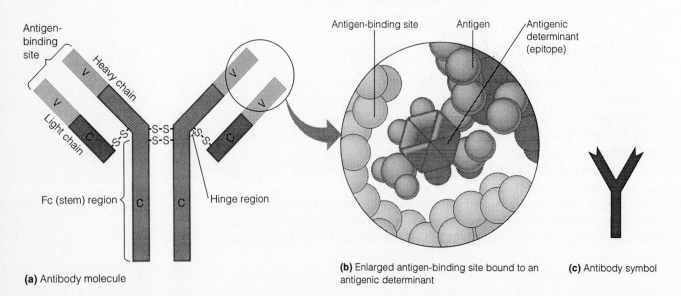

(a) Antibody molecule

(b) Enlarged antigen-binding site bound to an antigenic determinant

(c) Antibody symbol

FIGURE 17.1 **Structure of a typical antibody molecule.**   **(a)** The Y-shaped molecule is composed of two *light chains* and two *heavy chains* linked by disulfide bridges (S—S). Most of the molecule is made up of *constant regions* (C), which are the same for all antibodies of the same class. The amino acid sequences of the *variable regions* (V), which form the two *antigen-binding sites,* differ from molecule to molecule. **(b)** One antigen-binding site is shown enlarged and bound to an antigenic determinant.

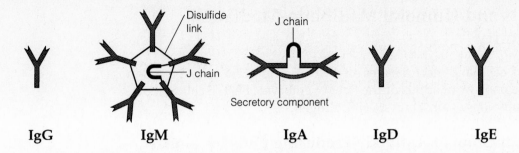

FIGURE 17.2 Human immunoglobulins. Structures of the five principal classes of human immunoglobulins are shown. Note that IgA and IgM are made of two and five monomers, respectively, in this drawing. In these cases, the monomers are held together by disulfide links, and some of these are joined by a polypeptide called the J (joining) chain. IgA, which may also occur as a monomer, is usually found attached to a protein called the secretory component.

the placenta to confer immunity to the fetus. IgG antibodies protect against circulating bacteria and viruses, neutralize bacterial toxins, trigger the complement system, and bind to antigens to enhance action of phagocytic cells.

IgM antibodies constitute 5% to 10% of antibodies in serum. IgM has a pentamer structure of five Y-shaped monomers held together by a **J (joining) chain.** IgM antibodies are the first to appear in response to an antigen, but their concentration declines rapidly. IgM antibodies generally do not enter surrounding tissue. IgM molecules are especially effective at cross-linking particulate antigens, causing their aggregation. The fact that IgM appears first in response to a primary infection and is relatively short-lived makes it valuable in disease diagnosis. Its presence in high concentrations in a patient makes it likely that the antibodies are associated with the disease pathogen. IgG is so long-lived that its presence may indicate immunity against a disease condition in the more distant past.

IgA antibodies account for about 10% to 15% of the antibodies in serum. It circulates in serum as a monomer, **serum IgA.** IgA may be joined by a J chain into dimers of two Y-shaped monomers called **secretory IgA.** IgA is found on mucosal surfaces and in body secretions such as colostrum. A **secretory component** protects the IgA from enzymes and may help it enter secretory tissues. The main function of IgA is preventing attachment of viruses and so on to mucosal surfaces.

IgD antibodies are only about 0.2% of the total serum antibodies. They are monomers and are found in blood and lymph cells and on B-cell surfaces. They do not fix complement or cross the placenta. IgD functions are little known, but IgD is present on the surface of B cells.

IgE antibodies are monomers slightly larger than IgG and constitute only 0.002% of the total serum antibodies. They bind by their Fc (stem) sites to mast cells and basophils. When an antigen reacts with IgE antibodies, the mast cell or basophil releases histamine and other chemical mediators involved in allergic reactions. These inflammatory reactions can be protective, attracting IgG and phagocytic cells.

B Cells and Humoral (Antibody-Mediated) Immunity

B cells, T cells, and macrophages all develop from **stem cells** located in the bone marrow of adults or the liver of embryos. Each B cell will produce antibodies against one specific antigen.

The Activation of Antibody-Producing Cells by Clonal Selection

IgM and IgD antibodies on the surface of B cells act as specific antigen receptors. The receptors on a given B cell will bind to only one specific antigen. Upon binding to an antigen, the B cell proliferates, a process called **clonal selection.** B cells activated by an antigen differentiate into **plasma cells** that secrete antibodies against the antigen. A population of **memory cells** that function in long-term immunity also are produced. B and T cells that interact with self antigens are destroyed, a mechanism called **clonal deletion.**

Antigen-Antibody Binding and Its Results.   A reaction between an antigen and an antibody results in an **antigen-antibody complex.** This may block active sites on toxins, neutralizing them, or inactivate viruses by preventing attachment to host cells. These complexes also can, in association with complement, lyse cells. Some antibodies are poorer matches for an antigen than others and are said to have less **affinity.**

Immunological Memory

Antibody titer is the amount of antibody present in the serum. First contact with antigen results in a **primary response** (see Figure 17.3). A second exposure to an antigen results in an intensified response called the **secondary response** (also called **memory** or **anamnestic response**). In cell-mediated immunity the memory is mainly in certain effector T cells that distinguish for an extended time between host cells and foreign cells. This is due to long-lived **memory cells** produced as a component of the primary response.

Molecular Basis of the Diversity of Antigen Receptors

An individual can respond to as many as 100 million or more different antigens. The mechanism that accounts for this is analogous to the generation of huge numbers of words from a limited alphabet. One set of genes (*V* genes) controls the variable regions of the light antibody chains, a single *C* gene controls the constant region, and four *J* genes code for the segments joining the variable and constant regions. The antibody *light chains* are assembled by selecting a *V* gene from the collection of hundreds and selecting one of the four *J* genes to connect it to the *C* gene.

  Heavy-chain genes are assembled by selecting from over 100 *V* genes, 12 diversity segments (*D* genes), and four *J* genes. The constant-region

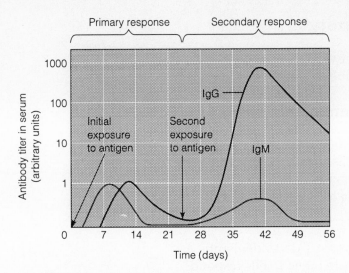

FIGURE 17.3 The primary and secondary immune responses to an antigen. IgM appears first in response to the initial exposure. IgG follows and provides longer-term immunity. The second exposure to the same antigen stimulates the memory cells formed at the time of initial exposure to rapidly produce a large amount of antibody. The antibodies produced in response to this second exposure are mostly IgG.

genes on heavy chains code for the differentiation of the different antibody classes. Selection from these various gene sets allows assembly of antibodies that can recognize a very large number of antigens. (See Figure 17.11 in the text.)

Monoclonal Antibodies and Their Uses

Recent techniques have been developed that allow the production of large volumes of antibodies in vitro—an important factor in many new diagnostic tests. In these techniques, a cancerous cell (considered immortal in the sense that it can be propagated indefinitely) and an antibody-secreting plasma cell (B cell), taken from a mouse that has been immunized with a particular antigen, are fused. This hybrid cell is called a **hybridoma.** Grown in culture, such a hybridoma will continue to produce the type of antibody characteristic of the ancestral B cell. Because the antibodies produced by such a hybridoma are identical, they are called **monoclonal antibodies.** We will see a number of applications of monoclonal antibodies later. One possibility is as a treatment for cancer. A monoclonal antibody programmed to react with a cancer can be combined with a toxin **(immunotoxin)** against the cancer cells and kill them. Endotoxins causing septic shock can be neutralized by monoclonal antibodies. A problem in therapeutic use of monoclonal antibodies is that they are produced from mouse cells; these are considered foreign, and human immune systems react against their presence. Monoclonal antibodies constructed with variable regions from mouse cells and constant regions from human sources **(chimeric monoclonal antibodies)** would be more compatible with the human immune system.

T Cells and Cell-Mediated Immunity

Some forms of immunity can be transferred between animals only by transferring certain lymphocytes—hence the name cell-mediated immunity.

The Components of Cell-Mediated Immunity

T cells are the key component of cell-mediated immunity. After developing from stem cells, they differentiate into T cells within the **thymus.** They then migrate to the lymph nodes, spleen, or other lymphoid organs. When stimulated by an antigen, they differentiate into effector T cells called **specialized T cells.**

Types of T Cells. Helper T cells (T_H) help present T-dependent antigens (discussed shortly) to B cells. They also help other T cells respond to antigens. **Cytotoxic T cells (T_C)** destroy target cells upon contact and can kill repeatedly. They are important in action against cancer, rejection of transplanted tissue, and protection against intracellular bacterial and viral infections. T_C cells recognize viral antigens on the host cell and cause destruction of that cell. Other helper T cells, called **delayed hypersensitivity T cells (T_D),** are associated with allergic reactions, such as to poison ivy, and with rejection of transplanted tissues. T_D cells produce substances that recruit defensive cells such as macrophages. **Suppressor T cells (T_S)** are not completely understood. They regulate the immune response when an antigen is no longer present. They also suppress the activity of some other T cells, so that immune responses are not developed against host antigens. T_S cells usually carry CD8 receptors; most T_H cells carry CD4 receptors.

Natural Killer Cells. Certain lymphocytes called **natural killer (NK) cells** are capable of destroying other cells such as virus-infected cells and tumor cells. They are not immunologically specific; they do not require antigenic stimulation. They contact and lyse target cells.

Chemical Messengers of Immune Cells: Cytokines (Lymphokines). When T cells are stimulated by an antigen, they release proteins called **cytokines** (also known as **lymphokines**). The best known cytokines are the **interleukins (IL).** Several representative interleukins and their functions are listed in Table 17.3 in the text. Cytokines, in summary, are soluble chemical messengers by which cells of the immune system communicate with each other.

The Cell-Mediated Immune Response

T cells have specificity for only a single antigen, one that recognizes receptors on the T cell. T cells are unable to bind with soluble antigens. They usually respond only to antigens processed by an antigen-presenting cell (APC), and also there must be a close association with **major histocompatibility complex (MHC) antigens** on the APC. MHC antigens are unique to each individual and are an expression of "self." This requirement for a close association between the two types of antigens is called

associative recognition. The advantage is that this minimizes the chances of a T cell mounting an immune attack on the body's own tissues.

Cell-Mediated Cytotoxicity

T_C cells activated by an antigen on a target cell cause the lysis of other such target cells. They bind to the cell's surface and release a protein called **perforin,** which forms a pore in the target cell, causing its lysis.

The Role of Helper T Cells. T_H cells are prolific cytokine producers. Binding of a T_H cell to an antigen-MHC complex on an APC stimulates the APC to secrete interleukin-1 (IL-1) (see Figure 17.4). The IL-1 in turn activates the T_H cell, which begins to synthesize interleukin-2 (IL-2). The IL-2 secreted by the T_H cell returns to receptors on the surface of the same cell, which proliferates and differentiates into mature T_H cells. These activated T_H cells produce IL-2 and other cytokines and in turn stimulate other T_H cells specific to that antigen to proliferate and mature. The cytokines also activate macrophages, cytotoxic T cells, and B cells.

Some Interactions of the Cell-Mediated Immune System with the Humoral Immune System

Helper T Cells' Role in the Activation of B Cells. **T-dependent antigens** require the cooperation of helper T cells before they can produce antibodies. T-dependent antigens are mainly proteins such as viruses, bacteria, foreign red blood cells, and hapten-carrier combinations.

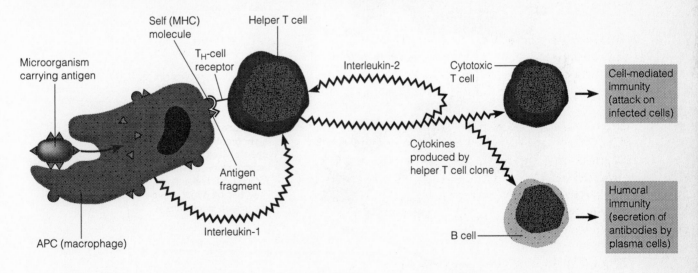

1. An antigen-presenting cell (APC) encounters and processes an antigen, forming MHC-antigen complexes on its surface.

2. A helper T (T_H) cell receptor binds to the complex, stimulating the APC to secrete interleukin-1.

3. This interleukin-1 stimulates the helper T cell to produce interleukin-2, which then stimulates that helper T cell to form a clone of helper T cells.

4. The cells of this clone in turn produce cytokines, stimulating cells of both immune systems.

FIGURE 17.4 **The central role of helper T cells.**

When a T_H cell encounters a processed T-dependent antigen (Figure 17.5) combined with an MHC on the APC, it is primed to activate the appropriate B cell. This interaction with a T_H cell stimulates the B cell to differentiate into an antibody-producing plasma cell.

   T-independent antigens, usually composed of polysaccharides or lipopolysaccharides with repeating subunits, are able to bind to multiple B cell receptors (Figure 17.6). They do not require T_H cell assistance. Their immune response is relatively weak.

Antibody-Dependent Cell-Mediated Cytotoxicity

If a target cell is coated with antibodies, leaving their F_C regions pointing outward, they can be killed by NK cells and cells of the nonspecific defense system. This process is **antibody-dependent cell-mediated**

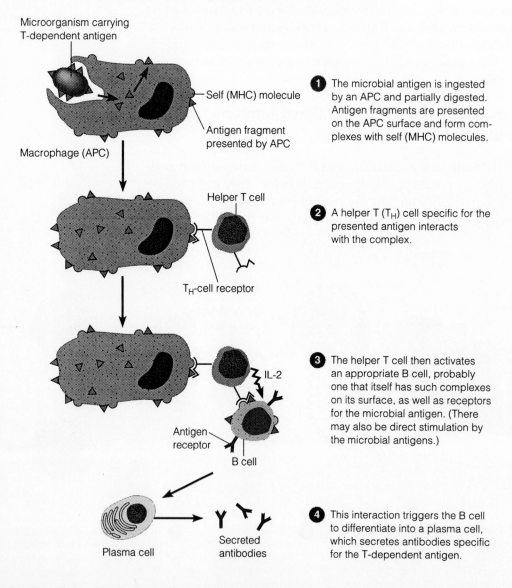

Microorganism carrying T-dependent antigen

Self (MHC) molecule

Antigen fragment presented by APC

Macrophage (APC)

Helper T cell

T_H-cell receptor

IL-2

Antigen receptor

B cell

Plasma cell

Secreted antibodies

1 The microbial antigen is ingested by an APC and partially digested. Antigen fragments are presented on the APC surface and form complexes with self (MHC) molecules.

2 A helper T (T_H) cell specific for the presented antigen interacts with the complex.

3 The helper T cell then activates an appropriate B cell, probably one that itself has such complexes on its surface, as well as receptors for the microbial antigen. (There may also be direct stimulation by the microbial antigens.)

4 This interaction triggers the B cell to differentiate into a plasma cell, which secretes antibodies specific for the T-dependent antigen.

FIGURE 17.5   **How helper T cells may activate B cells to make antibodies against T-dependent antigens.** B cells require the help of antigen-presenting cells (APCs) and specialized helper T cells to produce antibodies against T-dependent antigens.

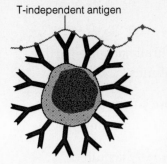

T-independent antigen

FIGURE 17.6 T-independent antigens. T-independent antigens have repeating units that are able to cross-link several antigen receptors on the same B cell. These antigens stimulate the B cell to make antibodies without the aid of helper T cells.

cytotoxicity (ADCC). By ADCC the immune system can successfully attack many relatively large organisms such as helminths.

Self-Tests

In the matching section, there is only one answer to each question; however, the lettered options (a, b, c, etc.) may be used more than once or not at all.

I. Matching

1. Antigen converts these into plasma cells.

2. Involved in cell-mediated immunity.

3. Directed against transplanted tissue cells and cancer cells.

4. Have been influenced by the thymus.

5. Defend mainly against bacteria and viruses circulating in blood and lymph.

6. Responsible for rejection of foreign tissue transplants.

a. B cells

b. T cells

II. Matching

1. Based on antibodies produced as a result of recovery from a disease.

2. Passed to a fetus across the placenta.

3. Passed to an infant in human colostrum.

4. Passed to a recipient by injection of gamma globulin blood fraction from other people.

5. Based on production of antibodies from the injection of a toxoid.

a. Naturally acquired active immunity

b. Artificially acquired passive immunity

c. Naturally acquired passive immunity

d. Artificially acquired active immunity

III. Matching

1. An incomplete antigen that will react with antibodies but will not stimulate their formation.

2. The number of determinant sites on an antigen or antibody.

3. The source of B cells and T cells.

4. Soluble chemical messengers by which cells of the immune system communicate with each other.

5. The result of the fusion of plasma cells and cancer cells.

a. Hapten

b. Valence

c. Stem cells

d. Cytokines

e. Hybridomas

f. Memory cells

IV. Matching

1. A pentamer; the first antibody class to appear, though comparatively short-lived.

2. The most abundant immunoglobulin in serum.

3. Functions of this immunoglobulin class are not well defined, but it is found on surface of B-cells.

4. Involved in allergic reactions, such as hay fever.

5. Often forms dimers of two immunoglobulin monomers.

a. IgA

b. IgG

c. IgD

d. IgE

e. IgM

V. Matching

1. Synonym for antigens.

2. T cells that interact with self antigens are destroyed.

3. Protein bound to IgA immunoglobulins.

4. Blood fraction that contains most of the serum immunoglobulins.

5. Antigenic; will stimulate the production of antitoxins.

a. Gamma globulin

b. Immunogens

c. Clonal deletion

d. Toxoid

e. Secretory component

VI. Matching

1. Regulate immune response when antigen is no longer present.

2. Needed to present T-dependent antigens to B cells.

3. Associated with allergic reactions such as to poison ivy.

a. Delayed hypersensitivity cells (T_D)

b. Helper T cells (T_H)

c. Cytotoxic T cells (T_C)

d. Suppressor T cells (T_S)

VII. Fill-in-the-Blanks

1. Resistance present at birth that does not involve humoral or cell-mediated immunity is _____ immunity.

2. A(n) _____ site is a specific chemical group on an antigen that combines with the antibody.

3. The stem region on an immunoglobulin monomer that allows the molecule to attach to a host cell is known by the letters _____ .

4. The five monomers that constitute the IgM molecule are held together by a _____ .

5. The antibody _____ is the measured amount of antibody in the serum.

6. The first breast milk secretions are called _____ ; they provide some immunity to the newborn.

7. IgA is protected from enzyme activity by a _____ component.

8. The fluid remaining when blood has been clotted and when blood cells and other matter have been removed is called _____ .

9. Certain lymphocytes called _____ cells kill virus-infected cells and tumor cells, but are not immunologically specific. They contact and lyse the target cells.

10. When an antigen-presenting cell is stimulated by an antigen, it secretes a cytokine called _____ .

11. Helper T cells carry _____ receptors on their surface.

12. Low-molecular-weight substances such as penicillin that do not (by themselves) cause formation of antibodies are known immunologically as _____ .

13. When a B cell is stimulated by an antigen, it becomes a _____ that produces humoral antibodies.

14. Antigens that are made up of repeating polysaccharide or protein subunits are more likely to be _____ antigens.

15. The second time we encounter an antigen, our response is faster and more intense; this is termed the _____ response.

16. Monoclonal antibodies with variable regions from mouse cells and constant regions from human sources are called _____ monoclonal antibodies.

17. The study of antibodies and antigens in antiserum is called _____ .

18. Some antibodies are poorer matches for an antigen than others; they are said to have less _____ .

19. A monoclonal antibody programmed to react with a cancer can be combined with a toxin, _____ , against the cancer cells and kill them.

20. Cytotoxic T cells cause the death of target cells by release of a protein called _____ that forms a pore in the target cell membrane.

VIII. Label the Art

I.

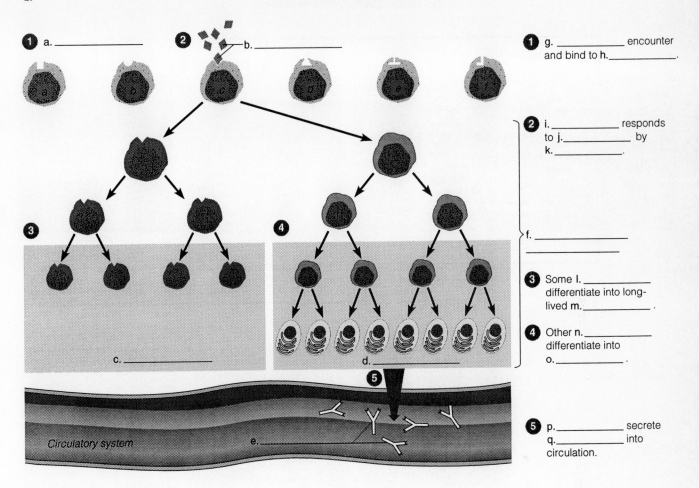

1 a. _____

2 b. _____

1 g. _____ encounter and bind to h. _____.

2 i. _____ responds to j. _____ by k. _____.

f. _____ _____

3 Some l. _____ differentiate into long-lived m. _____.

4 Other n. _____ differentiate into o. _____.

c. _____

d. _____

Circulatory system

e. _____

5 p. _____ secrete q. _____ into circulation.

(continued)

II.

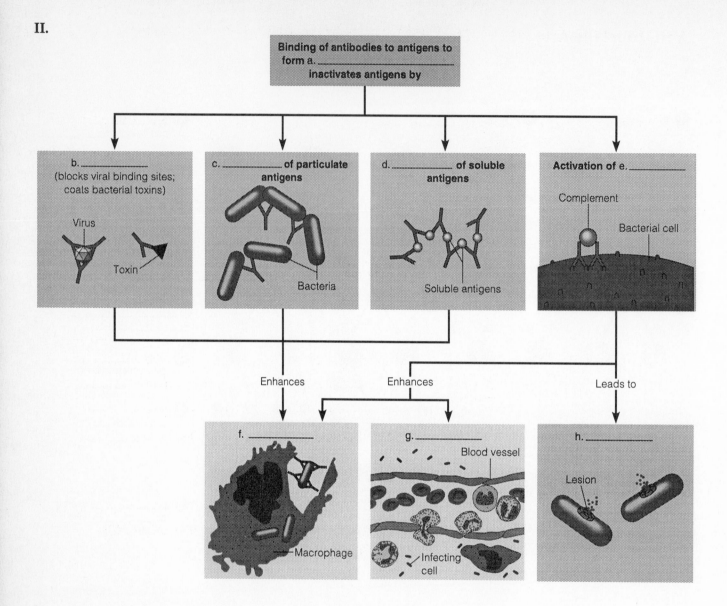

IX. Critical Thinking

1. Differentiate between innate resistance and acquired immunity.

2. An infant's mother had diphtheria prior to pregnancy. Is it necessary to immunize the infant for diphtheria? Why?

3. How are haptens related to allergies?

4. Draw and label the parts of an antibody molecule. What is the significance of the hinge region?

5. How do antibodies of the same class differ from each other? How do antibodies of different classes differ from each other?

6. Complete the following table:

Class of Ig	% in serum	Transplacental transfer?	Complement system?	Example of function

7. Antibodies of what class are most effective at cross-linking antigen, resulting in their aggregation? Why?

8. What are myelomas? Why are cells from myelomas used in the production of monoclonal antibodies?

9. Briefly discuss the two systems used to classify the cells of the cell-mediated immune system.

10. What are cytokines? Give two examples of cytokines and explain how each contributes to a vigorous response from the cell-mediated immune system.

CHAPTER 18

Practical Applications of Immunology

Learning Objectives

After completing this chapter, you should be able to:

- Define vaccination, herd immunity, attenuated, toxoid, and acellular vaccine.
- Explain why vaccination works.
- Differentiate between whole-agent vaccines and subunit vaccines. Provide an example of each.
- Differentiate between direct and indirect diagnostic tests.
- Explain how precipitation, immunodiffusion, agglutination, neutralization, viral hemagglutination, immunofluorescence, and ELISA can be used to diagnose diseases.
- Explain how countercurrent immunoelectrophoresis, hemagglutination, viral hemagglutination inhibition, complement fixation, and radioimmunoassay can be used to diagnose a disease or detect a compound.
- Explain how genetic engineering has been revolutionizing vaccine development and production.

Vaccines

A **vaccine** is a suspension of microorganisms, or some part or product of them, that will induce immunity when it is administered to the host. Vaccines had their origin in the practice of **variolation,** in which material from smallpox scabs was inoculated into the bloodstream to provide immunity to the disease. We now call this strategy **vaccination** or **immunization.**

A disease can be controlled if most, but not all, of the population is immune. This is called **herd immunity.**

Types of Vaccines and Their Characteristics

The microorganisms used in vaccines may be **inactivated** (killed) or **attenuated** (weakened). Attenuated vaccines tend to mimic actual infection and usually provide better immunity.

Subunit Vaccines. A **subunit vaccine** uses only those antigenic fragments of a microorganism that are best suited to stimulate an immune response. These subunits may be produced by bacteria or yeasts, for example, by use of genetic-engineering techniques; such vaccines are called **recombinant vaccines. Toxoids** are inactivated bacterial toxins.

Vaccines can be fragmented and the desired antigens separated out, rather than using the complete cell; these are called **acellular vaccines.** Some polysaccharide vaccines have enhanced effectiveness when combined with tetanus or diphtheria toxoids. These are **conjugated vaccines.**

Diagnostic Immunology

We cannot see antibodies and must infer their presence indirectly by a variety of reactions.

Precipitation Reactions

Precipitation reactions involve the reaction of *soluble* antigens with IgG or IgM antibodies to form large interlocking aggregates called **lattices.** Precipitation can be easily arranged by allowing the antigen and antibody to diffuse toward each other. A cloudy line of precipitation forms in the area in which the **optimal ratio** has been reached (the **zone of equivalence**). In a capillary tube this is called a **ring test.** In a Petri dish the antigens and antibodies can be placed into wells cut into the agar medium. This is the basis of the **Ouchterlony test** (see Figure 18.4 in the text). If passive diffusion of the reagents is too slow, the test can be speeded up by applying an electric current, a procedure called **immunoelectrophoresis,** used in research to separate proteins in human serum. The **countercurrent-immunoelectrophoresis (CIE) test** is used in diagnosis of bacterial meningitis. The pH is adjusted so the antigens and antibodies have opposite charges, and an electrical current causes them to move toward each other rapidly.

Agglutination Reactions

Agglutination reactions involve *particulate* antigens that can be linked together by antibodies in a process called agglutination.

Direct Agglutination Tests. **Direct agglutination tests** detect antibodies against relatively large *cellular* antigens, such as red blood cells, bacteria, and fungi. The amount of antigen in each well of a microtiter plate is the same. The serum containing the antibodies is sequentially diluted out in a series of wells. The higher the concentration of antibodies in the serum, the more dilutions are required to dilute it to the point at which no reaction occurs with the antigen. This is a measure of **titer,** or concentration of antibody. For diagnostic purposes a **rise in titer** during the course of a disease is very significant.

Direct ELISA

Indirect ELISA

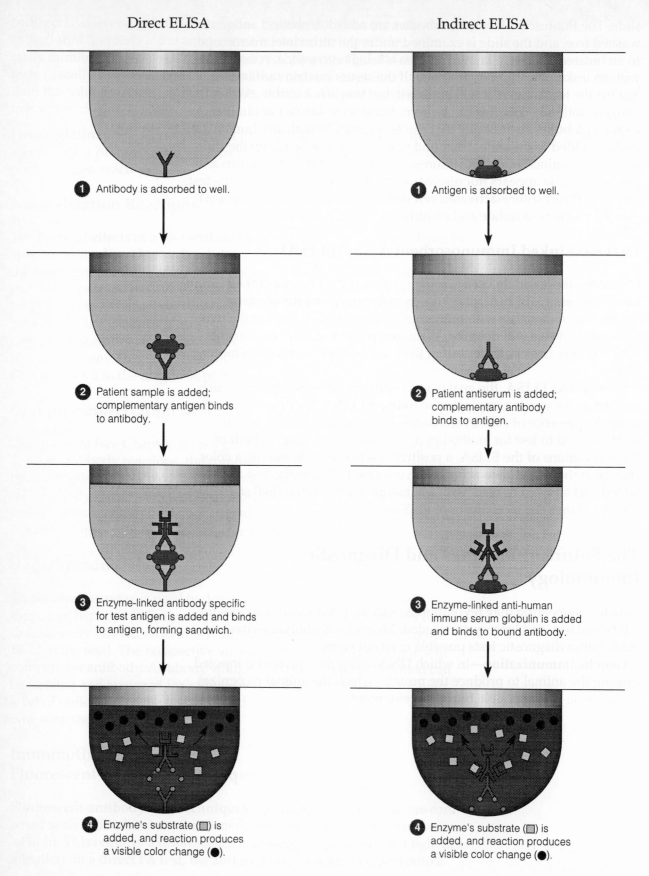

1 Antibody is adsorbed to well.

1 Antigen is adsorbed to well.

2 Patient sample is added; complementary antigen binds to antibody.

2 Patient antiserum is added; complementary antibody binds to antigen.

3 Enzyme-linked antibody specific for test antigen is added and binds to antigen, forming sandwich.

3 Enzyme-linked anti-human immune serum globulin is added and binds to bound antibody.

4 Enzyme's substrate (▨) is added, and reaction produces a visible color change (●).

4 Enzyme's substrate (▨) is added, and reaction produces a visible color change (●).

FIGURE 18.1 The ELISA, or EIA, method. The components are usually contained in small wells of a microtiter plate.

Self-Tests

In the matching section, there is only one answer to each question; however, the lettered options (a, b, c, etc.) may be used more than once or not at all.

I. Matching

1. Makes use of fact that certain viruses will cause agglutination of red blood cells.

2. The pH is adjusted so that the antigens and antibodies have opposite charges; then an electric current is applied to the solution.

3. The absence of complement is indicated by hemolysis.

4. A precipitation-type test in which wells are cut into the agar on a Petri dish.

5. Soluble antigens are detected by binding them to small latex particles, for example, and causing their agglutination.

6. The ELISA used to screen for AIDS antibodies in serum.

a. Ouchterlony test

b. Countercurrent immunoelectrophoresis test

c. Indirect agglutination test

d. Complement-fixation test

e. Direct ELISA

f. Indirect ELISA

g. Hemagglutination-inhibition test

II. Matching

1. Subunit vaccine using genetically engineered organisms to produce it.

2. Unwanted components are removed from a whole cell vaccine.

3. An inactivated toxin.

a. Recombinant vaccine

b. Toxoid

c. Acellular vaccine

d. Antitoxin

III. Fill-in-the-Blanks

1. Before the invention of modern vaccines, material from smallpox scabs was inoculated into the bloodstream to give immunity to the disease; this was called _____ .

2. The measure of the concentration of antibody in serum is called _____ .

3. Fluorescein-labeled antihuman gamma globulin would be used in the _____ fluorescent antibody test.

4. A vaccine using a living, weakened organism is called _____ .

5. For diagnostic purposes, a rise in _____ during the course of a disease is very significant.

6. A disease can be controlled if most, but not all, of the population is immune; this is called _____ immunity.

7. Polysaccharide vaccines can be enhanced in effectiveness by adding toxoids such as diphtheria; these are so-called _____ vaccines.

IV. Label the Art

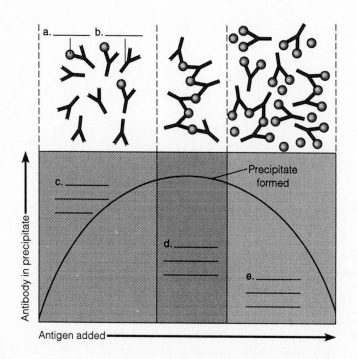

V. Critical Thinking

1. Discuss herd immunity as it relates to the control of disease.

2. Compare and contrast attenuated and inactivated vaccines. Which type of vaccination usually provides better immunity? Why?

3. Discuss two advantages of recombinant vaccines. Are there any disadvantages?

4. Why is the vaccinia virus a good choice for making subunit vaccines using recombinant methods?

5. What immunological technique is used to determine the relatedness of antigens? How is their relatedness indicated in this test?

6. How is the measured antibody titer used to diagnose disease? Which of the following titers indicates greater immunity: 1:94 or 1:312? Why?

7. Briefly outline how the indirect agglutination procedure is used to diagnose strep throat. What advantages (if any) does this procedure offer over culturing a throat swab on blood agar?

8. Discuss the advantages or disadvantages associated with each of the following procedures:

    a. Complement fixation

    b. Radioimmunoassay

    c. ELISA

9. A researcher needs to separate the helper T cells from the other types of cells in a sample. How might this be achieved?

10. What is genetic immunization, and how may it change the future of vaccines?

CHAPTER 19

Disorders Associated with the Immune System

Learning Objectives

After completing this chapter, you should be able to:

- Define hypersensitivity.
- Define shocking dose, sensitizing dose, and desensitization.
- Describe the mechanism of anaphylaxis.
- Describe the mechanism of cytotoxic reactions and how drugs can induce them.
- Describe the basis of the human blood group systems and their relationship to blood transfusions and hemolytic disease of the newborn.
- Describe the mechanism of immune complex reactions.
- Describe the mechanism of cell-mediated reactions.
- Describe a mechanism for self-tolerance.
- Give an example of a type II, type III, and type IV autoimmune disorder.
- Define HLA complex and explain its importance in disease susceptibility and tissue transplants.
- Explain how rejection of a transplant occurs and how rejection is prevented.
- Discuss the causes and effects of immune deficiencies.
- Describe the effects of HIV on the immune system and the stages of HIV infection.
- Describe how HIV can be transmitted.
- Discuss current and possible future treatment and prevention of HIV infection.
- Describe the immune responses to cancer and immunotherapy.

Antigens such as the staphylococcal enterotoxin are called **superantigens.** They indiscriminately activate many T cells at once, causing a harmful immune response.

Hypersensitivity

The term **hypersensitivity (allergy)** refers to sensitivity beyond what is considered normal. It occurs in people who have been previously sensitized by exposure to an antigen, called in this context an **allergen**. Once sensitized, another exposure to the antigen triggers an immune response that damages host tissue. There are four principal types of hypersensitivity reactions.

Type I (Anaphylaxis) Reactions

Systemic Anaphylaxis

Anaphylaxis is from the Greek words for "against" and "protection." These reactions result from combining antigens with IgE antibodies; they may be **systemic,** producing sometimes fatal shock and breathing difficulties, or **localized,** such as hayfever, asthma, or hives.

IgE antibodies bind to the surfaces of mast cells and basophils. **Mast cells** are prevalent in the connective tissue of skin, the respiratory tract, and surrounding blood vessels. **Basophils** circulate in the blood. When an antigen combines with antigen-combining sites on two adjacent IgE antibodies and bridges the space between them, the mast cell or basophil undergoes **degranulation,** releasing chemicals called mediators. The best-known mediator is **histamine,** which affects the blood vessels, causing edema (swelling), erythema (redness), increased mucus secretion, and smooth-muscle contractions resulting in breathing difficulty. Other mediators are **leukotrienes** (which tend to cause contractions, such as the spasms of asthmatic attacks) and **prostaglandins** (which tend to cause increased secretions of mucus). Collectively, mediators attract neutrophils and eosinophils to the site and cause inflammatory symptoms.

When an individual sensitized to an injected antigen, such as an insect sting or penicillin, receives a subsequent injection, the release of mediators can result in a drop in blood pressure (shock) that can be fatal in a few minutes. This is termed **systemic anaphylaxis,** or **anaphylactic shock.**

Localized Anaphylaxis

Localized anaphylaxis usually is associated with antigens that are ingested or inhaled, rather than injected. Examples are hay fever, for which **antihistamine** drugs often are useful to treat symptoms, and asthma, for which **epinephrine** is usually administered. It is sometimes difficult to distinguish between food hypersensitivity and food intolerance. Hives are often a characteristic of true food allergy.

Prevention of Anaphylactic Reactions

If contact with the allergen cannot be avoided, **desensitization** might be attempted. This consists of injections of a series of small doses of the antigen. The idea is to induce IgG antibodies to serve as **blocking antibodies** that intercept and neutralize antigens before they can react with cell-bound IgE.

Type II (Cytotoxic) Reactions

Immunological injury resulting from type II reactions is caused by antibodies that are directed at antigens on the host's blood cells or tissue cells. The host cell plasma membrane may be damaged by antibody and complement, or macrophages may attack antibody-coated cells. *Transfusion reactions,* such as those involving the ABO and Rh blood group systems, are of this type.

are often not apparent for a day or more, during which time participating cells migrate to and accumulate near the foreign antigens.

Causes of Type IV Reactions

Usually in type IV reactions the foreign antigens are phagocytized by macrophages and then presented to receptors on the T-cell surface. The T cells are primarily T_D cells but may include T cells. A principal factor is the release of lymphokines by T cells reacting with the target antigen.

Cell-Mediated Hypersensitivity Reactions of the Skin

The skin test for tuberculosis is a reaction by a sensitized individual to protein components of tuberculosis bacteria injected into the skin. A day or two is required for the reaction to appear. Cases of **allergic contact dermatitis** are usually caused by haptens that combine with proteins in the skin. Typical foreign antigens are poison ivy, cosmetics, and metals such as nickel in jewelry. The *patch test,* in which samples of suspected material are taped to the skin, may determine the offending environmental factor.

Autoimmune Disorders

Loss of Immunological Self-Tolerance

In our discussion of the different types of hypersensitivity reactions, we have mentioned several autoimmune diseases. These occur when there is a loss of **self-tolerance,** the immune system's ability to discriminate between self and nonself. It is believed that some clones of lymphocytes (forbidden clones) having the potential to respond to self antigens may be produced during fetal life but are destroyed **(clonal deletion)** or inactivated **(clonal anergy).**

Type II (Cytotoxic) Autoimmune Reactions

Myasthenia gravis is a disease in which antibodies coat the acetylcholine receptor junctions, preventing nerve impulses from reaching the muscles. In **Graves' disease,** antibodies attach to receptors on the thyroid gland and cause excessive production of thyroid-stimulating hormones.

Type III (Immune Complex) Autoimmune Reactions

Systemic lupus erythematosus is a systemic autoimmune disease in which individuals produce antibodies directed at components of their own cells, including DNA. Immune complexes damage the kidney glomeruli. **Rheumatoid arthritis** results when immune complexes are deposited in the joints. Immune complexes called *rheumatoid factors* may be formed by IgM binding to the Fc region of normal IgG. Chronic inflammation causes joint damage.

Type IV (Cell-Mediated) Autoimmune Reactions

Hashimoto's thyroiditis is a result of the destruction of the thyroid gland, primarily by T cells. **Insulin-dependent diabetes** is caused by immunological destruction of insulin-secreting cells of the pancreas by the cell-mediated immune system.

DISEASES AND REACTIONS RELATED TO THE HUMAN LEUKOCYTE ANTIGEN (HLA) COMPLEX

One inherited genetic characteristic is differences in histocompatibility antigens on cell surfaces. The genes controlling these antigens are the **major histocompatibility complex (MHC) antigens;** in humans these genes are also called the **human leukocyte antigen (HLA) complex.** For successful transplant surgery, **tissue typing** is used to match donor and recipient. Matching for class I antigens (HLA-A, -B, and -C) has long been standard procedure, but matching for class II antigens (HLA-DR, -DP, and -DQ) might be more important. The donor and recipient must be of the same ABO blood type.

Transplantation Reactions

The cornea and brain are examples of **privileged sites;** antibodies do not circulate to these regions. **Privileged tissue,** such as pig heart valves, are not antigenic and do not stimulate an immune response. The transfer of tissue such as skin from one part of an individual to another on the same individual is an **autograft. Isografts** are transplants between identical twins. Such transplants are not rejected. **Allografts**—transplants between related people—represent most transplants. **Xenografts** are transplants of tissue from animals other than humans, which tend to be strongly rejected. When bone marrow is transplanted to people, the transplanted tissue may carry cells capable of mounting an immune response. This is called **graft versus host (GVH) disease.**

Immunosuppression

People receiving transplants require suppression of their immune system **(immunosuppression)** to prevent rejection of the new tissue. The drug **cyclosporine** suppresses cell-mediated immunity, but at the cost of some liver and kidney toxicity. New immunosuppressant drugs such as tacrolimus (**FK 506**) and rapamycin have a mode of action similar to cyclosporine. Both block the synthesis of cytotoxic T cells.

Immune Deficiencies

Occasionally people are born with defective immune systems **(congenital immune deficiencies).** Hodgkin's disease, a form of cancer, lowers cell-mediated immunity, as does removal of the spleen. Several drugs, cancers, or infectious agents can result in such **acquired immune deficiencies.**

Acquired Immunodeficiency Syndrome (AIDS)

Acquired immunodeficiency syndrome (AIDS) is a type of immunodeficiency disease. A retrovirus known as **human immunodeficiency virus (HIV)** destroys helper T cells and is the cause of AIDS. The loss of an effective immune system leaves the victim susceptible to diseases that the immune system would normally prevent.

The HIV Infection

HIV is a *retrovirus* and requires the enzyme *reverse transcriptase* to form DNA from its RNA genome. The envelope of HIV has spikes of gp120 that allow the virus to attach to the CD4 receptors found on helper T cells. Attachment is followed by entry into the cell, where the viral DNA is integrated into the DNA of the host cell. It may cause new HIVs to bud from the T cell, or it may remain latent as a *provirus*. HIV is capable of a very rapid rate of antigenic changes.

A period of several months passes before seroconversion, when antibodies to HIV appear. The progression of the infection is divided into categories:

Category A. Asymptomatic or persistent lymphadenopathy (swollen lymph nodes). Serum concentration of CD4 T cells is 500 per mm^3 or above.

Category B. CD4 T cell counts fall into the 200–499 per mm^3 range and the immune system weakens. Certain infections appear, such as *Candida albicans* infections of mouth or vagina; shingles; persistent diarrhea and fever; whitish patches on the oral mucosa (hairy leukoplakia); and certain precancerous growths.

Category C. This is clinical AIDS. The CD4 T cell count is 200 per mm^3 or lower. Many AIDS indicator conditions appear, such as more extensive *Candida albicans* infections, cytomegalovirus eye infection, tuberculosis, *Pneumocystis* pneumonia, toxoplasmosis, and Kaposi's sarcoma.

The average time from infection to development of AIDS is about 10 years. Infants who acquire HIV infections from their mothers often die within a year.

Modes of HIV Transmission

Transmission primarily requires transfer of infected bodily fluids such as blood or semen. Routes of transmission include sexual contact, breast milk, blood-contaminated needles, organ transplants, and blood transfusions. Vaginal intercourse is more likely to transmit HIV from male to female, and transmission is more likely if genital lesions are present.

AIDS Research: Treatments and Vaccines

Chemotherapy.   Drugs such as **Zidovudine (AZT), ddI,** and **ddC** inhibit reverse transcriptase.

The Possibility of Preventive Vaccines.   Obstacles are the lack of a suitable animal host for the virus and the high mutation rate of the virus.

Immune Response to Cancer

Immunological Surveillance

The immune system's patrolling the body for cancer cells is called **immunological surveillance.** Individual cancer cells, before they become established, are recognized as foreign and are destroyed by an effective immune system.

Immunological Escape

Once established, the cancer becomes resistant to immune rejection. This is called **immunological escape.** One possible mechanism is **antigen modulation,** in which tumor cells shed their associated antigens.

Immunotherapy

A future approach to cancer treatment is **immunotherapy,** such as the use of monoclonal antibodies with immunotoxins, which we have discussed previously. There is considerable interest in tumor necrosis factor, interleukin-2, and interferons for cancer therapy.

Endotoxins from gram-negative bacteria stimulate production of **tumor necrosis factor** by macrophages. It interferes with the blood supply of the cancer.

Self-Tests

In the matching exercises there is only one answer to a question; however, the lettered options (a, b, c, etc.) may be used more than once or not at all.

I. Matching

1. Hypersensitivity.

2. Hypersensitivity specifically involving the interaction of humoral antibodies of the IgE class with mast cells.

3. A skin graft from a brother to a sister.

4. The heart of a baboon transplanted to a human.

5. A term used for an antigen causing hypersensitivity reactions.

6. A skin graft transferred from the thigh to the nose of the same person.

a. Allergen

b. Anaphylaxis

c. Xenograft

d. Allergy

e. Autograft

f. Allograft

g. Autoimmunity

h. Degranulation

II. Matching

1. A drug used for transplantation surgery.

2. A drug that suppresses cell-mediated immunity.

3. The reason why transplantation of a cornea is usually successful.

4. The mediator of a type I reaction that affects the blood capillaries and results in swelling and reddening.

5. The development of blocking antibodies by repeated exposure to small doses of the antigen.

a. Histamine

b. Leukotrienes

c. Prostaglandins

d. Cyclosporine

e. Privileged site

f. Privileged tissue

g. Desensitization

III. Matching

1. The naturally learned ability of the body not to respond immunologically against its own antigens.

2. The ability of an established cancer to avoid destruction by the immune system.

3. Inhibition of the immune response by drugs, radiation, and so on.

4. The treatment of cancer or other disease conditions by using monoclonal antibodies with which toxic compounds have been combined.

a. Immunological escape

b. Immunological surveillance

c. Immunosuppression

d. Immunological tolerance

e. Immunotherapy

IV. Matching

1. A mediator released from an antigen-triggered mast cell.

2. Zidovudine.

3. The release of mediators from mast cells or basophils during an anaphylactic reaction.

4. The destruction of Rh⁺ red blood cells by antibodies of maternal origin in a newborn infant; the antibodies are derived from the mother.

5. Individuals in whom ABO antigens are present in body fluids such as saliva and semen.

a. Leukotrienes

b. Erythroblastosis fetalis

c. Degranulation

d. Inhibition of reverse transcriptase

e. Secretors

V. Matching

1. Tuberculin test.

2. Asthma.

3. Glomerulonephritis.

4. Poison ivy dermatitis.

5. Graves' disease.

6. Reaction to an insect sting.

a. Type I (anaphylaxis) reaction

b. Type II (cytotoxic) reaction

c. Type III (immune complex) reaction

d. Type IV (cell-mediated) reaction

VI. Matching (HIV categories)

1. Persistent lymphadenopathy.

2. Full-blown AIDS.

3. CD4 T-cell count below 200/mm³.

a. Category A

b. Category B

c. Category C

VII. Fill-in-the-Blanks

1. Endotoxins from gram-negative bacteria stimulate macrophages to produce the cancer-inhibiting _____ factor.

2. The type of anaphylaxis that develops very rapidly after an antigen is presented to a sensitized host and that may result in life-threatening shock is _____ anaphylaxis.

3. In the ABO system, an absence of antigens makes a person blood type _____ .

4. A graft between identical twins is a(n) _____ .

5. The acronym MHC stands for _____ .

6. The acronym HLA stands for _____ .

7. One mechanism by which cancers escape the immune system is to shed associated antigens, which is called _____ modulation.

8. One result of immunosuppression could be development of graft _____ disease.

9. The name of the disease in which an immunological reaction against the thyroid gland receptor sites causes an excessive production of thyroid hormones is _____ disease.

10. The treatment for systemic anaphylaxis is to administer _____ promptly.

11. T cells are involved in _____ hypersensitivity; the reaction takes one or more days to develop.

12. Prevention of hemolytic disease of the newborn in response to Rh incompatibility can be done by administration of _____ .

13. An autoimmune condition caused by coating with antibodies the receptor sites at which nerve impulses reach muscles is called _____ .

14. A characteristic of true food allergy is the appearance of _____ on the skin.

15. Destruction of some clones of lymphocytes having the potential to respond to self antigens during fetal life is called _____ .

16. The cornea does not reject transplants; it is an example of a _____ .

17. The drug (name one) _____ revolutionized transplantation medicine by selectively suppressing cell-mediated immunity.

18. Pig heart valves are not antigenic and are an example of _____ .

19. About 85% of the population is Rh _____ .

20. Immune-caused destruction of white blood cells is called _____ .

VIII. Label the Art

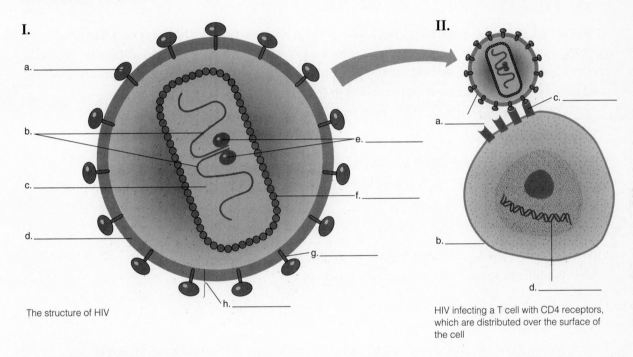

The structure of HIV

HIV infecting a T cell with CD4 receptors, which are distributed over the surface of the cell

IX. Critical Thinking

1. Six-year-old Susie is found playing in a patch of poison ivy by her mom. Although Mom was quite concerned, Susie didn't develop any symptoms. Five months later, on a camping trip, Susie again comes in contact with poison ivy. This time she develops a severe contact dermatitis. Why didn't Susie have any symptoms with her first exposure to poison ivy? Why did she have a reaction the second time? What substance in the poison ivy caused the reaction?

2. What are superantigens? Give an example of a superantigen and explain the reaction that it causes.

3. Explain the mechanism that triggers mast cells and basophils to undergo degranulation in anaphylactic reactions.

4. Discuss the signs and symptoms caused by each of the following substances:

    a. Histamines

    b. Prostaglandins

    c. Leukotrienes

5. A 4-year-old patient is prescribed penicillin and develops a rash without hives. Upon receiving this information the physician issues a prescription for a different antibiotic and tells the mother that it would be best for the child to avoid penicillin in the future. Is the child allergic to penicillin? Why avoid penicillin in the future?

6. Why would a person with type A blood have a reaction against type B blood upon the first exposure, whereas an Rh$^-$ person wouldn't have a reaction to Rh$^+$ blood until the second exposure?

7. What is serum sickness?

8. Define clonal deletion and clonal anergy. Discuss how they relate to immunological self-tolerance.

9. Briefly describe and give an example of each of the following:

    a. Type II autoimmune reaction

    b. Type III autoimmune reaction

    c. Type IV autoimmune reaction

10. List and briefly discuss some of the obstacles that must be overcome to produce an AIDS vaccine.

CHAPTER 20

Antimicrobial Drugs

Learning Objectives

After completing this chapter, you should be able to:

- Define a chemotherapeutic agent, and distinguish between a synthetic drug and an antibiotic.
- Identify the contributions of Paul Ehrlich, Alexander Fleming, René Dubos, and Selman Waksman to chemotherapy.
- Define the following terms: spectrum of activity, superinfection, broad spectrum.
- Describe the problems of chemotherapy for viral, fungal, protozoan, and helminthic infections.
- Identify five methods of action of antimicrobial agents.
- Describe the methods of action of each of the commonly used antibacterial drugs.
- Explain the actions of currently used antiviral drugs.
- Explain the actions of currently used antifungal, antiprotozoan, and antihelminthic drugs.
- Describe two tests for microbial susceptibility to chemotherapeutic agents.
- Describe the mechanisms of drug resistance.
- Compare and contrast synergism and antagonism.

Chemotherapy is the treatment of disease with chemicals (drugs) taken into the body. Drugs used for chemotherapy are **chemotherapeutic agents.** The class of chemotherapeutic agents used to treat infectious diseases is **antimicrobial drugs;** unlike disinfectants, they must act within the host where they kill the harmful organism without damaging the host, called **selective toxicity. Synthetic drugs** are synthesized in the laboratory; others, **antibiotics,** are produced by microorganisms.

History of Chemotherapy

During the early part of this century, Dr. Paul Ehrlich of Germany speculated about a "magic bullet," which would destroy pathogens but not harm the host. Eventually, he found an arsenic derivative, salvarsan, that was useful against syphilis. Prior to this discovery, the only chemotherapeutic agent available was quinine, for the treatment of malaria. Sulfa drugs were discovered during the 1930s. Penicillin, an antibiotic, was first discovered in 1928 but was not available in a useful form until after 1940.

Spectrum of Antimicrobial Activity

It is comparatively easy to find antimicrobials against procaryotes (bacteria) because procaryotes differ substantially from the eucaryotic cells of humans. Fungi, protozoa, and helminths are eucaryotic, which makes selective toxicity for the pathogen (without affecting the host) more difficult. It is difficult to find antimicrobials against viruses, which exist inside a host cell and interact with the host cell to synthesize new viruses.

If an antimicrobial drug affects relatively few bacteria, it has a narrow **spectrum of activity**, as opposed to drugs with a broad spectrum of activity. Antibiotics may eliminate normal microbiota and allow opportunistic pathogens to flourish **(superinfection)**.

Action of Antimicrobial Drugs

Inhibition of Cell Wall Synthesis

The cell walls of most bacteria contain peptidoglycans; human cell walls do not. Therefore, interference with the synthesis of bacterial cell walls usually does not harm the host. Antibiotics using this mode of action include penicillins, cephalosporins, bacitracin, and vancomycin. Because the peptidoglycan layer of gram-positive bacteria is more accessible than that of gram-negative, these bacteria are the most susceptible to such agents.

Inhibition of Protein Synthesis

Ribosome structure differs greatly between procaryotic and eucaryotic cells. Many antibiotics such as chloramphenicol, gentamicin, erythromycin, tetracyclines, and streptomycin interfere with protein synthesis by reacting with the ribosomes of bacteria.

Injury to the Plasma Membrane

Antibiotics, especially such polypeptides as polymyxin B, can adversely affect the membrane permeability of microbial cells. Loss of important metabolites occurs from these changes in permeability. Similarly, the effectiveness of nystatin, miconazole, ketoconazole, and amphotericin B against fungi is based on their combining with sterols to disrupt fungal plasma membranes. The activity of these drugs is selective because animal cell sterols are mostly cholesterol, whereas fungal cells are mostly ergosterol.

Inhibition of Nucleic Acid Replication and Transcription

Similarities between microbial and host cell DNA and RNA are so close that drugs that act by interfering with the nucleic acid synthesis of microbial cells have only limited clinical application. Drugs acting on this principle are rifampin, the quinolones, and the antiviral idoxuridine.

Inhibition of Synthesis of Essential Metabolites

In Chapter 5 we saw how the resemblance of sulfa drugs to para-aminobenzoic acid (PABA) was responsible for the antimicrobial activity of the former, based upon *competitive inhibition*. PABA is an essential metabolite for the production of folic acid, which humans do not synthesize. Trimethoprim and sulfones also are antimetabolites.

A summary of the actions of antimicrobial drugs is given in Figure 20.1.

Survey of Commonly Used Antimicrobial Drugs

Antibacterial Synthetics

Isoniazid (INH).   **Isoniazid,** used in the treatment of tuberculosis, is believed to inhibit synthesis of mycolic acids, which are part of the cell wall of mycobacteria.

Ethambutol.   **Ethambutol** is effective only against mycobacteria and is used in chemotherapeutic treatment of tuberculosis. It inhibits the incorporation of mycolic acid into the cell wall.

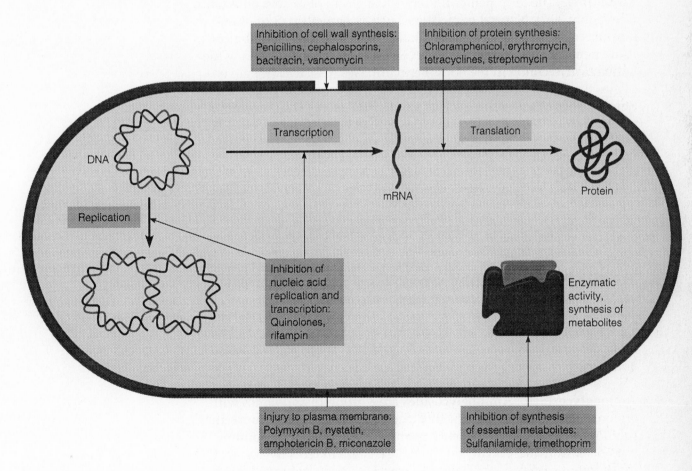

FIGURE 20.1 Summary of the major modes of action of antimicrobial drugs. This illustration shows these actions in a highly diagrammatic composite of a microbial cell.

Sulfonamides. The **sulfonamides** (sulfa drugs) act by competitive inhibition of folic acid, a precursor to nucleic acids. **Silver sulfadiazine** is used on burn patients. The most widely used sulfa-containing preparation is the combination of **trimethoprim** and **sulfamethoxazole.** These are structural analogs that inhibit synthesis of DNA at different stages.

Quinolones and Fluoroquinolones

The first quinolone was nalidixic acid, which selectively inhibits the enzyme DNA gyrase needed for DNA replication. This led to the development of the **fluoroquinolone group. Norfloxacin** and **ciprofloxacin** are used the most.

Antifungal Synthetics

The antifungal drug **flucytosine** is an antimetabolite of the base cytosine and interferes with protein synthesis. It is taken up only by fungal cells.

Antibacterial Antibiotics

Penicillins. The term **penicillin** refers to a group of related antibiotics (see Figure 20.2). **Natural penicillins,** such as **penicillin G** or **V,** are products of *Penicillium* mold growth. **Procaine penicillin** and **benzathine penicillin** combine penicillin G with other drugs to prolong the antibiotic's retention in the body. **Penicillinases** (β-*lactamases*) are enzymes that cleave the β-*lactam* ring of penicillins, causing resistance. **Semisynthetic penicillins** such as **ampicillin, amoxicillin, ticarcillin,** and **carbenicillin** have a broader spectrum of activity. Penicillinase resistance is a feature of several penicillins, such as **methicillin. Potassium clavulanate** (*clavulanic acid*), a noncompetitive inhibitor of penicillinase, has been combined with broad-spectrum penicillins such as **amoxicillin (Augmentin®)** and **ticarcillin (Timentin®).** Another penicillin variant has only a single-ring structure, the monobactams. One of these is **aztreonam,** which affects only gram-negative bacteria.

Cephalosporins. The structural nucleus of **cephalosporins** resembles that of penicillin. Cephalosporins, such as **cephalothin, cefamandole,** and **cefotaxime,** are often used as substitutes for penicillin.

Carbapenems. **Carbapenems** are a class of very broad-spectrum antibiotics. **Primaxin** is a combination of a β-lactam antibiotic *imipenem* and *cilastatin sodium*, which prevents degradation.

Aminoglycosides. **Aminoglycosides** are a group of antibiotics with amino sugars and an aminocyclitol ring. Examples are **streptomycin** (used for tuberculosis treatment), **neomycin** (used in topical ointment with bacitracin and polymyxin B), and **gentamicin** (effective against most gram-negatives, especially *Pseudomonas*). Aminoglycosides can affect the auditory nerve or the kidneys.

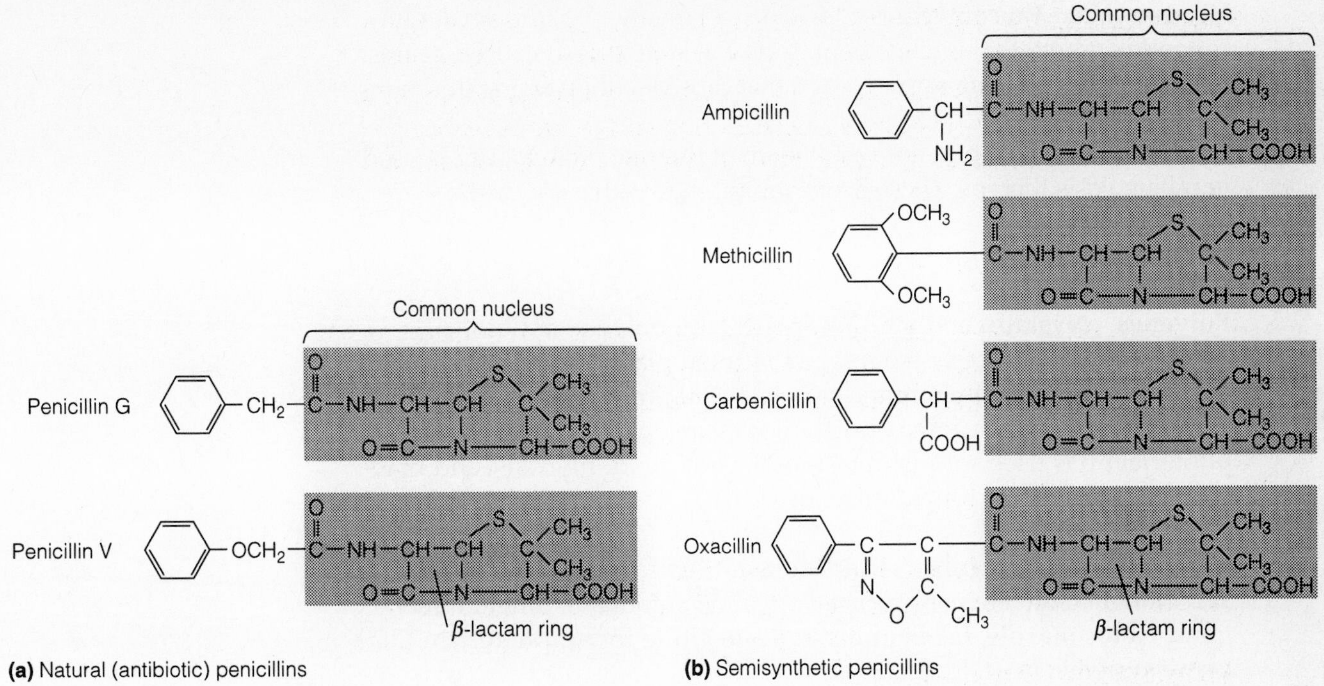

FIGURE 20.2 Structure of penicillins, antibacterial antibiotics. The portion that all penicillins have in common—their nucleus, which contains the β-lactam ring—is shaded. The unshaded portions represent the side chains that distinguish one penicillin from another.

Tetracyclines. **Tetracyclines** are broad-spectrum antibiotics that are also effective against chlamydias and rickettsias. They inhibit protein synthesis. They produce such toxic side-effects as tooth discoloration and liver damage. Commonly encountered are **tetracycline, oxytetracycline (Terramycin®),** and **chlortetracycline (Aureomycin®).** Some newer semisynthetic versions, such as **doxycycline** and **minocycline,** are retained in the body longer.

Chloramphenicol. **Chloramphenicol** is a broad-spectrum antibiotic that affects protein synthesis. Structurally simple, it is often synthesized chemically. It may cause a blood disorder, aplastic anemia. It is the drug of choice for typhoid fever and certain types of meningitis, for which the risk is considered justified.

Macrolides. **Macrolides** are named for their *macrocyclic lactone ring* and are especially effective against gram-positive bacteria. **Erythromycin** inhibits protein synthesis and is used in treating infections resistant to penicillins, as well as legionellosis and mycoplasmal pneumonia.

Polypeptides. **Bacitracin** inhibits cell wall synthesis, especially in gram-positives. It is toxic to kidneys and is usually applied topically. **Polymyxin B** injures plasma membranes. It is useful against gram-negative bacteria, especially *Pseudomonas* infections. Because of its toxicity to the kidneys and the brain, most applications are topical.

Vancomycin.   **Vancomycin** is a toxic drug, a member of the small gly-copeptide group, that inhibits peptidoglycan synthesis. It is used against penicillinase-producing staphylococci that cause life-threatening infections.

Rifamycins.   **Rifampin,** the best known of the rifamycin family, is used in tuberculosis therapy. These drugs inhibit the synthesis of mRNA.

Antifungal Drugs

Polyenes.   **Nystatin** and **amphotericin B** are polyene antibiotics whose activity is based on damage to fungal plasma membranes by combining with the membrane sterols. Nystatin is used for local fungal infections to the vagina and skin, such as *Candida* infections. Amphotericin B is used for histoplasmosis, coccidioidomycosis, and blastomycosis (systemic fungal infections).

Imidazoles and Triazoles.   **Imidazole** antifungals such as **miconazole** and **clotrimazole** are used topically against cutaneous fungal infections. **Ketoconazole,** taken orally, is a substitute for amphotericin B for many systemic fungal infections.

Griseofulvin.   **Griseofulvin** is a fungistatic drug that interferes with mitosis. Although taken orally, this drug binds selectively to keratin in skin, hair, and nails, preventing fungal growth at these sites.

Other Antifungal Drugs.   **Tolnaftate** is a topical agent used as an alternative to miconazole for athlete's foot infections. **Undecylenic** acid is a fatty acid with antifungal activity.

Antiviral Drugs

Amantadine was the first licensed antiviral drug in the U.S.A. How it works is not precisely known, but it has limited use in preventing influenza. **Ribavirin** also has activity against influenza; it is an analog of guanine and interferes with nucleic acid synthesis. Acyclovir, ganciclovir, and Zidovudine (AZT) are not active until acted upon by cell enzymes. **Acyclovir** is an analog of guanine and is active against herpesviruses. **Ganciclovir** is an analog of guanine. **Idoxuridine** and trifluridine are thymidine analogs, and vidarabine is an analog of adenine. **Zidovudine (AZT)** is a thymidine analog used to treat HIV infections. **Didanosine (ddI)** and **dideoxycytosine (ddC)** are nucleoside analogs also used to treat HIV, all by interfering with reverse transcriptase. **Interferon,** a protein secreted by some host cells infected by viruses, inhibits the virus from infecting new cells. Now available in large amounts from genetic engineering techniques, it is coming into use as an antiviral and anticancer drug.

Antiprotozoan and Antihelminthic Drugs

Antiprotozoan Drugs.   **Quinine** still has limited use against malaria, but it has usually been replaced with synthetic derivatives, such as **chloroquine** and **mefloquine. Quinacrine,** used against giardiasis, functions similarly. **Diiodohydroxyquin** (iodoquinol) is an amoebicide. **Metronidazole** is used for treatment of many protozoan diseases and also is effective against certain anaerobic bacteria. It probably causes disruption of DNA under anaerobic conditions. **Pentamidine isethionate** is mostly used for treatment of *Pneumocystis* pneumonia. **Nifurtimox** is used to treat Chagas' disease, caused by a trypanosome.

Antihelminthic Drugs. **Niclosamide** inhibits ATP production in tapeworms. **Praziquantel** also is effective against tapeworms and several fluke-caused diseases. **Mebendazole** is used to treat ascariasis. **Piperazine** paralyzes pinworms, which are then passed out of the body.

Tests to Guide Chemotherapy

Disk-Diffusion Method

The **disk-diffusion method (Kirby–Bauer test)** uses a dish of agar medium seeded uniformly with a test organism. Filter paper disks impregnated with known concentrations of chemotherapeutic agents are placed on the agar surface. If the chemotherapeutic agent is effective, a zone of inhibition (no growth) is observed around the disk. The diameter of the zone can be used to calculate the susceptibility of the organisms to the agent.

Broth Dilution Tests

A series of dilutions of an antibiotic can be placed in tubes (shallow wells in a plastic plate usually are used in practice) and inoculated with test bacteria. After incubation they are examined for turbidity. The **minimum inhibitory concentration (MIC)** of the antimicrobial is defined as the lowest concentration that prevents growth. Subculturing from the tubes that show no growth will determine if the bacteria have been killed or only inhibited. The lethal concentration that actually kills the bacteria is called the **minimum bactericidal concentration (MBC).** Many of these tests are highly automated and use light scattering to determine bacterial growth.

Effectiveness of Chemotherapeutic Agents

Drug Resistance

Resistance to drugs may be based on production of an enzyme such as penicillinase. Another serious threat to chemotherapy is drug resistance of various kinds that is carried on **plasmids,** called **resistance (R) factors.** Plasmids may transfer genes for resistance to several antibiotics at a time and to closely related bacterial species.

Self-Tests

In the matching exercises there is only one answer to a question; however, the lettered options (a, b, c, etc.) may be used more than once or not at all.

I. Matching

1. Plasmids that carry antibiotic resistance.

2. Chemotherapeutic agents produced by microorganisms.

3. Sulfa drugs are an example of this type of antibiotic.

   a. Antibiotics

   b. Synthetic drugs

   c. Chemotherapy

   d. Kirby–Bauer test

   e. R factors

II. Matching

1. Combines with sterols in plasma membrane.

2. Inhibition of protein synthesis.

3. Inhibition of RNA synthesis.

4. Inhibition of synthesis of cell wall peptidoglycans.

5. Inhibition of DNA synthesis.

6. Inhibition of cell wall mycolic acids.

   a. Cephalosporins

   b. Chloramphenicol

   c. Nystatin

   d. Isoniazid

   e. Sulfonamides

   f. Rifampin

III. Matching

1. Similar structurally to penicillins.

2. A synthetic drug used in tuberculosis chemotherapy.

3. Antifungal.

4. Causes plasma membrane leakage; useful against *Pseudomonas*.

5. An antiviral drug; a nucleoside analog.

   a. Cephalosporin

   b. Polymyxin B

   c. Ethambutol

   d. Idoxuridine

   e. Nystatin

IV. Matching

1. Used in treating diseases caused by protozoa.

2. An antifungal drug taken orally that concentrates in keratin.

3. Useful against tapeworms.

4. A drug that is useful against symptoms of genital herpes.

5. An antifungal drug of the polyene type.

6. A macrolide antibiotic.

a. Erythromycin

b. Griseofulvin

c. Amphotericin B

d. Niclosamide

e. Metronidazole

f. Acylovir

V. Matching

1. Inhibits ATP production in helminths.

2. Used in the treatment of malaria.

3. Used in the treatment of *Pneumocystis* pneumonia.

4. Paralyzes pinworms.

5. Used in the treatment of Chagas' disease.

a. Chloroquine

b. Niclosamide

c. Pentamidine isethionate

d. Piperazine

e. Nifurtimox

VI. Matching

1. Used in the treatment of HIV infections.

2. A synthetic antimicrobial that blocks the synthesis of nucleic acids.

3. A derivative of penicillin G designed to be retained for a longer time in the body.

4. A penicillin designed to be resistant to penicillinase.

5. A synthetic fluoroquinolone that acts against the gyrase enzyme.

6. An amoebicide.

7. Very broad spectrum; carbapenem group.

a. Zidovudine

b. Diiodohydroxyquin

c. Methicillin

d. Ampicillin

e. Procaine penicillin

f. Trimethoprim

g. Norfloxacin

h. Primaxin

VII. Fill-in-the-Blanks

1. Cell walls of most bacteria contain _____ , the target of activity by penicillins.

2. The treatment of disease with chemicals taken into the body is called _____ .

3. Nystatin and amphotericin B disrupt fungal plasma membranes by combining with _____ .

4. Nalidixic acid inhibits _____ synthesis.

5. Flucytosine is a structural analog of _____ .

6. The first "magic bullet" was an arsenical derivative called _____ , which was useful against syphilis.

7. Many bacteria develop resistance to penicillin by producing the enzyme _____ .

8. The usual principle of antibiotic activity is _____ , meaning it kills the harmful organism without damaging the host.

9. The drug of choice in treating typhoid fever, chloramphenicol, may sometimes cause a blood condition called _____ .

10. A rifamycin-type drug used in therapy of tuberculosis is _____ .

11. An aminoglycoside antibiotic used in tuberculosis treatment, but that may cause deafness and kidney damage, is _____ .

12. The term *penicillin* is applied to a group of antibiotics that all have a _____ in their structure.

13. The aminocyclitol ring and amino sugars are found in the _____ group of antibiotics.

14. A very toxic antibiotic, _____ , a member of the small glycopeptide group, is used only against life-threatening staphylococcal infections resistant to penicillin.

15. The lowest concentration of chemotherapeutic agent that will prevent growth is the _____ .

16. The lowest concentration of a chemotherapeutic agent that will kill the pathogen (as contrasted to inhibition) is called the _____ .

17. Miconazole and ketoconazole are examples of _____ -type antifungals.

18. Many antibiotics inhibit protein synthesis by reacting with the _____ of the bacterium, which differ greatly between procaryotic and eucaryotic cells.

19. When antibiotics eliminate much of the natural microbiota, there may be an overgrowth of resistant pathogens; this is called _____ .

20. Aztreonam is a penicillin variant with a single-ring structure; these antibiotics are called

   _____ .

21. Zidovudine, or _____ , is an analog of thymidine and is used in the treatment of HIV infections.

VIII. Critical Thinking

1. List and briefly discuss four criteria used to evaluate antimicrobial drugs.

2. Antibiotics as a supplement in animal feed has been linked to *Salmonella* infections in humans. Why did this practice begin? Why has it continued? What risks are associated with the continued use of antibiotics as an animal feed supplement?

3. What advantages and disadvantages are associated with the use of broad spectrum antibiotics?

4. Chloramphenicol, tetracycline, erythromycin, and streptomycin all inhibit protein synthesis in procaryotic cells, yet they do so by different mechanisms. Briefly describe the action of each of these drugs.

5. Both animal cells and fungal cells have sterols in their cell membranes. Why are drugs like amphotericin more toxic to fungal cells than to human cells?

6. List the advantages and disadvantages associated with each of the following drugs.

   a. Ethambutol

   b. Sulfonamides

   c. Carbapenems

   d. Nitrofurans

7. Nystatin is an antifungal drug that is used for clinical treatment of systemic fungal infections but is especially toxic to the kidneys. Why is it used to treat local vaginal infections without many problems?

8. Briefly discuss the action of acyclovir.

9. What problems are associated with antiprotozoan drugs? Give two examples of antiprotozoan drugs, indicating their action and an application.

10. Define synergism. For what purposes should combinations of antimicrobial drugs be used?

CHAPTER 21

Microbial Diseases of the Skin and Eyes

Learning Objectives

After completing this chapter, you should be able to:

- Describe the structure of the skin and mucous membranes and the ways pathogens can invade the skin.
- Provide examples of normal skin microbiota, and state their locations and ecological roles.
- Differentiate between staphylococci and streptococci, and list several skin infections caused by each.
- List the etiologic agent, method of transmission, and clinical symptoms of the following skin infections: otitis externa, acne, warts, smallpox, measles, rubella, cold sores, and scabies.
- Differentiate between cutaneous and subcutaneous mycoses, and provide an example of each.
- List the etiologic agent, method of transmission, and clinical symptoms of the following eye infections: conjunctivitis, trachoma, herpetic keratitis, and *Acanthamoeba* keratitis.
- Describe the epidemiologies of neonatal gonorrheal ophthalmia and inclusion conjunctivitis.

The first lines of defense of the body are the skin and the mucous membranes.

Structure and Function of the Skin

The **epidermis** is the outer, thinner portion of the skin; it is composed of several layers of epithelial cells. The outermost epidermal layer, the **stratum corneum,** consists mainly of dead cells containing the protein **keratin.** The inner layer of the skin is the **dermis,** composed of connective tissue with numerous blood and lymph vessels, nerves, hair follicles, and sweat and oil glands. *Perspiration* provides the moisture necessary for microbial growth; however, salts interfere with the growth of many organisms, and lysozyme in perspiration digests the cell walls of many bacteria.

Sebum from oil glands is mainly a mixture of unsaturated fatty acids, proteins, and salts. The fatty acids inhibit the growth of certain pathogens. Sebum helps prevent drying of the skin and hair but is nutritive to certain microorganisms, as will be discussed later. In the lining of body cavities such as the mouth, nasal passages, urinary and genital tracts, and gastrointestinal tract, there is a layer of specialized *epithelial cells,* which are attached at their bases to a layer of extracellular material called the *basement membrane.* Some cells produce **mucus,** which traps particles, including microbes, hence the name **mucous membrane (mucosa).** Other cells are ciliated, functioning to sweep particles and mucus from the body. The eyes are washed by tears, and lysozyme in tears destroys many bacteria.

Normal Microbiota of the Skin

Bacteria tend to clump on the skin in small groups. Many are gram-positive spherical bacteria such as staphylococci and micrococci. Gram-positive pleomorphic rods, **diphtheroids,** such as *Propionibacterium acnes,* metabolize the oil secretions in hair follicles. Acid produced by them keeps the skin pH between 3 and 5, a condition that tends to be bacteriostatic. Other diphtheroids, such as *Corynebacterium xerosis,* are aerobic. The yeast *Pityrosporum* grows on oily skin secretions. High moisture regions such as the armpit and between the legs are moist enough to support the growth of microorganisms, and populations there are high compared to dry regions like the scalp.

Bacterial Diseases of the Skin

Vesicles are small, fluid-filled lesions on the skin. Skin vesicles larger than about 1 cm in diameter are **bullae.** Flat, reddened lesions are **macules.** Raised lesions are called **papules,** or **pustules** when they contain pus.

Staphylococcal Skin Infections

Staphylococci are spherical gram-positive bacteria that form irregular grapelike clusters of cells. *Staphylococcus aureus* is the most pathogenic of the staphylococci. It forms golden yellow colonies on agar, and almost all pathogenic strains produce **coagulase,** an enzyme that coagulates the fibrin in blood. The coagulase test is used to distinguish *S. aureus* from other species of *Staphylococcus.* The predominant species of coagulase-negative staphylococci is *S. epidermidis. S. aureus* produces a number of toxins, such as enterotoxins, that affect the gastrointestinal tract (which will be discussed in Chapter 25), **leukocidins,** which destroy phagocytic leukocytes, and **exfoliative toxin,** which is responsible for scalded skin syndrome. *S. aureus* is found primarily in nasal passages and is a common hospital problem. It is universally present, and constant exposure to antibiotics causes rapid development of resistance. The species, therefore, causes many infections of surgical wounds and other artificial breaches of the skin barrier.

   Pimples are infections of hair follicles, which are natural openings in the skin barrier; an eyelash follicle infection is a **sty. Furuncles (boils)** are a type of **abscess** usually arising from a hair follicle infection. In furuncles, a region of pus is surrounded by inflamed tissue; more extensive invasion of tissue is termed a **carbuncle.** Established abscesses do not respond well to antibiotic therapy because the bloodstream does not reach them.

Impetigo of the newborn is troublesome in hospital nurseries, and prevention of this staphylococcal infection is based largely on the use of hexachlorophene-containing skin lotions. The circulation of bacterial toxins is called **toxemia.** When the outer skin layers peel away in reaction to staphylococcal toxins, it is known as the **scalded skin syndrome.** This syndrome is one of the characteristics of **toxic shock syndrome,** caused by growth of staphylococcal bacteria.

Streptococcal Skin Infections

Streptococci are gram-positive spherical bacteria that tend to grow in chains. They are aerotolerant anaerobes that do not use oxygen and are catalase-negative. They produce several toxins, as well as enzymes such as **hemolysins.** β-hemolytic group A streptococci, such as *Streptococcus pyogenes,* are the most important from the standpoint of human disease. The *M protein* distributed on the surface of their fibrils (which is also the basis for subdividing them into over 55 immunologic types) contributes to pathogenicity by its antiphagocytic properties and as an aid in adherence. Also produced by these cells are *erythrogenic toxins* (scarlet fever rash), *deoxyribonucleases* (enzymes degrading DNA), *NADase* (an enzyme that breaks down NAD), *streptokinases* (enzymes dissolving blood clots), *hyaluronidase* (an enzyme that dissolves hyaluronic acid, the cementing substance of connective tissue), and *leukocidins* (enzymes that kill white blood cells).

S. pyogenes is the most common cause of **erysipelas,** a disease in which the dermis is affected and the skin erupts into reddish patches. Usually it is preceded by a streptococcal infection, such as a sore throat, elsewhere in the body. Penicillin and erythromycin are used in treatment.

Impetigo is a superficial skin infection affecting grade-school-age children, characterized by pustules that become crusted and rupture. The bacteria enter through minor lesions. Staphylococci often are present but are probably secondary invaders. Penicillin and erythromycin are effective in treatment.

Infections by Pseudomonads

Pseudomonads, particularly *Pseudomonas aeruginosa,* may cause opportunistic skin infections. These bacteria are common in soil, water, and plants, and grow on minimal organic material. Resistance to antibiotics often makes them a hospital problem. *P. aeruginosa* produces several exotoxins and endotoxins. It may cause respiratory infections in compromised hosts (especially cystic fibrosis) or in burn patients, in whom infection may produce a characteristic blue-green pus from **pyocyanin** pigments. **Otitis externa,** or swimmer's ear, is a pseudomonad infection of the outer ear. Skin infections, collectively called *Pseudomonas* **dermatitis,** are contracted in such places as whirlpool baths and the like; the organisms are relatively resistant to chlorines. Gentamicin and carbenicillin, often used in combination, are effective; silver sulfadiazine is used to treat *P. aeruginosa* infections of burn patients.

schenckii. The nodule formed by the growing fungus may spread along the lymphatic vessels.

Candidiasis

Candida albicans is a yeastlike fungus that causes infections of the mucous membranes of the genitourinary tract (**vaginitis**) and oral mucosa (**thrush**). These infections, collectively called **candidiasis,** often appear when bacterial populations are suppressed by antibiotics that do not affect fungi. For topical use, nystatin, clotrimazole, or miconazole are recommended.

Parasitic Infestation of the Skin

Scabies

The tiny mite *Sarcoptes scabiei* burrows under the skin to lay its eggs. This sets up an inflammation that itches intensely. Treatment is by topical application of permethrin insecticide or gamma benzene hexachloride.

Microbial Diseases of the Eye

Conjunctivitis is an inflammation of the **conjunctiva,** the mucous membrane that lines the eyelid and covers the outer surface of the eyeball. **Neonatal gonorrheal ophthalmia** is a dangerous eye inflammation transmitted to the newborn during passage through the birth canal. Newborn infants have their eyes washed with 1% silver nitrate or antibiotics to prevent this infection.

Inclusion conjunctivitis is caused by the same organisms but involves only the conjunctiva, not the cornea. This disease may be spread (to newborns) by an infection in the birth canal or by unchlorinated water. Tetracycline is used topically in treatment.

Trachoma is the greatest single cause of blindness in the world. It is an infection of the epithelial cells of the eye caused by *Chlamydia trachomatis.* Scar tissue forms on the cornea. It is transmitted by contact with fomites. Tetracycline is useful in treatment.

Herpetic keratitis is caused by herpes simplex type 1 virus, with an epidemiology similar to cold sores. Corneal ulcers occur and may lead to blindness. Trifluridine is the drug of choice against this viral infection.

A freshwater amoeba, *Acanthamoeba,* causes a keratitis (**Acanthamoeba keratitis**) in the eyes of contact lens wearers. The damage is severe and may require a corneal transplant to correct.

Self-Tests

In the matching section, there is only one answer to each question; however, the lettered options (a, b, c, etc.) may be used more than once or not at all.

I. Matching

1. The inner layer of the skin, composed of connective tissue.

2. The lining of the inner eyelid and the surface of the eyeball.

3. Some of these specialized epithelial cells are ciliated.

4. The outermost epidermal layer; consists largely of dead cells containing the protein keratin.

5. Extracellular material to which epithelial cells of mucous membrane are attached.

a. Mucous membrane

b. Stratum corneum

c. Dermis

d. Conjunctiva

e. Basement membrane

II. Matching

1. Vesicles.

2. Papules.

3. Bullae.

4. Macules.

a. Flat, reddened

b. Small, fluid-filled

c. Fluid-filled lesions larger than about 1 cm

d. Raised lesions

III. Matching

1. *Streptococcus pyogenes.*

2. *Staphylococcus aureus.*

3. Tinea.

4. *Propionibacterium acne.*

5. Papillomavirus.

a. Scalded skin syndrome

b. Acne

c. Erysipelas

d. Warts

e. Ringworm

IV. Matching

1. Variola.

2. Varicella.

3. Herpes zoster.

4. Rubeola.

5. Rubella.

6. Shingles.

a. Impetigo

b. Smallpox

c. Measles

d. Chickenpox

e. German measles

V. Matching

1. Ringworm.

2. Tinea pedis.

3. Dermatophytes.

4. Thrush.

5. Sporotrichosis.

a. Cutaneous mycoses

b. Superficial mycoses

c. Subcutaneous mycoses

d. Candidiasis

VI. Matching

1. Swimmer's ear, usually caused by pseudomonads.

2. Boils.

3. Idoxuridine is an effective chemotherapeutic treatment.

4. Chlamydia-caused disease.

a. Herpetic keratitis

b. Trachoma

c. Otitis externa

d. Furuncles

VII. Matching

1. Treatment of cystic acne.

2. The location of M protein of streptococci.

3. Causes birth defects.

4. Prevention of neonatal gonorrheal ophthalmia.

5. Scabies.

a. Teratogenic

b. Fibrils

c. Isotretinoin

d. Mite

e. Silver nitrate

VIII. Fill-in-the-Blanks

1. The eyes are washed by tears, and the enzyme _____ in tears destroys many bacteria.

2. Gram-positive pleomorphic rods such as *Propionibacterium acne* and *Corynebacterium xerosis* are called _____ .

3. An occasional serious complication of chickenpox and influenza is _____ syndrome.

4. If a boil undertakes a more extensive invasion of the surrounding tissue, it is termed a _____ .

5. *Streptococcus pyogenes* is an example of _____ hemolytic group A streptococci.

6. The blue-green pus caused by opportunistic infections of burn patients is due to *Pseudomonas aeruginosa* forming water-soluble _____ pigment.

7. Benzoyl peroxide is useful in treating _____ .

8. Koplik spots are diagnostic for cases of _____ .

9. A fungal infection of the body is referred to as a _____ .

10. The presence of viruses in the bloodstream is termed _____ .

11. The greatest single cause of blindness in the world is the disease _____ .

12. Herpetic keratitis is caused by herpes simplex type _____ virus.

13. An eyelash follicle infection is called a _____ .

14. The M protein distributed on the surface on some streptococci contributes to pathogenicity by its _____ properties, and as an aid in adherence.

15. *Propionibacterium acne* metabolizes _____ , forming free fatty acids that cause the inflammation leading to acne scarring.

16. The first major disease to be deliberately eliminated on Earth is _____ .

17. Congenital rubella syndrome is caused by rubella infection during the first _____ of pregnancy.

18. Ringworm of the scalp is tinea _____ .

19. The _____ test is used to distinguish pathogenic *Staphylococcus aureus* from other species of *Staphylococcus*.

20. Scarlet fever rash is caused by an _____ -type toxin.

21. Ringworm of the feet, athlete's foot, is called tinea _____ .

22. A protozoan disease of the eye is caused by the amoeba _____ .

IX. Label the Art

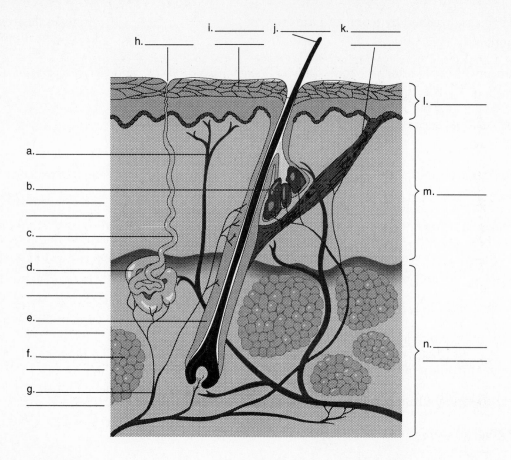

X. Critical Thinking

1. A surgical patient acquires a nosocomial infection. The primary organism isolated from the surgical wound is coagulase positive and resistant to penicillin. What is the probable etiologic agent? Why is this organism a common cause of nosocomial infections?

2. List three infections caused by *Pseudomonas.* What is the route of transmission in each example? What factors contribute to the transmission of *Pseudomonas* in each case?

3. Discuss the role of *Propionibacterium* in acne. Does *Propionibacterium* provide any benefits to its host?

4. Why are the symptoms of shingles different from those of chicken pox?

5. Compare and contrast herpes simplex type I and type II.

6. What factors have contributed to the increase in measles cases since 1983?

7. Discuss the factors that may result in development of a vaginal yeast infection.

8. A 43-year-old patient consulted his physician concerning mild inflammation of his eyes. The patient was instructed to avoid wearing his contact lenses for one week. The symptoms worsened over the next few days. After eliminating the possibility of a bacterial agent, the physician ordered a corneal scraping. What etiologic agent does the physician suspect? What evidence of this agent is the physician looking for in the scraping?

9. Why were efforts to eliminate smallpox successful?

10. Give three examples of how normal microbiota of the skin can contribute to the discomfort of the host.

CHAPTER 22

Microbial Diseases of the Nervous System

Learning Objectives

After completing this chapter, you should be able to:

- Define the following terms: central nervous system, peripheral nervous system, cerebrospinal fluid, blood-brain barrier.
- Differentiate between meningitis and encephalitis.
- Discuss the epidemiology of meningitis caused by *H. influenzae, S. pneumoniae, N. meningitidis,* and *L. monocytogenes.*
- Discuss the epidemiology of tetanus and leprosy, including method of transmission, etiology, disease symptoms, and preventive measures.
- Provide the etiologic agent, symptoms, suspect foods, and treatment for botulism.
- Discuss the epidemiology of poliomyelitis, rabies, and arbovirus encephalitis, including method of transmission, etiology, disease symptoms, and preventive measures.
- Compare the Salk and Sabin polio vaccines.
- Compare the preexposure and postexposure treatments for rabies.
- Identify the causative agent, vector, symptoms, and treatment for cryptococcosis.
- Identify the causative agent, vector, symptoms, and treatment for African trypanosomiasis and *Naegleria* microencephalitis.
- List the characteristics of diseases thought to be caused by prions.

Structure and Function of the Nervous System

The **central nervous system (CNS)** consists of the **brain** and **spinal cord.** It is the control center that picks up sensory information from the environment and, after interpreting it, sends out impulses to coordinate body activities. The **peripheral nervous system** consists of all the nerves branching from the brain and spinal cord. These nerves are the communication lines between the CNS and the body, and the external environment. The skull protects the brain and the backbone protects the spinal cord; both the brain and spinal cord are covered by the **meninges** (Figure 22.1). The meninges consist of the dura mater (outermost layer), the arachnoid (middle layer), and the pia mater (innermost layer). **Cerebrospinal fluid** circulates in the space between the pia mater and the arachnoid layers (*subarachnoid space*). The **blood–brain barrier** consists of capillaries that

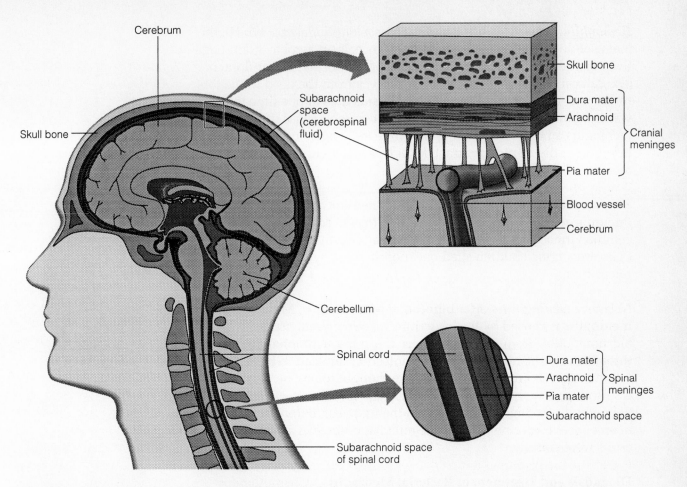

FIGURE 22.1 The meninges and cerebrospinal fluid. The meninges, whether cranial or spinal, consist of three layers: dura mater, arachnoid, and pia mater. Between the arachnoid and the pia mater is the subarachnoid space, in which cerebrospinal fluid circulates. Note how the cerebrospinal fluid is vulnerable to contamination by microbes carried in the blood that are able to penetrate the blood–brain barrier at the walls of the blood vessels.

permit certain substances to pass from the blood to the brain but that prevent others from passing. Microorganisms can gain access to the CNS by trauma (accidental, or medical procedures such as spinal taps). Also, microorganisms may enter by movement along peripheral nerves or by the bloodstream and lymphatic system. An infection of the meninges is **meningitis;** an infection of the brain is **encephalitis.** Antibiotics often are unable to cross the blood–brain barrier.

Bacterial Diseases of the Nervous System

Bacterial Meningitis

The three major types of bacterial meningitis are meningococcal meningitis, caused by *Neisseria meningitidis;* pneumococcal meningitis, caused by *Streptococcus pneumoniae;* and *Hemophilus influenzae* meningitis. Nearly 50 other bacteria can cause meningitis, as can fungi, viruses, and protozoa. Meningitis patients complain of headache and have symptoms of nausea and vomiting. This may proceed to convulsions, coma, and even death.

Hemophilus influenzae **Meningitis.** *Hemophilus influenzae* type b, the main serological type of medical importance, is an aerobic, gram-negative pleomorphic bacterium. A member of the normal throat flora, it also can cause meningitis. The name derives from the need for factors from lysed red blood cells and the erroneous idea at one time that it was the cause of influenza. It is the most common cause of bacterial meningitis among children under four years of age. Following a viral infection of the respiratory tract, *H. influenzae* enters the blood and then invades the meninges.

Streptococcus pneumoniae **Meningitis (Pneumococcal Meningitis).** **Pneumococcal meningitis** is caused by *Streptococcus pneumoniae,* a common inhabitant of the nasopharyngeal system. About half the cases are among children aged one month to four years. The mortality rate is high.

Neisseria meningitidis **(Meningococcal Meningitis).** **Meningococcal meningitis** is caused by the bacterium *Neisseria meningitidis,* an aerobic, nonmotile, gram-negative coccus. It is a frequent inhabitant of the throat behind the nose, and a throat infection can lead to bacteremia followed by meningitis. Symptoms are thought to be due to endotoxins produced by the bacteria. The disease primarily affects the very young, with the highest incidence in the first year. Before development of an effective vaccine used by the military, outbreaks among young recruits were common.

Diagnosis and Treatment of Bacterial Meningitis. Chemotherapy must be prompt; broad-spectrum third-generation cephalosporins are the first choices. Diagnosis requires a sample of cerebrospinal fluid obtained by a spinal tap. Gram stains and cultures can be made. Serological tests such as rapid latex agglutination tests are used.

Listeriosis

Listeria monocytogenes, the cause of **listeriosis,** is widely distributed, mainly in animal feces. Listeriosis is a mild disease in most adults but is more serious in the immunosuppressed, including cancer patients. When infecting a pregnant woman it often causes abortion or serious damage or death to the fetus. The organism proliferates within phagocytic cells and grows at refrigeration temperatures.

Tetanus

The cause of **tetanus** is an obligately anaerobic, endospore-forming, gram-positive rod, *Clostridium tetani.* It is a common soil organism, particularly soil contaminated with animal fecal wastes. Growth of *C. tetani* in wounds releases **tetanospasmin,** a neurotoxin that blocks the nerve pathway that signals muscle relaxation. Spasms result as opposing muscle sets contract (spastic paralysis). Contraction of the jaw muscles prevents opening of the mouth **(lockjaw),** and death from respiratory failure results in many cases. Most people in the U.S.A. have been immunized by the tetanus toxoid included in the DPT (diphtheria, pertussis, tetanus) vaccine. A booster of toxoid may be given when a dangerous wound is received; this booster causes the body to renew its immunity level in a rapid anamnestic response. If the patient has had no previous immunity, toxoid may not cause a rapid enough appearance of antibodies, and a temporary immunity can be provided by tetanus immune globulin (TIG) pooled from immunized humans. Table 22.1 summarizes wound management for tetanus.

TABLE 22.1 Wound Management for Prevention of Tetanus

	MINOR WOUNDS		EXTENSIVE WOUNDS	
HISTORY OF TETANUS IMMUNIZATION	ADMINISTER TOXOID	ADMINISTER TETANUS IMMUNE GLOBULIN	ADMINISTER TOXOID	ADMINISTER TETANUS IMMUNE GLOBULIN
Unknown or 0 to 2 doses	Yes	No	Yes	Yes
3 or more doses	No*	No	No†	No

*Yes, if more than 10 years since last dose.
†Yes, if more than 5 years since last dose.

Botulism

The cause of **botulism** is an obligately anaerobic endospore-forming bacterium, *Clostridium botulinum,* found in soil and freshwater sediments. The exotoxin produced is a neurotoxin that blocks the transmission of nerve impulses across the synapses, causing progressive *flaccid paralysis.* Symptoms typically appear in a day or two. The word *botulism* is derived from the Latin *botulus,* meaning "sausage," a common vehicle of the disease at one time. Most botulism results from attempts at preservation by heat that fail to eliminate the *C. botulinum* endospore but provide anaerobic conditions for its growth. There are several toxin types. **Type A toxin** is found mostly in the western United States. It is proteolytic and probably the most virulent. **Type B toxin** is less virulent, and both proteolytic and nonproteolytic strains occur. Type B is responsible for most outbreaks in the eastern United States and Europe. **Type E toxin** is produced by a strain found in wet soils and sediments and often involves seafood. This organism is nonproteolytic, its endospores are less heat-resistant, and the toxin can be produced at refrigerator temperatures. The botulinal toxin is heat-labile; that is, it is destroyed by boiling. It is not formed in acid foods below pH 4.7. Nitrites are included in some meat products to prevent bacterial growth following germination of endospores. For treatment, antitoxins of trivalent ABE type are available, but they do not affect attached toxins. Toxin identification can be done by inoculating mice with the suspected toxin; if the mice who survive the inoculation are protected by, for example, type A antitoxins, then the suspected toxin is considered to be type A. **Wound botulism** can occur from *C. botulinum* growth in wounds, and **infant botulism** may occur from growth in the infant GI tract, which is rarely a factor in adults.

Leprosy

Leprosy, sometimes called **Hansen's disease** for the person who first iso-lated the organism, is caused by an acid-fast rod, *Mycobacterium leprae.* It is probably the only bacterium that grows primarily in the peripheral nervous system. The microorganisms have not been grown on artificial media but can be grown in armadillos. There are two main forms of lep-rosy. The **tuberculoid (neural) form** is characterized by regions of the skin that have lost sensation and are surrounded by a border of nodules. The **lepromin test,** similar to the tuberculin test in design, usually is positive for this form. In the **lepromatous (progressive) form** of leprosy, skin cells are infected, and disfiguring nodules form over the body. Mucous mem-branes of the nose (one sign is a lion-faced appearance) and the hands are particularly affected; the organism prefers cooler regions of the body. Pro-gression to this stage of leprosy indicates a less-effective immune system.

Leprosy is not very contagious. Death is usually a result of complica-tions from diseases such as tuberculosis. Laboratory diagnosis is by iden-tification of acid-fast rods in lesions. The sulfone drugs, such as dapsone, are the most effective treatment. Rifampin and a fat-soluble dye, clofaz-imine, also are used.

Viral Diseases of the Nervous System

Poliomyelitis

Poliomyelitis is informally called **polio.** Adults are more likely to get the disease when living in regions of good sanitation, where immunity is not acquired as an infant; therefore, in adults, the disease is more likely to reach the paralytic stage. Symptoms usually are a few days of headache, sore throat, and fever; only a few cases are paralytic. The poliovirus oc-curs in three different serotypes. The primary mode of transmission is in-gestion of fecal-contaminated water. In paralytic cases, the virus pene-trates capillary walls and enters the CNS, where it displays a high affinity for nerve cells. The motor nerve cells in the upper spinal cord, called the anterior horn cells, are particularly affected.

The **Salk vaccine** (inactivated polio vaccine, IPV) uses viruses inacti-vated by formalin. This vaccine requires a series of injections and peri-odic booster shots. The newer **enhanced polio vaccine** (E-IPV, grown on human diploid cells) is available; it is often used for immunosuppressed individuals. The **Sabin vaccine** (oral polio vaccine, OPV) contains three live attenuated strains. The immunity from this vaccine resembles that ac-quired naturally, but on rare occasions the vaccine itself may cause the disease, probably by a mutation to virulence. Diagnosis is usually based on isolation of the virus and observation of cytopathic effects on cell cul-tures.

Rabies

The **rabies virus,** a rhabdovirus with a bullet shape, is typically transmit-ted in the saliva of a biting rabid animal. The virus travels along periph-eral nerves to the CNS, where it produces encephalitis. Symptoms include spasms of the muscles of the mouth and pharynx when swallowing liquids. This painful reaction causes an aversion to water, which is the basis of the term **hydrophobia** (fear of water) that is applied to rabies.

Animals with **furious rabies** are restless, snapping at anything. Some animals suffer from **paralytic rabies,** in which there is only minimal excitability, although they may snap irritably if handled.

Laboratory diagnosis is usually made by a fluorescent antibody test. In early years, the Pasteur treatment was standard for prevention of rabies in exposed individuals. The vaccination procedure consisted of injections of dried spinal cords of rabbits infected with rabies virus. Currently, treatment of exposed people begins with the administration of human rabies immune globulin (RIG), followed by active immunization. The latter is done with human diploid cell vaccines (HDCV) grown in human diploid cell lines. HDCV is administered in five or six injections over a 28-day period. Indications for antirabies treatment are unprovoked bites from dogs, cats, skunks, and similar animals. Bites from rodents and rabbits generally do not require antirabies treatment. It also is possible to contract rabies by inhalation of aerosols of the virus.

Arthropod-Borne Encephalitis

Encephalitis, caused by arboviruses that are transmitted by a mosquito, is common in the U.S.A. Horses frequently are infected, accounting for terms such as **Eastern equine encephalitis (EEE)** and **Western equine encephalitis (WEE). St. Louis encephalitis (SLE)** and **California encephalitis (CE),** which do not infect horses, also occur. These diseases are summarized in Table 22.2. In the Far East, **Japanese B encephalitis** is a similar disease. All cause a percentage of neurological damage; however, EEE is the most severe in the U.S.A., causing a high incidence of brain damage, deafness, and similar neurological problems. It also has the highest mortality rate. Diagnosis is usually made by, for example, complement-fixation tests. Control of mosquitoes is an essential preventive measure.

TABLE 22.2   Types of Arthropod-Borne Encephalitis

DISEASE	MOSQUITO VECTOR	HOST ANIMALS	GEOGRAPHIC DISTRIBUTION IN U.S.A.	COMMENT
St. Louis encephalitis	Culex	Birds	Throughout country	Mostly urban outbreaks; affects mainly adults over 40.
California encephalitis	Aedes	Small mammals	North Central states, New York State	Affects mostly 5–18 age group in rural or suburban areas; La Crosse strain medically most important. Mainly in upper Midwest.
Western equine encephalitis	Culex	Birds, horses	Throughout country	Severe disease; frequent neurological damage, especially in infants.
Eastern equine encephalitis	Aedes, Culiseta	Birds, horses	East coast	More severe even than WEE; affects mostly young children and younger adults; relatively uncommon in humans.

Fungal Disease of the Nervous System

Cryptococcus neoformans Meningitis (Cryptococcosis)

Cryptococcus neoformans is a yeastlike fungus widely distributed in soil, especially in soil enriched by pigeon droppings. The disease **cryptococcosis** is thought to be transmitted by inhalation of dried infected pigeon droppings. Often the disease does not proceed beyond an infection of the lungs, but it may spread by the bloodstream to the brain and meninges. Laboratory diagnosis is by latex agglutination tests of cerebrospinal fluid. The drugs of choice for treatment are amphotericin B and flucytosine.

Protozoan Diseases of the Nervous System

African Trypanosomiasis

African trypanosomiasis is a disease caused by protozoa that affects the nervous system. The flagellates *Trypanosoma brucei gambiense* and *Trypanosoma brucei rhodesiense* enter the body through a tsetse fly bite. Domestic and wild animals are reservoirs for the disease. Later they move into the cerebrospinal fluid, where the resulting symptoms are responsible for the informal name of sleeping sickness. Suramin, pentamidine isethionate, and a toxic arsenical called melarsoprol are used in chemotherapy. A new drug, eflornithine, blocks an enzyme needed for proliferation. It is very effective against *T. b. gambiense* but variable against *T. b. rhodesiense*.

Naegleria Microencephalitis

An amoebic protozoa found in ponds and streams, *Naegleria fowleri* can infect the nasal mucosa and from there reach and proliferate in the brain. The fatality rate of *Naeglaria* **microencephalitis** is nearly 100%.

Nervous System Diseases Thought to Be Caused by Prions

There are a number of fatal diseases of the human central nervous system caused by prions. **Sheep scrapie** is a typical prion disease. **Creutzfeldt–Jakob disease** and **kuru** are similar diseases in humans. A current international public health concern is **bovine spongiform encephalopathy,** or mad cow disease, which affects many herds of cattle in the British Isles. The agents causing these diseases are unknown; they do not have any detectable nucleic acid, and one hypothesis is that they are composed entirely of protein (prions). Chapter 13 has a further discussion of this subject.

Self-Tests

In the matching section, there is only one answer to each question; however, the lettered options (a, b, c, etc.) may be used more than once or not at all.

I. Matching

1. A membrane layer covering the brain and spinal cord.
2. A prion-caused disease.
3. The site of attack of the poliomyelitis virus.
4. Hansen's disease.
5. Human diploid cell vaccine.

a. Meninges
b. Anterior horn cells
c. Rabies
d. Kuru
e. Viroid
f. Leprosy

II. Matching

1. Innermost layer of the meninges.
2. Outermost layer of the meninges.
3. Middle layer of the meninges.

a. Dura mater
b. Ventricles
c. Subarachnoid space
d. Arachnoid
e. Pia mater

III. Matching

1. Treated by the Pasteur treatment.
2. Treated by human diploid cell vaccine after exposure.
3. Caused by a bullet-shaped rhabdovirus.
4. Also known by the term *hydrophobia*.
5. Thought to be transmitted by inhalation of the pathogen in dried pigeon droppings.
6. Caused by the bacterium *Neisseria meningitidis*.
7. Protozoan disease.

a. Rabies
b. Meningococcal meningitis
c. *Hemophilus influenzae* meningitis
d. Cryptococcosis
e. Poliomyelitis
f. Pneumococcal meningitis
g. African trypanosomiasis

IV. Matching

1. A prion-caused disease.

2. A mosquito-borne virus.

3. The drugs of choice for treatment are amphotericin B and flucytosine.

4. Opposing muscles contract, causing spastic paralysis.

5. Pathogen grows at refrigerator temperatures.

a. Creutzfeldt–Jakob disease

b. Meningococcal meningitis

c. Listeriosis

d. *Crytococcus neoformans* meningitis

e. Tetanus

f. California encephalitis

V. Matching

1. Uses live viruses.

2. On rare occasions, the vaccine has caused the disease by a mutation to virulence.

a. Salk polio vaccine

b. Sabin polio vaccine

VI. Matching

1. An amoebic protozoa found in ponds and streams that causes a lethal brain infection.

2. Spread by the bite of a tsetse fly.

3. An important cause of bacterial meningitis.

4. The cause of African sleeping sickness.

a. *Naegleria fowleri*

b. *Trypanosoma brucei gambiense*

c. *Cryptococcus neoformans*

d. *Streptococcus pneumoniae*

VII. Matching

1. Probably the most virulent; the most common type in western United States.

2. Outbreaks often involve seafoods; nonproteolytic.

3. Toxin can be produced at refrigerator temperatures.

a. Type A botulism

b. Type B botulism

c. Type E botulism

VIII. Fill-in-the-Blanks

1. An infection of the brain is called _____ .

2. The _____ nervous system consists of the brain and spinal cord.

3. An infection of the meninges is called _____ .

4. The _____ consists of capillaries that permit certain substances to pass from the blood to the brain but that prevent others from passing.

5. The _____ nervous system consists of all the nerves branching from the brain and the spinal cord.

6. The polio vaccine that uses living attenuated viruses is the _____ vaccine.

7. The bacterium causing _____ can be successfully grown in armadillos.

8. Of the several types of arthropod-borne encephalitis that occur in the U.S.A., the most severe in its effects is _____ .

9. The T in the DPT vaccine stands for _____ .

10. Tetanospasmin is a toxin affecting the _____ system.

11. The word *botulism* is derived from the Latin for _____ .

12. _____ fluid circulates in spaces within the brain (ventricles).

13. The most effective rabies vaccine, and the one with the fewest adverse effects such as allergic reactions, is the _____ cell vaccine.

14. _____ are used in some foods to prevent the growth of *Clostridium botulinum.*

15. People with a less-effective cell-mediated immune system are more likely to develop the _____ form of leprosy.

IX. Critical Thinking

1. Discuss the signs and symptoms associated with tetanus. What factors contribute to transmission of tetanus?

2. People who recover from botulism are not immune to the disease. Why? Why are antibiotics not indicated in botulism cases?

3. Why is it recommended that honey not be fed to infants under 1 year of age?

4. Why do many people infected with polio develop a sore throat and nausea? How do these symptoms relate to transmission of the polio virus?

5. People that have been infected with rabies may be able to develop immunity from a vaccination given after exposure to the virus. Why is this possible?

6. Explain the difficulties associated with the development of a vaccine effective against African trypanosomiasis.

7. List and briefly discuss at least three examples of slow viral infections.

8. What is Hansen's disease? How is it transmitted? How is it diagnosed?

9. Differentiate between the Salk and Sabin polio vaccines.

10. Describe the progression of infection with cryptococcosis.

CHAPTER 23

Microbial Diseases of the Cardiovascular and Lymphatic Systems

Learning Objectives

After completing this chapter, you should be able to:

- Identify the role of the cardiovascular system in spreading infections and in eliminating infections.
- Identify the role of the lymphatic system in spreading infections and in eliminating infections.
- List the signs and symptoms of septicemia, and explain the importance of infections that develop into septicemia.
- Describe the epidemiologies of puerperal sepsis, bacterial endocarditis, and myocarditis.
- List the cause of and treatment and preventive measures for rheumatic fever.
- Discuss the epidemiologies of tularemia, brucellosis, anthrax, and gas gangrene.
- List four diseases acquired by animal bites.
- Compare and contrast the causative agents, vectors, reservoirs, symptoms, treatments, and preventive measures for plague, relapsing fever, Lyme disease, and typhus.
- Describe the epidemiologies of Burkitt's lymphoma and infectious mononucleosis.
- Compare and contrast the causative agents, vectors, reservoirs, symptoms, and treatments for yellow fever and dengue.
- Compare and contrast the causative agents, methods of transmission, reservoirs, symptoms, and treatments for toxoplasmosis, American trypanosomiasis, and malaria.
- Discuss the worldwide effects of these diseases on human health.
- Diagram the life cycle of *Schistosoma*. Show where the cycle can be interrupted to prevent human disease.

The **cardiovascular system** consists of the heart, blood, and blood vessels. The **lymphatic system** consists of lymph, lymph vessels, lymph nodes, and lymphoid organs such as the thymus, spleen, and tonsils. The cardiovascular system can serve as a vehicle to spread infections in the body, but is also—particularly the lymphatic system—the site of many of the body's defenses against infections.

Structure and Function of the Cardiovascular System

The **heart** is the center of the cardiovascular system (see Figure 23.1). **Blood** is pumped to the lungs where it exchanges carbon dioxide for oxygen and returns to the heart. This oxygenated blood is then pumped throughout the body through tubes called **blood vessels.** Circulating blood provides oxygen to tissue cells and picks up waste carbon dioxide. Blood passes from the heart in **arteries** and returns in **veins.** Arteries eventually branch into smaller vessels, **arterioles** (the equivalent in veins are **venules**), and then branch into even smaller vessels called **blood capillaries.** The body's main veins are the **superior vena cava** and the **inferior vena cava.** Capillary walls are only one cell thick. This thinness allows blood and tissue cells to exchange materials. The liquid of blood is **plasma.** Red blood cells, also called **erythrocytes,** carry oxygen. White blood cells, **leukocytes,** are important in defending against infection and include phagocytes and the B and T cells involved in acquired immunity.

Structure and Function of the Lymphatic System

Some plasma filters through blood capillary walls and, as **interstitial fluid,** bathes tissue cells. This fluid is picked up by **lymph capillaries** and is referred to as **lymph.** Microorganisms are also able to enter lymph capillaries easily. Lymph eventually rejoins venous blood just before it reenters the heart. **Lymph nodes,** small oval structures, are located at various points along the lymphatic system. Phagocytic and antibody-producing cells help remove foreign bodies in the lymph or cause antibodies against them to be produced. The tonsils, appendix, spleen, and thymus gland also are lymphatic organs.

Bacterial Diseases of the Cardiovascular and Lymphatic Systems

Septicemia

Uncontrolled proliferation of microbes in the blood is called **septicemia.** This condition causes fever and the drop in blood pressure known as **septic shock. Lymphangitis,** which becomes apparent as red streaks under the skin running up the arm or leg from an infection site, occurs when the lymph system becomes involved. Gram-negative rods are most commonly involved with septicemia today. The endotoxins released by the death and lysis of such cells are responsible for most of the symptoms. Gram-positive bacteria, such as the staphylococci, were a more common cause of such infections at one time.

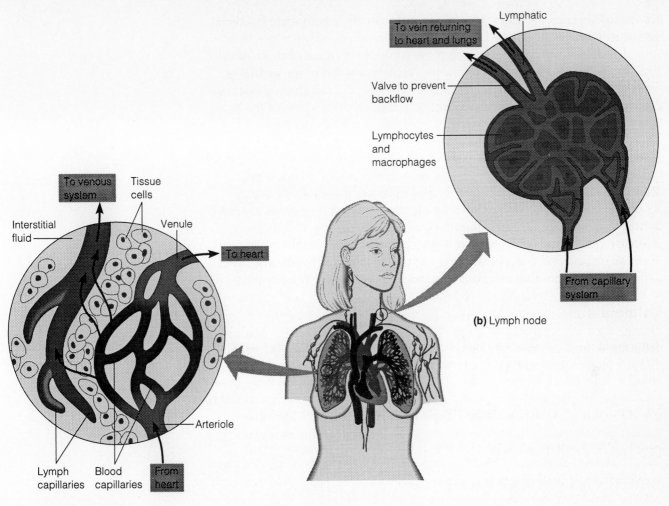

FIGURE 23.1 Relationship of the cardiovascular and lymphatic systems.   From the blood capillaries, some plasma filters into the surrounding tissue and enters the lymph capillaries, shown in the lung in part (a). This fluid, now called lymph, returns to the heart through the lymphatic circulatory system (darker gray), which channels the lymph to a vein. All lymph returning to the heart must pass through at least one lymph node, as shown in (b). (See also Figure 16.6 in the text.)

Puerperal Sepsis

An infection of the uterus that often leads to septicemia is called **puerperal sepsis,** also known as **puerperal fever** or **childbirth fever.** Group A β-hemolytic streptococci such as *Streptococcus pyogenes* are the most common cause. **Peritonitis,** an infection of the abdominal cavity, may be a final result. Semmelweiss and Holmes long ago clearly showed that poor hygienic practices by doctors and midwives were the primary cause, and that proper disinfection and sterilization could prevent transmission. Infections from abortions are mostly caused by anaerobic bacteria such as *Bacteroides* and *Clostridium* species.

Bacterial Endocarditis

The heart has three layers. The outer layer, called the **pericardium,** is a sac enclosing the heart; a middle layer of muscle tissue is called the **myocardium;** and the inner layer of epithelium is called the **endocardium.**

Endocarditis is an inflammation of the endocardium. **Subacute bacterial endocarditis** usually is due to α-hemolytic streptococci. Microbes from tonsillectomies and tooth extractions are a common cause of this disease. Blood clots are particularly dangerous. **Acute bacterial endocarditis** is usually caused by *Staphylococcus aureus*. Streptococci can cause **pericarditis** (inflammation of the pericardium). **Myocarditis** (infection of the heart muscle, or myocardium) is caused by a varied array of microbes.

Rheumatic Fever

Rheumatic fever is generally considered to be an autoimmune reaction by the body to repeated infections by *Streptococcus pyogenes,* usually from a streptococcal sore throat. Inflammations cause arthritis and damage heart valves. Group M proteins of the streptococci may be the main antigenic structure. Some people develop a complication of rheumatic fever called **Sydenham's chorea,** characterized by involuntary movements.

Tularemia

Tularemia is a disease of the lymph nodes caused by *Francisella tularensis,* a gram-negative, facultatively anaerobic rod. Hunters in contact with small animals such as rabbits are often infected, but arthropod bites (deer flies, for example) may also be involved. The infection results in a small ulcer at the infection site and later lymph node involvement. The microorganism grows intracellularly in phagocytes. This intracellular location is a problem in chemotherapy.

Brucellosis (Undulant Fever)

Several species of the genus *Brucella* cause **brucellosis,** also known as **undulant fever** because of the periodic spiking fever. The organisms are capable of intracellular growth in phagocytic cells and are difficult to treat. *Brucella abortus* and *Brucella melitensis* are most commonly transmitted by the ingestion of the unpasteurized milk of cattle and goats, respectively. *Brucella suis* is transmitted mostly by contact with swine carcasses and is the most common mode of transmission in the U.S.A. today. Brucella tend to multiply in the uterus of susceptible animals, favored by the presence of *mesoerythritol* in the fetus and in surrounding membranes.

Anthrax

Robert Koch first identified **anthrax** as being caused by a spore-forming rod now called *Bacillus anthracis.* The spores are found in soil and are ingested by animals while grazing. Humans can get anthrax by contact with spore-containing animal hides and other products. A pustule forms at the site of entry on the skin, and septicemia may result. Human gastrointestinal anthrax caused by ingestion of endospores has been reported. **Woolsorter's disease**—pneumonia caused by inhaling large numbers of the spores—is particularly dangerous. The vaccine is based on the toxin produced by the organisms.

Gangrene

Anaerobic conditions occur in tissue if a wound interrupts the blood supply *(ischemia)*. In turn, ischemia leads to death *(necrosis)* of tissue, or **gangrene.** Substances from dead cells provide nutrients for bacteria such as the endospore-forming anaerobes of the genus *Clostridium.* Gas gangrene, especially in muscle tissue, occurs when *Clostridium perfringens* ferments carbohydrates in wound tissue and produces carbon dioxide and hydrogen gases. This microorganism produces exotoxins and enzymes that further interfere with blood supply and otherwise favor the spread of the infection. This condition spreads the area of necrosis; ultimately severe toxemia and death may ensue. Treatments for gas gangrene include debridement (surgical removal of tissue) or amputation. *Hyperbaric chambers*, in which the patient is subjected to an oxygen-rich atmosphere that interferes with the growth of the obligately anaerobic clostridia, are particularly useful for abdominal gangrene. Antibiotics such as penicillin are effective.

Systemic Diseases Caused by Bites and Scratches

Animal bites inflicted on humans often become infected. *Pasteurella multocida,* which normally is more likely to cause septicemia in animals, can cause infections in humans. The infection may be localized at the site of the wound or may progress to pneumonia and life-threatening septicemia. Anaerobes such as *Clostridium, Bacteroides,* and *Fusobacterium* can also infect deep bites. Rat bites can cause recurring fever and arthritislike symptoms, collectively known as **rat bite fever.** Organisms of the genus *Streptobacillus* or *Spirillum* are the typical pathogens. Infections resulting from human bites are often troublesome. **Cat scratch disease,** usually from minor scratches, may cause swollen lymph nodes and prolonged fever. The pathogen may be a gram-negative rod, *Afipia felis,* or a rickettsialike *Rochalimaea henselae.*

Plague

In the Middle Ages, **plague** was known as the Black Death because of the characteristic blackish areas on the skin caused by hemorrhages. The name **bubonic plague** derives from swellings, called *buboes,* that form as the lymph nodes in the groin and armpit regions become enlarged. The disease is caused by a gram-negative rod, *Yersinia pestis,* that is able to survive and even increase in number inside phagocytic cells. Normally the disease is transmitted by the rat flea from rat reservoirs, but in the western U.S.A. the disease is endemic in wild rodents.

A particularly dangerous development is **pneumonic plague,** which arises when the bacteria are carried by the blood to the lungs. Because of airborne droplet infection, pneumonic plague is highly infective. Untreated, plague has a mortality rate of 50% to 75%, and for pneumonic plague the rate approaches 100%. Diagnosis is most commonly done by isolation and by identification of the isolated bacterium by fluorescent antibody tests or phage tests. Antibiotics such as streptomycin and tetracycline are useful for prophylactic protection of exposed people and for treatment.

Relapsing Fever

All members of the genus *Borrelia* (a spirochete) cause **relapsing fever.** The disease is transmitted by soft ticks that feed on rodents. A high fever is accompanied by jaundice and rose-colored spots. The fever subsides after three to five days, but several relapses, each caused by a different antigenic type, may occur. Diagnosis is made by observation of spirochetes in the patient's blood.

Lyme Disease (Lyme Borreliosis)

Lyme disease is caused by the bite of a tick, *Ixodes pacificus* on the Pacific coast and mostly *Ixodes scapularus* in the rest of the U.S.A. A skin lesion spreads from the site of the bite, clearing in the center. Later complications are arthritis and, occasionally, heart and neurological abnormalities. A spirochete, *Borrelia burgdorferi,* is the pathogen. The disease can be successfully treated with antibiotics.

Typhus

Several related diseases are caused by rickettsias that are spread by arthropod vectors such as mites, ticks, mosquitoes, fleas, and lice. **Epidemic typhus,** caused by *Rickettsia prowazekii,* is louse-borne. The pathogen is excreted in the louse feces and rubbed into the wound when the bitten host scratches the bite. Symptoms are a high and prolonged fever, stupor, and a rash of red spots caused by subcutaneous hemorrhaging. Mortality rates can be high, but tetracycline and chloramphenicol are effective treatments. Laboratory diagnosis is based on an indirect fluorescent antibody test and a positive agglutination test using latex beads. A vaccine is available. The related **endemic murine typhus** occurs sporadically rather than in epidemics. The most common hosts are rodents such as rats and squirrels. The pathogen, *Rickettsia typhi,* is transmitted by rat fleas. Diagnosis and treatment of both typhus diseases are similar.

Spotted Fevers

The best-known spotted fever in the United States is **tick-borne typhus,** or **Rocky Mountain spotted fever,** which is most prevalent in the southeastern U.S.A. and Appalachia. The rickettsial pathogens are transmitted to humans by ticks. In the east, dog ticks are mainly responsible, and in the Rocky Mountains, wood ticks. The rickettsias are often transmitted among ticks by *transovarian passage,* infecting tick eggs as they are produced. Clinically, the disease resembles the typhus fevers. Chloramphenicol and tetracycline are effective in treatment if administered early.

Viral Diseases of the Cardiovascular and Lymphatic Systems

Burkitt's Lymphoma

The Epstein–Barr virus (a herpesvirus) causes two forms of cancer: **Burkitt's lymphoma** and **nasopharyngeal carcinoma.** Fortunately, not many infected with the EB virus in this country develop these cancers.

Infectious Mononucleosis

Infectious mononucleosis is caused by the Epstein–Barr virus. The disease is characterized by enlarged and tender lymph nodes, enlarged spleen, fever, sore throat, headache, nausea, and general weakness. It probably is most commonly spread by saliva. As a result of the infection, mononuclear white blood cells proliferate in a manner resembling leukemia. It is rarely fatal. Diagnosis involves detection of nonspecific heterophil antibodies and by immunofluorescent techniques against the Epstein–Barr virus itself. Recovery results in a good immunity, but patients may continue to shed the virus because of latent infection.

Yellow Fever

Arboviruses (arthropod-borne viruses) reproduce in arthropods or in humans. **Yellow fever** is caused by the yellow fever virus of this type. Once injected by the mosquito vector, the virus multiplies in local lymph nodes; later the liver, spleen, kidneys, and heart are infected. Early stages of the disease are characterized by fever, chills, headache, nausea, and vomiting, followed by jaundice resulting from deposition of bile pigments in the skin and mucous membranes—a reflection of liver damage. Yellow fever is endemic in Central and tropical South America and Africa. Monkeys are a reservoir of the virus, but the mosquito vector *Aedes aegypti* can be controlled in limited areas. The vaccine is an attenuated live viral strain.

Dengue

Aedes aegypti is also the vector of **dengue,** milder than yellow fever and endemic in many tropical climates. Dengue is rarely fatal, but because of the muscle and joint pain experienced, the name of **breakbone fever** has been used. An Asian mosquito, *Aedes albopictus,* has been introduced into the U.S.A. It transmits the virus by transovarian passage and from person to person. It has a range that covers much of the country. For this reason, dengue may become more common.

Protozoan Diseases of the Cardiovascular and Lymphatic Systems

Toxoplasmosis

The protozoa *Toxoplasma gondii,* which causes the disease of the blood and lymph vessels called **toxoplasmosis,** forms spores (oocysts) much like the malarial parasite. Cats are essential for reproduction of the organisms, as the sexual phase appears to occur only in their intestinal tract. Shed in cat feces, the protozoa infects a new host, usually by ingestion. *Oocysts* contain *sporozoites* that invade host cells and form *tachyzoites,* which reproduce. In the chronic phase the parasite is in *tissue cysts* containing *bradyzoites.* Loss of immune function can result in reactivation. Eating undercooked meat of an animal infected in this manner or inhalation of dried cat feces may cause human infections. In pregnant women, infection of the fetus across the placenta can cause drastic change to the fetus. A serological test using fluorescent antibody or an ELISA can be used in diagnosis.

American Trypanosomiasis (Chagas' Disease)

An example of a protozoan disease of the cardiovascular system is **American trypanosomiasis (Chagas' disease).** The causative agent is *Trypanosoma cruzi,* a flagellated protozoa. The disease occurs in the southern U.S.A. and throughout Mexico, Central America, and South America. The arthropod vector is the reduviid bug. Reservoirs for *T. cruzi* are various wild animals. The disease is transmitted to humans when insect bites are contaminated by the insect's feces. Most damage is caused by inflammatory reactions after transport by the blood to the liver, spleen, heart, and so on. One symptom is loss of involuntary muscular contractions in the esophagus and gastrointestinal tract. These organs become grossly enlarged—megaesophagus and megacolon. The disease is most dangerous to children, in whom it damages the heart. Nifurtimox is the drug of choice but is not a cure.

Malaria

Malaria is characterized by chills and fever, and often vomiting and severe headaches. Symptoms typically occur in cycles of one to three days. Caused by the spore-forming protozoa of the genus *Plasmodium,* malaria is found anywhere the protozoan parasite is present in human hosts, and wherever the *Anopheles* mosquito is found. (It can also be spread by blood transfusions or contaminated syringe needles.) It is a serious and increasing problem in tropical Asia, Africa, Central America, and South America.

There are four forms of malaria: *Plasmodium falciparum* (most dangerous and geographically widespread), *Plasmodium vivax* (also widely distributed), *Plasmodium malariae,* and *Plasmodium ovale* (both of which have a lower infection rate and are geographically restricted, milder diseases). The infecting plasmodium, in the form of a *sporozoite,* enters liver cells, where it undergoes schizogony. Later, *merozoites* are released into the bloodstream to infect red blood cells. When more merozoites are released from rupturing red blood cells, paroxysms of chills and fever result. Anemia from loss of red blood cells and excessive enlargement of the liver and spleen are added complications. Some merozoites become male and female *gametocytes* and are picked up by feeding mosquitoes. These then pass through a sexual cycle in the mosquito to produce sporozoites.

Laboratory diagnosis is made by identifying the protozoa in blood smears. Sickle-cell anemia victims tend to be resistant to malaria. Quinine derivatives such as primaquine, mefloquine, and chloroquine are used in chemotherapy.

Helminthic Diseases of the Cardiovascular and Lymphatic Systems

Schistosomiasis

Schistosomiasis is caused by a flatworm parasite, a fluke. The disease is a major world health problem. Waters become contaminated with ova excreted with human wastes. The motile larval form of *Schistosoma*, the **miracidium,** is released from the ova and enters a species of snail. Eventually, the pathogen emerges from the snail as a fork-tailed **cercaria.** This penetrates the skin of the human host and is carried by the bloodstream to the liver or urinary bladder (depending on the species of fluke), where it matures into an adult form, producing a new supply of ova. Defensive reactions by the body to the presence of ova cause tissue damage (*granulomas*). The disease causes damage, such as abscesses and ulcers, to the liver and to other organs, such as the lungs or urinary system. Diagnosis is mainly by detection of flukes or ova in fecal or urine specimens. Praziquantel and oxamniquine are used in chemotherapy.

Swimmer's itch sometimes troubles swimmers in lakes in the northern U.S.A. This is an allergic reaction to cercaria of a similar wildfowl parasite that does not mature in humans.

Self-Tests

In the matching section, there is only one answer to each question; however, the lettered options (a, b, c, etc.) may be used more than once or not at all.

I. Matching

1. The largest vessels carrying oxygenated blood from the heart.

2. The walls are only one cell thick; they aid in the exchange of materials.

3. The venous equivalent of arterioles.

a. Capillaries

b. Venules

c. Arterioles

d. Veins

e. Arteries

II. Matching

1. Blood cells important in phagocytosis and antibody production.

2. A lymphoid organ.

3. Small, oval structures in the lymphatic system; sites of considerable defensive activity by the body.

4. Plasma that bathes tissue cells after their passage through capillary walls.

5. Blood cells that carry oxygen.

a. Platelets

b. Erythrocytes

c. Leukocytes

d. Plasma

e. Lymph

f. Interstitial fluid

g. Lymph nodes

h. Tonsils

III. Matching

1. Characterized by uncontrolled proliferation of bacteria in the blood.

2. Of autoimmune origin due to group M proteins of streptococci.

3. Inflammation of the inner layer of the heart linings.

4. A common cause is microorganisms from tonsillectomies and tooth extractions.

5. The middle layer surrounding the heart; mostly muscle tissue.

6. Characterized by red streaks from the site of infection.

a. Lymphangitis

b. Septicemia

c. Pericardium

d. Endocarditis

e. Myocardium

f. Rheumatic fever

g. Myocarditis

IV. Matching

1. Probably transmitted by saliva.

2. Childbirth fever; a form of septicemia.

3. Often transmitted by contact with small animals such as rabbits.

4. Undulant fever, at one time transmitted by ingestion of contaminated milk, is now transmitted by contact with animal carcasses.

5. Caused by a spore-forming rod that is present in the soil.

6. The cat is essential in the reproductive cycle and the transmission of the causative organisms.

7. Caused by a protozoa that forms oocysts.

8. Heterophil antibodies are used in diagnosis.

9. Caused by the Epstein–Barr virus.

a. Brucellosis

b. Toxoplasmosis

c. Anthrax

d. Rocky Mountain spotted fever

e. Tularemia

f. Puerperal sepsis

g. Infectious mononucleosis

V. Matching

1. The bite of a tick transmits a spirochete, *Borrelia burgdorferi*.

2. A swimming organism called a cercaria is an essential part of the life cycle of the pathogen.

3. A rickettsial disease transmitted by dog ticks or wood ticks.

4. A rickettsial disease transmitted by a louse.

5. A rickettsial disease transmitted by a rat flea.

6. Chagas' disease.

7. A spore-forming protozoa is the cause.

a. Gangrene

b. Lyme disease

c. Epidemic typhus

d. Rocky Mountain spotted fever

e. Endemic murine typhus

f. Yellow fever

g. Plague

h. Sydenham's chorea

i. American trypanosomiasis

j. Malaria

k. Schistosomiasis

VI. Fill-in-the-Blanks

1. The surgical removal of tissue is called _____ .

2. The fluid portion of the blood is called _____ .

3. A general name for a white blood cell is _____ .

4. When a gram-negative bacterium lyses, it releases harmful _____ .

5. The outer sac enclosing the heart is called the _____ .

6. Acute bacterial endocarditis usually is caused by _____ . (Give genus and species.)

7. Group M proteins are associated with the bacterial genus _____ .

8. Infection of the abdominal cavity is called _____ .

9. A bacterial species classified as group A β-hemolytic streptococcus is _____ .

10. *Francisella tularensis* causes _____ .

11. *Brucellis suis* is most likely to infect people coming into contact with animals such as

    _____ .

12. A characteristic of _____ is a pustule that forms at the site of entry through the skin.

13. Sydenham's chorea is a complication of _____ .

14. Burkitt's lymphoma is caused by the same virus that causes _____ .

15. Infections caused by obligate anaerobes such as *Clostridium perfringens* are sometimes treated by putting the patients in _____ chambers.

16. The causative organism of _____ grows intracellularly in phagocytes and is usually spread by contact with rabbits or the bite of deer flies.

17. Infections from abortions are mostly caused by anaerobic bacteria of the genera

    _____ and _____ .

18. A disease known to be transmitted by arthropods such as deer flies is _____ .

19. Many years ago, Semmelweiss showed how proper hygiene and disinfection of hands and instruments could prevent _____ .

20. Two diseases in which the microbial cell grows inside the host's phagocytic cells and is therefore relatively resistant to immune reactions and chemotherapy are

    _____ and _____ .

21. When malaria is transmitted by the bite of a mosquito, the parasite form injected is the

    _____ .

22. The most dangerous form of malaria is caused by *Plasmodium* _____ .

23. Snails are essential to the life cycle of the disease organism causing _____ .

24. The bite of the *Anopheles* mosquito causes _____ .

VII. Label the Art

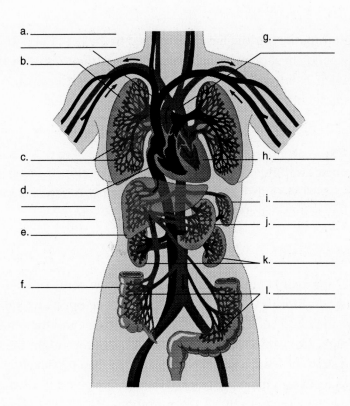

a. _____

b. _____

g. _____

c. _____

h. _____

d. _____

i. _____

j. _____

e. _____

k. _____

f. _____

l. _____

VIII. Critical Thinking

1. What organisms are most often associated with septicemia? What causes the symptoms of septicemia?

2. After examining a patient with acute bacterial endocarditis, her physician concludes that the probable cause was a chronic tooth infection. What is the probable etiologic agent? How can dental infections lead to endocarditis?

3. Why is it so important to treat cases of strep throat?

4. List and discuss three species of *Brucella*, how they are transmitted, and the signs and symptoms of the infection caused in each example.

5. What is a hyperbaric chamber, and what role does it play in the treatment of gas gangrene?

6. A recent immigrant from the Middle East is admitted to the hospital with a high fever, difficulty breathing, and chest pain. Examination of the patient reveals that he has septicemia. The primary organism isolated is gram-positive and forms endospores. What is the diagnosis? What is the etiologic agent? Speculate as to the patient's profession.

7. After returning from a camping trip, a 22-year-old man developed fever, jaundice, and rose-colored spots. The fever subsided three days later, only to return again. What is the etiologic agent? How is it transmitted? Why did the fever return?

8. In November 1993, approximately one month after clearing some brush from her rural property in Southern California, a 36-year-old woman developed an oval rash inside the bend of her arm. The weather had been unusually warm that year, and there had been several fires in the local mountains. Concerned about the rash, the woman sought the advice of her physician. The physician ordered a blood test and prescribed tetracycline. What is the suspected etiologic agent and the diagnosis? Although the blood test was negative, the physician instructed the woman to finish the antibiotics. Why? How did the weather and fires contribute to transmission of this organism?

9. How was it determined that EBV was the cause of infectious mononucleosis?

10. Human infection with *Toxoplasma* is usually subclinical or produces mild, flulike symptoms. Discuss two situations in which toxoplasmosis may be fatal.

CHAPTER 24

Microbial Diseases of the Respiratory System

Learning Objectives

After completing this chapter, you should be able to:

- Describe how microorganisms are prevented from entering the respiratory system.
- Characterize the normal microbiota of the upper and lower respiratory systems.
- Differentiate among pharyngitis, laryngitis, tonsillitis, and sinusitis.
- List the etiologic agent, symptoms, prevention, preferred treatment, and laboratory identification tests for streptococcal pharyngitis, scarlet fever, diphtheria, cutaneous diphtheria, otitis media, and the common cold.
- List the etiologic agent, symptoms, prevention, preferred treatment, and laboratory identification tests for pertussis and tuberculosis.
- Compare and contrast the eight bacterial pneumonias discussed in this chapter.
- List the etiologic agent, symptoms, prevention, and preferred treatment for viral pneumonia, influenza, and pulmonary syndrome *Hantavirus*.
- List the etiologic agent, method of transmission, preferred treatment, and laboratory identification tests for four fungal and protozoan diseases of the respiratory system.

Structure and Function of the Respiratory System

The **upper respiratory system** consists of the nose and throat, including the middle ear and auditory tubes (Figure 24.1). Mucus traps dust and microorganisms, and cilia assist in moving these to the throat for elimination. The **lower respiratory system** consists of the larynx, trachea, bronchial tubes, and lungs (Figure 24.2). **Alveoli** are air sacs in lung tissue where oxygen and carbon dioxide are exchanged, and the double-layered membrane around the lungs is the **pleura.** Macrophages in the alveoli destroy many pathogens, and IgA antibodies in mucus, saliva, and tears aid in resistance.

Cutaneous diphtheria, characterized by skin lesions, is fairly common in tropical countries. In the U.S.A. it affects mainly lower socioeconomic groups.

Otitis Media

Infections of the upper respiratory system may cause **otitis media,** or middle ear infection leading to earache. The symptoms are due to pus formation and pressure against the eardrum. Enlarged adenoids and the relatively small auditory tubes in many children contribute to susceptibility. *Staphylococcus aureus, Streptococcus pneumoniae, Hemophilus influenzae,* and *Moraxella (Branhamella) catarrhalis* are the most commonly identified causes of this disease. Broad-spectrum penicillins are the first choice in chemotherapy.

Viral Disease of the Upper Respiratory System

Common Cold

The **common cold** is caused mainly by viruses classified as coronaviruses or rhinoviruses. There are more than 200 viruses that cause colds, so immunity to any one is of limited use. Immunity develops in isolated populations due to IgA antibodies, and immunities accumulate with age. Laryngitis and otitis media are common complications. Symptoms can be alleviated, but there is no cure.

DISEASES OF THE LOWER RESPIRATORY SYSTEM

A lower respiratory tract infection affecting the bronchi (Figure 24.2) is **bronchitis** or **bronchiolitis.** A severe complication of bronchitis is **pneumonia.**

Bacterial Diseases of the Lower Respiratory System

Pertussis (Whooping Cough)

Infection by *Bordetella pertussis,* a gram-negative coccobacillus, can cause **pertussis (whooping cough).** Ciliary action is blocked by accumulations of dense masses of these bacteria in the trachea and bronchi. Eventually, *tracheal cytotoxin* causes loss of ciliated cells. The initial, **catarrhal,** stage resembles a common cold. The **paroxysmal** stage is characterized by attempts to clear accumulations by coughing. The bacteria may be cultured from a throat swab inserted through the nose on a wire while the patient coughs. Fluorescent antibody tests may confirm identification. Recovery results in good immunity. Vaccination is recommended, although there is some concern about adverse effects. (The P in DTP vaccine stands for pertussis.)

Tuberculosis

Mycobacterium tuberculosis, a slow-growing rod that sometimes forms filaments, is the cause of **tuberculosis.** The organism is *acid-fast,* meaning it retains the carbol fuchsin dye of the Ziehl–Neelson staining technique. The stain characteristically reflects the large lipid content of the cell wall, which also may make it resistant to drying, sunlight, and chemical disinfectants.

Tuberculosis is usually acquired by inhaling the tubercle bacillus. In the alveoli (Figure 24.2), the bacilli are phagocytized by macrophages. These may destroy the pathogen; if they do not, the bacilli multiply within the macrophages. New macrophages are attracted to the site and a **tubercle** (a small lump that is characteristic of tuberculosis) forms. Hypersensitivity reactions against bacillus-laden macrophages kill many of them, and a **caseous center** in the tubercle, surrounded by macrophages and lymphocytes, forms. As the caseous lesions heal they become calcified and are called **Ghon complexes,** which show up on X rays. If the disease is not arrested, the caseous area enlarges—called **liquefaction**—eventually forming an air-filled **tuberculous cavity.** The bacilli now multiply rapidly, eventually leading to release of the microbes into the circulatory system. The disseminated infection, affecting many organs, is called **miliary tuberculosis.**

Stress and genetic differences contribute to susceptibility to tuberculosis. The intracellular growth of tuberculosis shields it from antibiotics. Drugs such as streptomycin, rifampin, isoniazid (INH), and ethambutol are used as treatment, usually in combinations to avoid resistance. Skin tests such as the Mantoux test involve injecting a purified protein derivative (PPD) of the tuberculosis bacterium. T cells sensitized by exposure to tuberculosis cause a delayed hypersensitivity reaction at the injection site.

Mycobacterium bovis is the cause of **bovine tuberculosis,** which is usually spread by unpasteurized milk. This disease mainly affects bones and the lymphatic system. A vaccine, **BCG,** is a live attenuated culture of *M. bovis.*

Tuberculosis, which the body combats by cell-mediated immunity, is a frequent complication of AIDS. Other mycobacterial diseases affect AIDS patients. Most are a group of related organisms called the *M. avium-intracellulare* complex.

Bacterial Pneumonias

The term *pneumonia* is applied to many pulmonary infections.

Pneumococcal Pneumonia. **Pneumococcal pneumonia,** caused by the pneumococcus *Streptococcus pneumoniae,* is the classic form of pneumonia. The organism can be distinguished from other similar organisms by inhibition of its growth near a disk of optochin, or by bile solubility. Serological typing using the *quellung reaction* is useful. A vaccine is used to immunize elderly or debilitated individuals. Penicillin is the drug of choice in treatment.

Hemophilus influenzae **pneumonia.** Clinically, this is similar to pneumonia caused by *S. pneumoniae.* A Gram stain often helps in differentiation; *H. influenzae* is gram-negative.

Mycoplasmal Pneumonia (Primary Atypical Pneumonia). Pneumonias typically have a bacterial etiology. If a bacterial agent is not isolated, the pneumonia is considered *atypical.* The most common atypical type **(primary atypical pneumonia)** is caused by a bacterium with no cell wall, *Mycoplasma pneumoniae,* which does not grow on routine isolation media. The bacterium can, however, be grown and isolated on horse serum and yeast extract medium on which the colonies have a minute "fried-egg" appearance. The disease is relatively mild and is sometimes called **walking pneumonia.** Tetracyclines are often effective in treatment. Diagnosis may use a complement-fixation test to detect a rise in titer. ELISAs are also available.

Legionellosis. Legionellosis (Legionnaires' disease), characterized by high fever and cough, is caused by a gram-negative rod, *Legionella pneumophila.* The bacterium may grow in the water of air-conditioning cooling towers or in the hot-water lines of hospitals. Erythromycin and rifampin are the drugs of choice for treatment. *L. pneumophila* also causes a mild, self-limiting respiratory disease called **Pontiac fever.**

Chlamydial pneumonia. In recent years there have been outbreaks of pneumonia caused by the newly named *Chlamydia pneumoniae.* It resembles mycoplasmal pneumonia and is transmitted person to person.

Psittacosis (Ornithosis)

Psittacine birds, such as parakeets and other parrots, carry this disease, hence the name **psittacosis,** or **ornithosis.** *Chlamydia psittaci,* a gram-negative obligately intracellular bacterium, is the cause. It forms **elementary bodies,** which are the airborne infective agent. Within the cell they develop into larger **reticulate bodies** that reproduce by fission; eventually they become elementary bodies and leave the cell. The dried droppings of sick birds are the usual source of these. Diagnosis is made by isolation in embryonated eggs, mice, or cell culture. Fluorescent antibody testing is used for final identification of isolated organisms, and a complement-fixation test is used to detect serum antibodies.

Q Fever

Coxiella burnetii, an obligately intracellular rickettsia, causes **Q fever.** It is a parasite of cattle ticks that eventually contaminates dairy products. An occasional complication is endocarditis. Diagnosis is made by isolation of the organism in egg embryo or cell culture.

Other Bacterial Pneumonias

Staphylococcus aureus, M. (B.) catarrhalis, Streptococcus pyogenes, Klebsiella pneumoniae, and *Pseudomonas* spp. occasionally cause bacterial pneumonias.

Viral Diseases of the Lower Respiratory System

Viral Pneumonia

Viral pneumonia is seldom confirmed by culture but is often assumed if mycoplasmal and bacterial pneumonia are ruled out. **Respiratory syncytial virus** is the most common cause, especially in children.

Influenza (Flu)

The **influenza** virus has an envelope with two types of projections: *hemagglutinin* (H) *spikes* and *neuraminidase* (N) *spikes*. Antibodies against influenza are directed mainly at these spikes. Variation in their composition alters the antigenic type of the virus and helps it evade established resistance in the human population. *Type A* of the virus is more widespread and severe, and *type B* is milder and geographically limited. *Antigenic drift* is the result of minor variations of antigenic makeup; *antigenic shifts* are major changes involving recombination of the eight RNA segments of the genome. Vaccines are available to high-risk groups; they are usually multivalent, directed at the types then in circulation. These latter cause the designation—the Hong Kong virus of 1968, for example—to change and enable the virus to evade almost all established immunity. Amantadine, if taken very early, will significantly reduce symptoms.

Pulmonary Syndrome *Hantavirus*

In 1993 several cases of severe respiratory illness, mostly fatal, were reported. Death resulted from flooding of the lungs with blood plasma, a type of internal hemorrhaging. The viral cause was identified as a strain of *Hantavirus* (named for the Hantaan River in Korea). Infections are usually associated with contact with rodents, which carry the virus. The disease occurs in widely separated locales in the U.S.A., but most cases have been reported in the Southwest.

Fungal Diseases of the Lower Respiratory System

Histoplasmosis

Histoplasmosis is caused by the dimorphic fungus *Histoplasma capsulatum* and may resemble tuberculosis. It is, however, much milder. It is found mainly in the Mississippi and Ohio River valleys. Bird droppings in the soil encourage growth of the fungus, which is spread by airborne conidia. Amphotericin B or ketoconazole are used in chemotherapy.

Coccidioidomycosis

A dimorphic fungus, *Coccidioides immitis,* whose sources are found in the soil of the American Southwest, is the cause of **coccidioidomycosis.** A tuberculinlike skin test is used in screening. Amphotericin B and ketoconazole are used to treat the disease, but most cases are subclinical; about 1% are progressive, resembling tuberculosis.

Blastomycosis (North American Blastomycosis)

Another dimorphic fungus, *Blastomyces dermatitidis*, causes **North American blastomycosis.** It is found most often in the Mississippi Valley. Although most cases are subclinical, a rapidly progressive infection with cutaneous ulcers and abscess formation can occur. Amphotericin B is used in treatment.

Other Fungi Involved in Respiratory Disease

Other fungi, such as *Aspergillus fumigatis* and other species of *Aspergillus*, may cause **aspergillosis.** Compost piles are likely growth sites. Other mold genera also may cause pulmonary infections in susceptible individuals.

Protozoan Disease of the Lower Respiratory System

Pneumocystis pneumonia is caused by *Pneumocystis carinii.* It is uncertain if the organism is a protozoa or a fungus. The parasite may be transmitted by direct human contact, mostly to immunosuppressed individuals. Pneumocystis infection causes the alveoli to become filled with a frothy exudate; untreated infections are usually fatal. This disease is a common condition among the immunosuppressed, especially AIDS patients. Pentamidine isethionate and trimethoprim-sulfamethoxazole are effective for treatment.

Self-Tests

In the matching section, there is only one answer to each question; however, the lettered options (a, b, c, etc.) may be used more than once or not at all.

I. Matching

1. Mycoplasmal pneumonia.

2. The causative organism is suspected to grow in the water-cooling towers of air-conditioning systems.

3. Rodents are carriers of the pathogen.

4. Production of the harmful exotoxin involved in this disease is controlled by a lysogenic phage.

5. Pharyngitis.

6. DPT vaccine is useful in prevention.

a. Diphtheria

b. Legionellosis

c. Primary atypical pneumonia

d. Pulmonary syndrome *Hantavirus*

e. Streptococcal sore throat

II. Matching

1. The infectious agent causing the disease is an elementary body formed intracellularly by a bacterial pathogen.

2. Caused by *Coxiella burnetii.*

3. Caused by a dimorphic fungus growing in soil contaminated by bird droppings.

4. Caused by a dimorphic fungus widely distributed in airborne spores in the American Southwest.

5. Caused by rickettsial organisms parasitic of cattle ticks.

6. A fungal disease found with considerable frequency in the Ohio and Mississippi River valleys.

7. Also called ornithosis.

8. Caused mainly by viruses classified as coronaviruses or rhinoviruses.

a. Common cold

b. Histoplasmosis

c. Coccidioidomycosis

d. Q fever

e. Psittacosis

III. Matching

1. Caused by a lysogenized bacterium.

2. Pertussis.

3. The DPT vaccine helps prevent this disease.

4. Infection of the middle ear.

5. Ghon complexes are one possible characteristic.

6. The Mantoux test is useful in screening for this disease.

7. Caused by an acid-fast organism.

8. Ethambutol, isoniazid, and rifampin are useful in treatment.

9. A disease of the respiratory tract near the larynx caused by *Hemophilus influenzae.*

a. Scarlet fever

b. Otitis media

c. Whooping cough

d. Tuberculosis

e. Acute epiglottitis

IV. Fill-in-the-Blanks

1. The disease agent causing _____ is characterized by designations H and N.

2. The air sacs in lung tissue where oxygen and carbon dioxide are exchanged are the
    _____ .

3. The double-layered membrane around the lungs is the _____ .

4. Among diseases of the upper respiratory tract are pharyngitis, laryngitis, tonsillitis, sinusitus,
    and epiglottitis. Of these, only _____ is likely to be life-threatening.

5. *Mycobacterium avium-intracellulare* complex is a frequent opportunistic infection of
    _____ .

6. Immunity to the common cold is based mainly on antibodies of the _____
    type.

7. Tuberculosis spreading progressively to new sites is termed _____
    tuberculosis.

8. The vaccine used to help prevent tuberculosis is called _____ .

9. The quellung reaction is used in diagnosis or serological typing of the respiratory disease
    _____ .

10. In tropical areas or among people in lower socioeconomic groups in the U.S.A. we often see a
    nonrespiratory form of diphtheria, called _____ .

11. The Ziehl–Neelson staining technique is valuable in diagnosis of _____ .

12. Tuberculosis bacteria entering the lungs may be ingested by alveolar _____ .

13. Minor year-to-year variations in the antigenic makeup of the influenza virus is called
    antigenic _____ .

14. Major changes in the antigenic makeup of the influenza virus that allow it to evade almost all
    previous immunity is called antigenic _____ .

15. The trachea contains few bacteria, and the lower respiratory tract is usually
    _____ .

16. Laboratory diagnosis of _____ can be made by bacterial culture from a swab
    passed through the nose on a wire while the patient coughs.

17. The terms *catarrhal stage* and *paroxysmal stage* refer to the disease _____ .

18. The influenza virus has an envelope with two types of projections: _____
    spikes and _____ spikes.

19. Diagnosis for streptococcal sore throat involves streaking a throat swab of bacteria onto blood
    agar to observe the type of _____ .

20. A disease for which administration of antitoxin is essential in treatment is

    _____ .

21. A protozoa-caused (or fungal; the taxonomy is uncertain) pneumonia often seen in AIDS patients is _____ .

22. The rickettsial organism *Coxiella burnetii* is the cause of the disease _____ .

23. The terms *liquefaction* and *caseous center* are associated with a progressive case of

    _____ .

V. Label the Art

INFLUENZA VIRUS

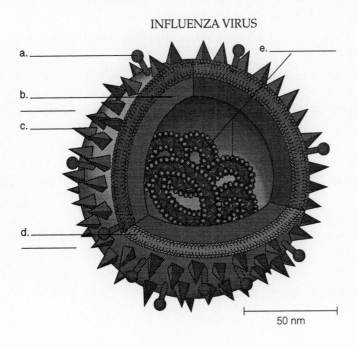

50 nm

VI. Critical Thinking

1. Discuss the new approach that has been suggested to develop a vaccine to prevent the common cold.

2. What are the consequences of *Bordetella pertussis* growing on cilia of the trachea?

3. Outline the procedure used to confirm an active case of tuberculosis when a person has a positive skin reaction.

4. Why do certain ethnic groups such as Native Americans, Asians, and Hispanics tend to have higher rates of tuberculosis than do people of European descent? (Hint: See Chapter 17.)

5. List four organisms that cause bacterial pneumonia. How is each transmitted? What is the drug of choice for each?

6. A pet-store owner is admitted to the hospital with pneumonia. The physician orders fecal samples to be taken from the birds at the pet shop. What is the suspected etiologic agent? How was the pet-shop owner infected? What antibiotic will the physician probably prescribe?

7. Explain the relationship between Q fever and endocarditis.

8. What are H spikes and N spikes, and how do they relate to reproduction of the influenza virus?

9. Why is there sometimes an increased incidence of histoplasmosis and coccidioidomycosis after earthquakes in California?

10. Discuss the difficulties in the diagnosis of atypical pneumonia.

CHAPTER 25

Microbial Diseases of the Digestive System

Learning Objectives

After completing this chapter, you should be able to:

- List the structures of the digestive system that contact food.
- List examples of microbiota for each part of the gastrointestinal tract.
- Describe the events that lead to dental caries and periodontal disease.
- List the etiologic agents, suspect foods, signs and symptoms, and treatments for staphylococcal food poisoning, salmonellosis, typhoid fever, bacillary dysentery, cholera, peptic disease syndrome, and gastroenteritis.
- List the etiologic agents, methods of transmission, sites of infection, and symptoms for mumps, CMV inclusion disease, and viral gastroenteritis.
- Differentiate among hepatitis A, hepatitis B, hepatitis C, hepatitis D, and hepatitis E.
- Identify causes of ergot and aflatoxin poisoning.
- List the etiologic agents, methods of transmission, symptoms, and treatments for giardiasis, amoebic dysentery, and cryptosporidiosis.
- List the etiologic agents, methods of transmission, symptoms, and treatments for tapeworms, hydatid disease, pinworms, hookworms, ascariasis, and trichinosis.

Structure and Function of the Digestive System

The first group of organs of the essentially tubelike **gastrointestinal (GI) tract,** or **alimentary canal,** is the mouth, pharynx (throat), esophagus (food tube), stomach, small intestine, and large intestine (see Figure 25.1). **Accessory structures** consist of the teeth, tongue, salivary glands, liver, gallbladder, and pancreas. Except for the teeth and tongue, these lie outside the GI tract and produce secretions released into the tract. The action of secretions such as bile, pancreatic enzymes, saliva, and stomach and intestinal enzymes converts ingested foods into their end-products of digestion. By the time the liquefied food leaves the small intestine as *feces*, the absorption of sugars, fatty acids, and amino acids produced by digestion is almost complete. Water, vitamins, and nutrients are absorbed in the large intestine.

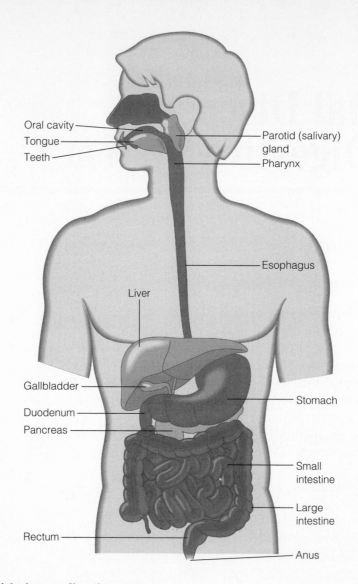

Oral cavity
Tongue
Teeth
Parotid (salivary) gland
Pharynx
Esophagus
Liver
Gallbladder
Duodenum
Pancreas
Stomach
Small intestine
Large intestine
Rectum
Anus

FIGURE 25.1 Anatomy of the human digestive system.

Normal Microbiota of the Digestive System

In the mouth, saliva contains millions of bacteria per milliliter. The stomach and small intestine have relatively few because of acidity. The large intestine has enormous numbers of bacteria, mostly anaerobes or facultative anaerobes.

Bacterial Diseases of the Mouth

Dental Caries (Tooth Decay)

The hard surface of the teeth accumulates masses of microorganisms and their products. This is called **plaque** and is an important factor in **dental caries,** or tooth decay. The most important *cariogenic* (caries-causing) bacterium is *Streptococcus mutans.* This, and certain other bacteria, convert

tic of the disease is **carriers,** which c[...]
after recovery. Antimicrobials such a[...]
effective treatments.

Bacillary Dysentery (Shigello[...]

Bacteria of the genus *Shigella—S. son[...]
boydii—all cause **bacillary dysentery**[...]
common type in the U.S.A., causes a[...]
dysenteriae, not common in the U.S.A[...]
It is known as the Shiga bacillus (its[...]

Cholera

Vibrio cholerae, the cause of **cholera,** i[...]
organism produces an enterotoxin in[...]
of body fluids and electrolytes, as w[...]
0:1, biotype eltor, of *V. cholerae* cause[...]
Vibrio cholerae not belonging to serog[...]
disease that does not have the epide[...]
latter organisms are indigenous in m[...]
ters. They are more likely to be invas[...]

Treatment of any form of cholera i[...]
lost fluids and electrolytes; antibioti[...]
the death rate can be as high as 50%[...]
treated.

Vibrio parahaemolyticus Gast[...]

Vibrio parahaemolyticus causes a gastr[...]
milder. Because it is present in coasta[...]
ciated with seafood. It is the most co[...]
organism has a requirement for sodi[...]
2% to 4% sodium chloride.

Escherichia coli Gastroenteriti[...]

One of the most numerous microorg[...]
Escherichia coli normally is harmless,[...]
teritis. Pathogenic strains possess ad[...]
testinal mucosa. These strains may p[...]
E. coli), invade the lining of the large[...]
have no well-understood mechanism[...]
isms may be the major cause of trave[...]
breaks of epidemic diarrhea in nurse[...]
increasing importance. Serotype 015[...]
toxin like the Shiga toxin and is knov[...]
as *hemorrhagic colitis* and *hemolytic u*[...]

sucrose sugar into a gummy polysaccharide called *dextran,* which local-
izes the bacteria and their acid production, forming a hole in the *enamel* of
the tooth (Figure 25.2). Some antimicrobial protection is provided by
lysozyme in saliva and the tissue exudate called *crevicular fluid.* Old
plaque deposits are called *dental calculus* or *tartar.* Once the enamel has
been penetrated, other bacteria are involved in decay of the *dentin* and
the *pulp.*

Periodontal Disease

Many who avoid tooth decay later lose their teeth to **periodontal** (surrounding the tooth) **disease.** As gums recede with age, they expose the *cementum.*

Gingivitis. The most common periodontal disease is **gingivitis,** infection of the gingivae, or gums, a condition characterized by bleeding of
the gums while brushing the teeth.

Periodontitis. A chronic form of gingivitis is **periodontitis.** The gums
are inflamed and bleed. Pus often forms in pockets surrounding the
teeth. Tooth loss arises from destruction of bone and tissue supporting
the teeth.

**Acute necrotizing ulcerative gingivitis (Vincent's disease or trench
mouth)** is a serious mouth infection. Chewing causes pain and the breath
is foul.

Bacterial Diseases of the Lower Digestive System

Intoxication is a disease resulting from the ingestion of toxins preformed
in food by microbial growth. **Infection** is a disease resulting from micro-

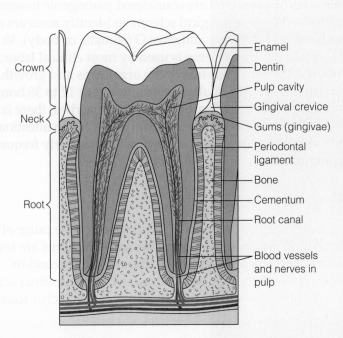

Crown

Neck

Root

Enamel
Dentin
Pulp cavity
Gingival crevice
Gums (gingivae)
Periodontal
ligament
Bone
Cementum
Root canal
Blood vessels
and nerves in
pulp

**FIGURE 25.2 A healthy human
tooth.**

bial growth in the tissues of the host.
blood and mucus is termed **dysenter**
describes inflammation of the stoma
symptoms are *abdominal cramps, naus*

Staphylococcal Food Poisonin
(Staphylococcal Enterotoxicos)

Staphylococcus aureus produces an exc
digestive system, is called an **enterot**
and diarrhea about one to six hours a
terotoxin is responsible for most of th
sistant and withstands boiling for 30
lococci produce **coagulase,** an enzym
is usually associated with production
tant to drying, radiation, and osmotic
ham are foods with a high incidence
the presence of sugar concentrations
nonstaphylococcal competition.

Food can be contaminated by stapl
sions. If allowed incubation time for
Pathogenic staphylococci ferment m
agulase-positive, form yellow coloni
when growing on foods. The source
determined by **phage typing** the bac
ble way to test for the presence of en
an enzyme, **thermostable nuclease, i**
probable toxin production in a food.

Salmonellosis (*Salmonella* Ga

Salmonella are not assigned to specie
of serovars. All serovars of *Salmonella*
degree. The Kauffmann–White serolo
signs numbers and letters to different
(capsular), and H (flagellar). This sch
ries. **Salmonellosis** results from inges
the cells in the intestinal tract. An inc
is normal. Symptoms are moderate fe
low mortality rate. Salmonellae are c
tract of many animals; poultry and eg
sources of food contamination.

Typhoid Fever

Salmonella typhi is a very virulent sero
typhoid fever. Incubation is about tw
and malaise. Diarrhea appears only la
third week. In severe cases there may
and dissemination of the organism in

most commonly transmitted by blood but can be transmitted by many body fluids, such as saliva, breast milk, and semen. Three distinct structures with the hepatitis B surface antigen (HB$_s$Ag) can be found in patients' blood. **Dane particles** probably represent intact hepatitis B virions, whereas **filamentous** and **spherical particles** are probably unassembled components of Dane particles (Figure 25.3). Many cases are asymptomatic, but the disease is generally more severe than hepatitis A. About 10% enter a chronic phase and carry the virus indefinitely. Liver cancer is often a subsequent complication of chronic infections. A vaccine is available.

Hepatitis C (Non-A, Non-B). Screening of donated blood for HBV antibodies became so effective that hepatitis of this type became rare. However, another hepatitis named non-A, non-B appeared, which has since been named **hepatitis C (HCV).** A serological test has been developed, in turn, to detect this form. Many cases are subclinical or mild, but the disease is likely to become chronic and cause liver damage. There is a possibility that HCV is not the only blood-transmitted hepatitis remaining to be discovered.

Hepatitis D. A hepatitis virus (HDV) discovered in 1977 was originally called the **delta antigen;** in people infected with HBV, this led to a high incidence of severe liver damage and mortality. HDV can occur as an acute (*coinfection form*) or a chronic (*superinfection form*) hepatitis. The severe symptoms are associated with superinfection. The HDV is a single strand of RNA with a capsid; it cannot cause infection. Only

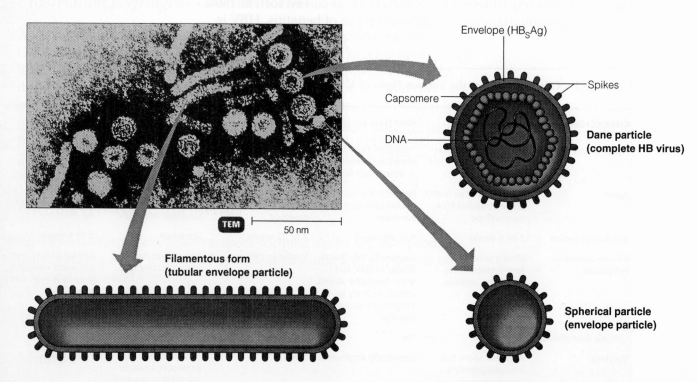

FIGURE 25.3 Hepatitis B virus (HBV). Micrograph and illustrations depict the distinct types of HBV particles discussed in the text.

when it obtains an external envelope of HB_sAg from coinfection with HBV is it infectious.

Hepatitis E.   Another form of non-A, non-B hepatitis is spread by fecal–oral transmission, much like hepatitis A, which it clinically resembles. The agent is **hepatitis E virus (HEV).** It does not cause chronic liver disease but is responsible for a very high mortality rate in pregnant women.

Viral Gastroenteritis

About 90% of viral gastroenteritis is caused by rotavirus or Norwalk agent. **Rotavirus** (*rota* means wheel) is the most common, mostly causing disease in small children. It causes low-grade fever, diarrhea, and vomiting. Major outbreaks of gastroenteritis caused by the **Norwalk agent** have occurred. Infected individuals suffer from gastroenteritis for one to three days.

Fungal Diseases of the Digestive System

Ergotism is caused by the toxin ergot, which is produced by the fungus *Claviceps purpurea,* which grows on grains. **Aflatoxin** is produced by the mold *Aspergillus flavus,* which grows on grains and so on. It is toxic to livestock and is suspected to be a human carcinogen.

Protozoan Diseases of the Digestive System

Giardiasis

A flagellated protozoa, *Giardia lamblia,* causes the prolonged diarrheal disease in humans called **giardiasis.** The disease is also characterized by malaise, nausea, weakness, weight loss, and abdominal cramps. Treatment is by metronidazole and quinacrine hydrochloride. The general population has many carriers who may excrete the cyst stage, which is relatively resistant to the chlorine used to disinfect drinking water.

Amoebic Dysentery (Amoebiasis)

The cysts of *Entamoeba histolytica* are ingested with contaminated food and water and can cause severe **amoebic dysentery.** Severe infections can cause perforation of the intestinal wall and abscesses. For treatment, metronidazole plus iodoquinal are the drugs of choice.

Cryptosporidiosis

A protozoa, *Cryptosporidium,* causes **cryptosporidiosis,** a severe, prolonged diarrhea. Formerly recognized only as a pathogen of calves, in recent years *Cryptosporidium* has been observed in immunosuppressed humans, especially those with AIDS. The oocysts are resistant to chlorination and must be removed from water supply systems by filtration.

Helminthic Diseases of the Digestive System

Tapeworm Infestations

Cysticerci, larval forms of the **tapeworm** that become encysted in muscles of the intermediate host, are ingested by humans and develop into adult tapeworms. They attach to the intestinal wall of the human host and shed eggs with the host's feces. *Taenia saginata* (the beef tapeworm), *Taenia solium* (the pork tapeworm), and *Diphyllobothrium latum* (the fish tapeworm) are the best known forms in the U.S.A. Tapeworms may be 6 or 7 meters in length. Niclosamide is used in treating tapeworm infestations. The pork tapeworm may produce the larval stage in humans, called **cysticercosis.** The infection arises from ingestion of tapeworm eggs. Symptoms can be serious, especially if they develop in the eyes or in the brain.

Hydatid Disease

Echinococcus granulosus, a tapeworm only a few millimeters in length, may infect humans by way of dog feces. The dog becomes infected by eating the flesh of infected sheep. The tapeworm egg migrates in the human to various tissues and develops into a large **hydatid cyst.** These cysts are damaging because of their size and, if they rupture, possible anaphylactic shock.

Nematode Infestations

Pinworms.   The most familiar nematode infection in humans is by *Enterobius vermicularis* **(pinworm).** The tiny worm migrates out of the anus of the human host and lays its eggs, causing local itching. Drugs such as pyrantel pamoate or mebendazole are effective treatments.

Hookworms.   **Hookworm** infestations were once common in the southeastern states, mostly caused by *Necator americanus.* The hookworm attaches to the intestinal wall and feeds on blood and tissue. Blood loss can lead to craving for starch or clay soils (pica), which is a symptom of iron deficiency anemia.

Ascariasis.   One of the most widespread helminthic diseases is **ascariasis,** caused by *Ascaris lumbricoides.* The worms can be up to about a foot in length but cause few symptoms, as they live on partially digested food in the intestinal tract. Ingested eggs hatch into small, wormlike larvae that pass from the bloodstream into the lungs. In the lungs they migrate into the throat and are swallowed, developing into adults in the intestinal tract. Mebendazole is an effective treatment.

Trichinosis.   The small roundworm *Trichinella spiralis* causes **trichinosis.** The disease is acquired by ingestion of undercooked meat, usually pork, that contains the encysted larvae. In the intestine of a human the ingested cyst matures into the adult form, producing larvae that invade tissue, especially muscle of the diaphragm and eye. Thiabendazole is an effective drug.

Self-Tests

In the matching section, there is only one answer to each question; however, the lettered options (a, b, c, etc.) may be used more than once or not at all.

I. Matching

1. Produced from sucrose by *Streptococcus mutans*.

2. Pertaining to the inflammation and degeneration of gums and bones supporting the teeth.

3. An enzyme produced by pathogenic strains of *Staphylococcus aureus*.

4. A disease resulting from ingestion of a toxin.

5. A disease resulting from microbial growth in tissues of the host.

6. Loss of appetite.

7. Vincent's disease.

8. Chronic gingivitis.

a. Periodontal disease

b. Coagulase

c. Intoxication

d. Infection

e. Dental plaque

f. Anorexia

g. Periodontitis

h. Acute necrotizing ulcerative gingivitis

II. Matching

1. Affects the parotid glands.

2. Causes diarrhea; same organism that causes gas gangrene.

3. Caused by a small roundworm that encysts its larval form in muscle tissue.

4. MMR vaccine used.

5. A virus commonly causing gastroenteritis.

6. Caused by *Salmonella typhi*.

7. A disease caused by growth of *Salmonella* bacteria in the intestinal tract; incubation time of 12 to 36 hours is normal.

8. Orchitis is a possible complication.

a. Hydatid disease

b. Salmonellosis

c. Norwalk agent

d. Trichinosis

e. Typhoid fever

f. *Clostridium perfringens* gastroenteritis

g. Mumps

III. Matching

1. Shigellosis.

2. *Streptococcus mutans.*

3. The most common food poisoning in Japan; sodium-requiring bacteria.

4. Prolonged diarrhea caused by a flagellated protozoa.

5. *Entamoeba histolytica.*

6. Primary infection during pregnancy can transmit to fetus and cause severe damage.

7. Coagulase-positive organisms.

a. Dental caries

b. Staphylococcal food poisoning

c. Bacillary dysentery

d. *Vibrio parahaemolyticus* gastroenteritis

e. Cytomegalovirus inclusion disease

f. Giardiasis

g. Amoebic dysentery

IV. Matching

1. Viral hepatitis, usually foodborne.

2. A term associated with the serological classification of *Salmonella.*

3. An infectious viral particle that causes disease only from coinfection with the hepatitis B virus.

4. Mycotoxin.

5. Associated with tapeworm infestation.

6. Formerly non-A, non-B hepatitis.

a. Aflatoxin

b. Hepatitis A

c. Hepatitis B

d. Non-A, non-B

e. HDV

f. Kauffmann–White

g. Bacteriocinogen

h. Hydatid cysts

i. Hepatitis C

V. Fill-in-the-Blanks

1. Cysticercosis occurs in humans from infection by *Taenia* _____ .

2. The epidemic form of cholera is caused by serogroup _____ of *Vibrio cholerae.*

3. The term for skin yellowing due to liver infection is _____ .

4. The term for the larval form of tapeworm that becomes encysted in the muscles of the intermediate host is _____ .

5. *Diphyllobothrium latum* is the name of the _____ tapeworm.

6. The name for the pork tapeworm is *Taenia* _____ .

7. Ergotism is caused by the mycotoxin _____ .

8. The species of *Shigella* most common in the U.S.A. that causes a relatively mild dysentery is *Shigella* _____ .

9. Cholera is caused by the organism named _____ *cholerae.*

10. _____ is the toxin produced by the mold *Aspergillus flavus*, which grows on grains.

11. _____ immunoglobulins in saliva tend to prevent microbial attachment.

12. Dental plaque is formed mostly from _____ produced from sucrose.

13. _____ disease is a collective term for a number of conditions characterized by inflammation and degeneration of the gums, supporting bone, and so on.

14. Serological type _____ enterotoxin is responsible for most cases of staphylococcal food poisoning.

15. Pathogenic staphylococci usually ferment _____ .

16. In the alimentary canal, the throat is the _____ and the food tube is the _____ .

17. The stomach and the _____ intestine have few microorganisms, but the _____ intestine contains as many as 100 billion bacteria per gram of feces.

18. The Dane particle probably represents an intact _____ virion.

19. *Trichinella spiralis* is a small roundworm that causes _____ .

20. For treatment of amoebiasis, one drug of choice is _____ ; it is also used to treat giardiasis.

VI. Label the Art

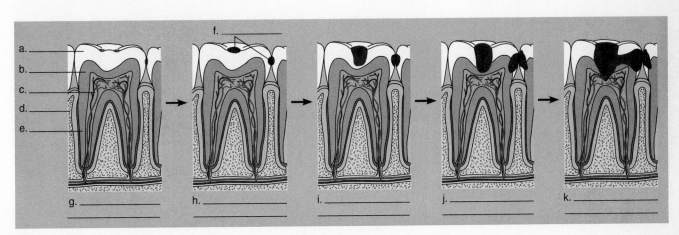

VII. Critical Thinking

1. Is food contaminated with *Staphylococcus aureus* safe to eat after reheating? Why? What about food contaminated with *Salmonella*?

2. Explain why there is so much variation in the incubation period associated with salmonella gastroenteritis.

3. Compare and contrast cholera and *Vibrio parahaemolyticus* gastroenteritis.

4. Discuss the importance of proper preparation (butchering and so on) and handling of meat in the control of food poisoning with *Clostridium perfringens*.

5. Discuss two complications of infection with CMV.

6. Discuss the signs and symptoms associated with rotavirus and Norwalk agent infections.

7. Why are outbreaks of infection with *Giardia* frequent in the U.S.A.?

8. What role does the cyst play in infection with *Entamoeba histolytica*?

9. Why must a person be infected with HBV in order to become infected with HDV?

10. List four species of pathogenic *Shigella*. What animal reservoirs are associated with transmission of *Shigella*? Which of the four species has the highest mortality rate? Which species is most common in the U.S.A.?

CHAPTER 26

Microbial Diseases of the Urinary and Reproductive Systems

Learning Objectives

After completing this chapter, you should be able to:

- List the antimicrobial features of the urinary system.
- Identify the portals of entry for microbes into the reproductive system.
- Describe the normal microbiota of the upper urinary tract, the male urethra, and the female urethra and vagina.
- Describe methods of transmission for urinary and reproductive system infections.
- List the microorganisms that cause cystitis, pyelonephritis, and leptospirosis, and name the predisposing factors for these diseases.
- Describe the cause and treatment of glomerulonephritis.
- List the etiologic agents, symptoms, methods for diagnosis, and treatments for gonorrhea, PID, syphilis, NGU, vaginitis, lymphogranuloma venereum, chancroid, candidiasis, and trichomoniasis.
- Discuss the epidemiology of genital herpes and genital warts.
- List genital diseases that can cause congenital and neonatal infections, and explain how these infections can be prevented.

The **urinary system** consists of organs that regulate the chemical composition of the blood and excrete waste products. The **reproductive system** consists of organs that produce gametes for reproduction of the species or to support the developing embryo.

Structure and Function of the Urinary System

The urinary system consists of two kidneys, two ureters, a urinary bladder, and the urethra (see Figure 26.1). The **urine** passes down the **ureters** from the **kidneys,** where wastes have been removed from the blood, to the **bladder.** Elimination eventually occurs through the **urethra.** The urethra in the male also conveys seminal fluid. Valves in ureters prevent backflow to the kidneys from the bladder, an aid in preventing infection. Also, the acidity of urine and the flushing action of excretion aid in preventing infection.

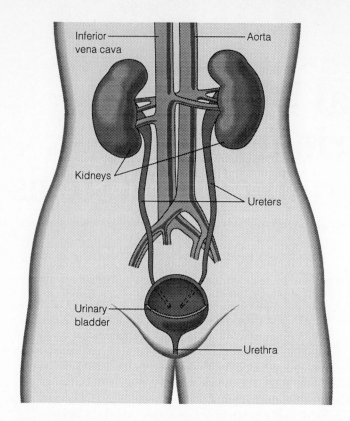

Inferior vena cava

Aorta

Kidneys

Ureters

Urinary bladder

Urethra

FIGURE 26.1 **Organs of the human urinary system, shown here in the female.**

Structure and Function of the Reproductive System

The female reproductive system consists of two **ovaries,** two **uterine (fallopian) tubes,** the **uterus,** the **vagina,** and the external genitals. The ovum (egg) is released from the ovary (a process called ovulation) and passes down the uterine tube. If it is fertilized in the uterine tube, it becomes implanted in the wall of the uterus, where it develops into an embryo and fetus. The vagina is a copulatory canal and part of the birth canal. The external genitals are the clitoris, labia, and glands for lubricating secretions. The male reproductive system consists of two **testes,** which make sperm and male sex hormones; ducts to transport seminal fluid and sperm; and the **penis.**

Normal Microbiota of the Urinary and Reproductive Systems

The urine in the bladder and the organs of the upper urinary tract are normally sterile. The urethra contains a resident microbiota. In infancy, the female vagina is populated by acid-forming lactobacilli, but the acidity becomes more neutral in childhood, and the microbiota then contains a variety of corynebacteria, cocci, and bacilli. At puberty, acid-forming lactobacilli again become the dominant flora. At menopause, the microbiota return to those of childhood and the pH becomes neutral.

DISEASES OF THE URINARY SYSTEM

Bacterial Diseases of the Urinary System

Infections of the urinary system are mostly opportunistic. Inflammation of the urethra is called *urethritis;* of the bladder, *cystitis;* and of the ureters, *ureteritis.* Microbial populations of as few as 1000/ml of any single bacterial type, or as few as 100/ml of coliforms such as *E. coli,* are considered an indication of infection. In females it is not unusual to have the urinary tract contaminated with intestinal bacteria such as *E. coli* and other enteric bacteria. About 35% of nosocomial infections occur in the urinary tract because of operations in this area or catheterization for draining urine.

Cystitis

Cystitis is an inflammation of the urinary bladder and is more common in females. Most cases are due to infection by *E. coli* or *Staphylococcus saprophyticus.*

Pyelonephritis

Pyelonephritis is an inflammation of the kidneys. Scar tissue can be formed that impairs kidney function. It is generally a complication of infections elsewhere in the body; in females, these usually are lower urinary tract infections. *E. coli* is responsible for about 75% of the cases.

Leptospirosis

Leptospirosis is caused by a spirochete, *Leptospira interrogans.* The disease is characterized by chills, fever, and headache. Localization of the pathogens in the kidneys may result in jaundice and possible kidney failure. Leptospirosis is mainly a disease of animals, which excrete the organism with their urine. Domestic dogs are commonly immunized against it. The disease is transmitted to humans and animals by contact with urine-contaminated water. The organism can pass through minor abrasions in mucous membranes. Serological agglutination tests can be used in diagnosis. Penicillin is the most satisfactory drug, but it is not always effective.

Glomerulonephritis

Glomerulonephritis, or **Bright's disease,** is an inflammation of the glomeruli (blood capillaries in the kidneys that assist in blood filtration). Glomerulonephritis is an immune-complex disease caused by soluble streptococcal antigens from infections by β-hemolytic streptococci. The complexes cause inflammation of the glomeruli and kidney damage. The disease is characterized by fever, high blood pressure, and protein and red blood cells in the urine.

DISEASES OF THE REPRODUCTIVE SYSTEM

Bacterial Diseases of the Reproductive System

Diseases of the reproductive system transmitted by sexual activity are called **sexually transmitted diseases (STDs).**

Gonorrhea

By far the most common reportable communicable disease in the U.S.A. is **gonorrhea,** which is caused by a gram-negative diplococcus, *Neisseria gonorrhoeae.* More than 600,000 cases a year are reported. The bacteria must attach to the epithelial wall of the urethra by means of fimbriae in order to be infective. The organism may invade spaces separating mucosal cells found in the oral-pharyngeal area and genitalia, as well as the eyes and rectum.

Males become aware of gonorrheal infection by painful urination and discharge of pus from the urethra. About 80% show these symptoms after a few days' incubation. Complications such as sterility resulting from partial blockage of reproductive ducts by scarring may occur.

Females usually are unaware of the infection, but pain can accompany the fairly common occurrence of a spread of the infection to the uterus and uterine tubes.

Complications of gonorrhea may involve the joints, heart (gonorrheal endocarditis), meninges (gonorrheal meningitis), eyes, and pharynx. Gonorrheal arthritis is caused by the growth of the gonococcus in fluids in the joints. **Ophthalmia neonatorum** is a condition resulting from infection of the eyes of an infant as it passes through the birth canal. Laws usually require administration of antibiotics at birth to prevent this. Gonorrheal infections also can be transferred by hand contact in adults. Gonorrhea can affect any area of sexual contact. **Pharyngeal gonorrhea** often resembles the symptoms of a sore throat, and **anal gonorrhea** can be painful, with the discharge of pus.

There is no immunity to reinfection. Penicillin generally is effective for treatment, but resistance is an increasing problem. Ceftriaxone is currently the drug of choice. Tetracycline often is administered because 20% to 25% of gonorrhea patients also have concurrent chlamydial infections. Diagnosis is made by identification of the organism in the infectious discharges of both men and women. Serological tests are coming into use. The organisms are observed as gram-negative cocci in pairs contained within phagocytic leukocytes. The organism survives poorly outside the body and requires special transporting media to keep it alive even for short intervals before cultivating.

Nongonococcal Urethritis (NGU)

Inflammations of the urethra not caused by *Neisseria gonorrhoeae* are called **nongonococcal urethritis (NGU),** or **nonspecific urethritis.** NGU may be the most common sexually transmitted disease in the U.S.A., but it is not reportable. Many of these infections are caused by chlamydias, obligate intracellular parasites not easily cultured. New tests are available to detect chlamydial infections. Tetracycline is an effective treatment. Other bacteria, such as the mycoplasmas *Ureaplasma urealyticum* or *Mycoplasma hominis,* may cause NGU.

Pelvic Inflammatory Disease

Pelvic inflammatory disease (PID) is a collective term for any extensive bacterial infection of the female pelvic organs. PID is often caused by *N. gonorrhoeae;* however, coinfection with chlamydial bacteria is frequent, and they are also an important agent of this disease.

Infection of the uterine tubes **(salpingitis)** is the most serious kind of PID. Scarring can block passage of ova from the ovary to the uterus. This can result in sterility or **ectopic** (or tubal) **pregnancy** (the fetus develops in the uterine tube rather than in the uterus).

Recommended treatment is administration of doxycycline and cefoxitin sodium, which are active against both the gonococcus and chlamydia.

Syphilis

The number of **syphilis** cases in the U.S.A. has remained fairly stable in recent years. The causative organism is a gram-negative spirochete, *Treponema pallidum.* Syphilis is transmitted mainly by sexual contact. The incubation period averages about three weeks.

There are several identifiable stages of the disease. The initial symptom of the **primary stage** is a **chancre,** or sore, which usually appears at the site of infection. The fluid formed in the chancre is highly infectious, and spirochetes can be seen with darkfield microscopy. In a few weeks this lesion disappears; many women are entirely unaware of it. Serological tests are positive in about 80% of patients in the primary stage. Bacteria enter the bloodstream and the lymphatic system and become distributed in the body.

Several weeks after the primary stage, the **secondary stage,** characterized by rashes, appears. The rash occurs on the skin and mucous membranes, and lesions are very infectious. Nonsexual transmission by kissing, dental work, and so on, is most likely at this stage. Serological tests become almost uniformly positive. In a few weeks the symptoms of secondary syphilis subside, and the **latent period** is entered. The majority of cases do not progress beyond the secondary stage, even without treatment, perhaps because of developing immunity.

After two to four years, the disease is not normally infectious but can be transmitted from mother to fetus. Such **congenital syphilis** can cause damage to the infant's mental development, as well as other neurological disturbances. In fewer than half the cases, even without treatment, the disease reappears in a tertiary stage; when it does, it is usually after an interval of at least ten years. Many of the lesions of the **tertiary stage,** called **gummas,** can cause extensive tissue damage. The symptoms may be due to hyperimmune reactions to surviving spirochetes. This stage of syphilis is no longer common.

The **VDRL (Venereal Disease Research Laboratory)** slide test and other rapid screening tests for diagnosis, such as the **rapid plasma reagin (RPR)** card test, detect antibodies, not against the spirochete itself, but reagin-type antibodies. The production of these is stimulated by lipid material formed as a response to infection by the spirochete. The antigen used in these slide tests is an extract of beef heart, which contains lipids similar to those formed by the body that originally stimulated the reagin production. Confirmation of a positive screening test can be made by **fluorescent treponemal antibody absorption (FTA-ABS)** tests. In these tests an avirulent cultivated strain of the spirochete is allowed to react with antibodies present in the patient's blood. Benzathine penicillin, a long-lasting formulation, has continued to be effective in the treatment of syphilis, especially in the primary stage.

Gardnerella Vaginosis

Vaginitis, infections of the vagina, are most often caused by the fungus *Candida albicans,* the protozoa *Trichomonas vaginalis,* or the bacterium *Gardnerella vaginalis. C. albicans* usually is an opportunist, whereas *T. vaginalis* and *G. vaginalis* usually are sexually transmitted. *G. vaginalis* infection, once called nonspecific vaginitis, involves interaction with anaerobic vaginal bacteria. There is no sign of inflammation, so the term **vaginosis** is preferred to vaginitis. Diagnosis is based on a fishy odor, a vaginal pH above 5, and "clue cells," sloughed-off vaginal epithelial cells covered with gram-negative rods, in the discharge. Metronidazole, which eliminates the anaerobic bacteria, is used for treatment.

Lymphogranuloma Venereum

Chlamydia trachomatis, the probable major cause of NGU and the cause of trachoma, also is responsible for **lymphogranuloma venereum.** The disease is found in much of the tropical world and in the southeastern U.S.A. Infection causes the regional lymph nodes to become enlarged and tender. Scarring of the lymph drainage ducts occasionally leads to massive enlargement of the male genitalia and rectal narrowing in females. Chlamydias, if properly stained, may be seen as inclusions in pus. Tetracycline antibiotics generally are effective in treatment.

Chancroid (Soft Chancre)

Chancroid (soft chancre) is mainly a tropical disease, but cases in the U.S.A. are steadily increasing. In chancroid, a painful ulcer forms on the genitalia and spreads to adjacent lymph nodes. The causative organism is *Hemophilus ducreyi.* Erythromycin and ceftriaxone are effective in treatment.

Viral Diseases of the Reproductive System

Genital Herpes

Herpes simplex virus type 2, and sometimes type 1, causes a sexually transmitted disease called **genital herpes.** The lesions cause a burning sensation, and vesicles appear. The lesions are infectious; there is a serious danger of infection to an infant at birth—**neonatal herpes.** Such an infection can be fatal or cause serious neurological damage.

The virus may enter a latent stage in nerve cells and reappear at intervals, much like cold sores. There is no cure at this time for genital herpes, although oral and topical administration of acylovir alleviates symptoms and decreases the frequency of recurrence in many cases.

Genital Warts

Warts are caused by viruses, and sexual transmission of genital warts is common. The papillomaviruses that cause warts also are sometimes factors in cervical cancer.

AIDS

AIDS is caused by a virus that is often sexually transmitted; it is discussed in Chapter 19 because it affects the immune system.

Fungal Disease of the Reproductive System

Candidiasis

Candida albicans is a yeastlike fungus that grows on mucous membranes. Infection of the mucous membrane of the vagina is **vulvovaginal candidiasis.** There is irritation; severe itching; a thick, yellow, cheesy discharge; and a yeasty odor. Predisposing conditions include pregnancy, diabetes, certain tumors, and treatment with immunosuppressive drugs or broad-spectrum antibiotics. Diagnosis is made by microscopic identification of the fungus, and treatment consists of topical application of clotrimazole and miconazole.

Protozoan Disease of the Reproductive System

Trichomoniasis

The protozoa *Trichomonas vaginalis* is a fairly normal inhabitant of the female vagina and the male urethra. Trichomoniasis results when the acidity of the vagina is disturbed and the protozoa outgrow the normal microbiota. Males rarely have any symptoms; females exhibit a purulent discharge with a disagreeable odor, accompanied by irritation and itching. Diagnosis is made by microscopic examination of the discharge. Metronidazole is used in treatment.

Self-Tests

In the matching section, there is only one answer to each question; however, the lettered options (a, b, c, etc.) may be used more than once or not at all.

I. Matching

1. Most common reportable communicable disease in the U.S.A.

2. The VDRL test useful in diagnosis.

3. Caused only by a protozoa.

4. Most of the cases probably are caused by chlamydias.

5. In diagnosis, the causative organism of this disease is seen as pairs of gram-negative cocci contained within phagocytic leukocytes.

6. *Treponema pallidum.*

7. Caused by a spirochete.

8. Metronidazole has been an effective treatment.

9. Caused by a virus.

a. Nongonococcal urethritis

b. Gonorrhea

c. Genital herpes

d. Syphilis

e. Trichomoniasis

II. Matching (Stages of Syphilis)

1. Characteristic symptoms are mainly rashes; serological tests at this stage almost uniformly positive.

2. The disease is not normally infectious during this stage, but it can be transmitted from mother to fetus.

3. The usual symptom characteristic of this stage is the chancre.

4. Lesions in this stage can cause extensive damage; symptoms may be due to hyperimmune reactions.

a. Primary

b. Latent

c. Tertiary

d. Secondary

III. Matching

1. The site of ectopic pregnancies.

2. Where the ovum is implanted to develop into a fetus.

3. The channel for eventual elimination of urine.

4. Carries urine from the kidneys to the urinary bladder.

5. The site of cystitis.

    a. Ureter

    b. Bladder

    c. Urethra

    d. Uterus

    e. Uterine tube

    f. Urine

IV. Matching

1. Caused by *Chlamydia trachomatis*.

2. Bright's disease.

3. An immune-complex disease caused by soluble streptococcal antigens.

4. Caused by a spirochete.

    a. Glomerulonephritis

    b. Leptospirosis

    c. Chancroid

    d. Lymphogranuloma venereum

V. Fill-in-the-Blanks

1. The ovum is released from the ovary in a process called _____ .

2. Inflammation of the urethra is called _____ .

3. Inflammation of the ureters is called _____ .

4. The spread of an infection such as gonorrhea to the uterus and uterine tubes is called _____ .

5. Infection of the eyes of the newborn by the gonococcus at the time of birth is _____ .

6. Cases of gonorrhea are often treated with tetracycline antibiotics that are directed at a probable coinfection with _____ .

7. Slide tests to screen for syphilis detect _____ - type antibodies.

8. *Hemophilus ducreyi* is the cause of _____ .

9. Herpes simplex virus type _____ is the most common cause of genital herpes.

10. The uterine tubes are also known as the _____ tubes.

11. Pelvic inflammatory disease leads to a higher incidence of _____ pregnancies.

12. Pyelonephritis is an inflammation of the _____ .

13. Domestic dogs are commonly immunized against a disease, _____ , usually spread by contact with urine-contaminated water.

14. Gonorrheal _____ is caused by growth of the gonococcus in fluids in the joints.

15. To be infective, the organism causing gonorrhea attaches to the epithelial walls of the urethra by _____ .

16. Syphilis transmitted by the mother to the fetus is known as _____ syphilis.

17. Infection of the mucous membrane of the vagina by *Candida albicans* is _____ .

18. Rubbery lesions of the tertiary stage of syphilis, called _____ , sometimes cause extensive tissue damage.

19. In females, the urinary tract can become contaminated with intestinal bacteria such as

_____ .

20. Another name for chancroid is _____ .

21. A drug useful in alleviating symptoms of genital herpes is _____ .

VI. Label the Art

I.

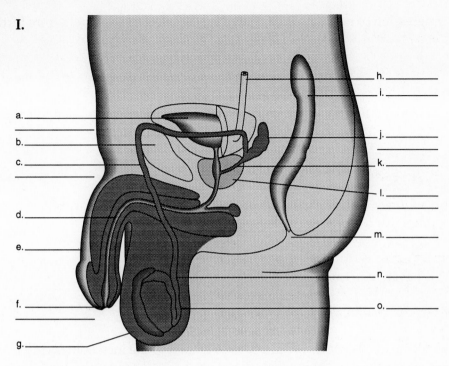

a. _____
_____
b. _____
c. _____
_____
d. _____
e. _____
f. _____
g. _____

h. _____
i. _____
j. _____
k. _____
l. _____
_____
m. _____
n. _____
o. _____

This side view of a male pelvis in cross section shows the reproductive organs.

II.   ### III.

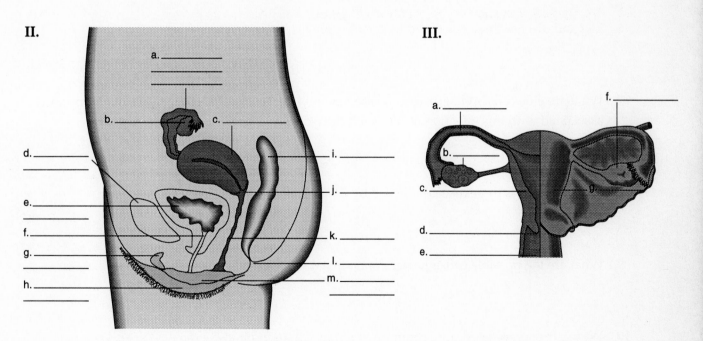

a. _____
_____
_____
b. _____
c. _____
d. _____
_____
e. _____
_____
f. _____
g. _____
_____
h. _____
_____

i. _____
j. _____
k. _____
l. _____
m. _____

a. _____
b. _____
c. _____
d. _____
e. _____

f. _____

Side view cross-section of female pelvis showing reproductive organs.   Front view of female reproductive organs, with left side of the structures shown sectioned.

VII. Critical Thinking

1. Discuss the progression of normal microbiota of the female genital system from birth to maturity.

2. Explain why bacterial infections of the urinary system are more common in females than in males.

3. What causes the kidney damage that is associated with glomerulonephritis?

4. What is PID? What are two causes of PID?

5. How is the widespread use of oral contraceptives related to the increased incidence of gonorrhea and other STDs?

6. List and discuss three tests used to detect syphilis.

7. What are clue cells? What disease is associated with clue cells? List the signs and symptoms associated with this infection. What is the drug of choice used to treat the disease?

8. List and discuss three diseases caused by *Chlamydia trachomatis*.

9. What are the possible consequences of neonatal herpes?

10. Why is trichomoniasis more common in females than in males?

CHAPTER 27

Soil and Water Microbiology

Learning Objectives

After completing this chapter, you should be able to:

- Explain how the components of soil affect soil microbiota.
- Outline the carbon and nitrogen cycles, and explain the roles of microorganisms in these cycles.
- Identify a role for microorganisms in removal of pollutants from soil.
- Describe the freshwater and seawater habitats of microorganisms.
- Discuss the causes and effects of eutrophication.
- Explain how water is tested for bacteriologic quality.
- Compare primary, secondary, and tertiary sewage treatments.
- List some of the biochemical activities that take place in an anaerobic sludge digester.
- Define BOD, septic tank, oxidation pond, activated sludge, and trickling filter.
- List three examples illustrating the use of bacteria to remove pollutants.

Some microbes are **parasites** that prey on others. **Mutualism** occurs when two organisms both benefit, such as the association between fungi and algae in lichen. A common relationship is **commensalism,** in which one organism benefits without affecting the other. One example of commensalism is partial degradation of cellulose by one organism so that the resulting glucose can be used by others. In **cometabolism** the microbes are unable to degrade a compound directly. However, if an energy source such as glucose is provided, the compound is degraded.

SOIL MICROBIOLOGY AND CYCLES OF THE ELEMENTS

The Components of Soil

Minerals

Most soil consists of a mixture of rock and mineral fragments formed by the weathering of preexistent rock.

Organic Matter

Organic matter is derived from the remains of various living organisms and their waste products. Much of organic matter is of plant origin; much comes from the vast numbers of microorganisms and small animals in the soil. The anaerobic environment of swamps and bogs leads to slower decomposition and therefore a higher organic matter content—about 95% compared to 2% to 10% found in agricultural soils. A considerable part of soil organic matter is **humus,** dark residual organic material relatively resistant to decay. Organic matter gives soil its spongy nature, prevents formation of heavy crusts, and increases the proportion of the pore spaces in soil, which then increases aeration and water retention.

Water and Gases

Water is found between the soil particles and adheres to them. Various inorganic and organic constituents of soil are dissolved in the soil water and made available to living inhabitants of the soil. Because of biological respiratory activity, soils contain a higher concentration of carbon dioxide and a lower proportion of oxygen than the atmosphere.

Organisms

Microorganisms in the soil exist in a condition of near starvation with low reproductive rates. When usable nutrients are added to soil, the microbial populations and their activity rapidly increase but return to normal levels when the nutrients are depleted. The most numerous organisms in soil are **bacteria.** A typical garden soil will have millions per gram, mainly in the top few centimeters. Numbers determined by the usual plate-count methods probably greatly underestimate the actual amount. **Actinomycetes,** although bacteria, are usually considered separately when enumerating soil bacteria. Found in soil in large numbers, they produce **geosmin,** the gas that gives soil its characteristic musty odor. The actual **biomass** (weight) of actinomycetes is probably about that of conventional bacteria. Plate counts may enumerate mainly the asexual spores and mycelial fragments of these filamentous organisms. The genus *Streptomyces* produces many of our antibiotics.

    Fungi are found in soil in fewer numbers than bacteria and actinomycetes. Estimates of the biomass of fungi, however, indicate that it probably equals that of bacteria and actinomycetes combined. Molds greatly outnumber yeasts in soil. **Algae** and **cyanobacteria** sometimes form abundant growth on the surface of moist soils but also are found in desert soils. They are located mainly in the surface layers. **Protozoan** populations in the soil tend to rise and fall with their food supply, which is mainly bacteria.

Pathogens in Soil. **Human pathogens** find the soil an alien, hostile environment. Even relatively resistant enteric pathogens such as *Salmonella* species survive only a few weeks or months. However, endospore-forming bacteria such as *Bacillus anthracis*, the cause of anthrax in animals, survive in soil and are ingested by grazing animals. *Clostridium tetani* (tetanus), *Clostridium botulinum* (botulism), and *Clostridium* *perfringens* (gas gangrene) are endospore-forming soil organisms that are pathogenic when introduced into anaerobic food or wounds, where they produce toxins.

Plant pathogens, the cause of rusts, smuts, blights, and wilts that affect plants, are often caused by fungi that pass part of their life cycle in soil. **Insect pathogens** such as *Bacillus thuringiensis* (whose spores and toxins are sold commercially to control some insect larvae) also are found in soil.

Microorganisms and Biogeochemical Cycles

Perhaps the most important role of soil microorganisms is their participation in **biogeochemical cycles,** the recycling of certain chemical elements so that they can be used over again.

The Carbon Cycle

The first step in the **carbon cycle** is the utilization of atmospheric carbon dioxide (about 0.03% of the atmosphere) in photosynthesis to produce organic matter. Eventually this organic matter, from green plants or creatures living directly or indirectly off plants, is changed by decay into carbon dioxide and released again to the atmosphere. Much carbon dioxide also is released by the burning of fossil fuels. There is considerable concern that accumulations of carbon dioxide in the atmosphere will cause world temperatures to increase—the **greenhouse effect.**

The Nitrogen Cycle

Nitrogen is needed by all organisms for the synthesis of protein, nucleic acids, and other nitrogen-containing compounds (Figure 27.1). Molecular nitrogen (N_2) constitutes about 80% of the earth's atmosphere. As nitrogen-containing organic matter enters the soil (much of the nitrogen is in protein form), it undergoes decay by microorganisms. Decomposition of proteins releases amino acids. Ammonia (NH_3) is formed by the removal of amino groups from amino acids, called **ammonification.** In **nitrification,** ammonia is oxidized to yield energy by autotrophic bacteria (Figure 27.1). During nitrification, *Nitrosomonas* bacteria convert ammonia to nitrites (NO_2), and *Nitrobacter* bacteria oxidize the nitrites to nitrates (NO_3), the form preferred by plants as a nitrogen source. **Denitrification** is an anaerobic process by which nitrates are used as an electron acceptor in place of oxygen (an example of anaerobic respiration). *Pseudomonas* species appear to be the most important of several groups involved. The result is, by a series of steps, formation of nitrogen gas, which returns to the atmosphere. This process can be an important means of nitrogen loss in soil.

Nitrogen fixation is the ability to use nitrogen directly from the atmosphere, an accomplishment of only a few bacteria or cyanobacteria. There are several types of **nonsymbiotic** (free-living) bacteria capable of nitrogen fixation. Among them are the aerobic *Azotobacter* and *Beijerinckia.* Some anaerobic species of *Clostridium* such as *Clostridium pasteurianum* and a number of species of the photosynthetic cyanobacteria fix nitrogen. The cyanobacteria are important in rice paddies of the Orient, where they form a nitrogen-fixing symbiosis with *Azolla* water plants. The **nitro-**

ples are typhoid fever, cholera, hepatitis A, and the protozoa *Giardia lamblia.* Many helminthic diseases, such as schistosomiasis, are spread among people coming into contact with contaminated water, not by ingestion; a larval form penetrates the skin.

Chemical Pollution. Many industrial and agricultural chemicals enter water and are resistant to degradation. Some become concentrated in biological organisms, as happens with fat-soluble pesticides and metallic mercury. Mercury is converted by benthic bacteria to a soluble form that is taken up by fish. Eating such fish can cause neurological damage.

    Eutrophication, a term meaning "well nourished," can occur when nitrogen and phosphorus, in particular, enter waters. Algae get carbon from atmospheric carbon dioxide and energy from light. Therefore, algae and non-nitrogen-fixing cyanobacteria require only relatively small amounts of nitrogen and phosphorus for heavy growth (blooms). Many cyanobacteria fix their own nitrogen; for them, phosphorus is the only ingredient lacking. The death of these organisms results in their degradation by bacteria, a process that depletes the oxygen in water and hastens the filling of the lake. Oil spills are often effectively treated with certain nutrients and specially adapted bacteria.

    Coal-mining wastes often are high in sulfur content, mostly iron sulfides (FeS_2). Bacteria such as *Thiobacillus ferrooxidans* obtain energy from the oxidation of the ferrous ion (Fe^{2+}) and convert the sulfides into sulfates. These enter the streams as sulfuric acid, damaging aquatic life and causing formation of insoluble yellow iron hydroxides. These organisms are also used in mining to leach such minerals as uranium from ores in a soluble form.

Tests for Water Purity

Microbiological tests for water safety are aimed at detection of **indicator organisms.** In the U.S.A. these are usually **coliform bacteria** (aerobic or facultative anaerobic, gram-negative, nonendospore-forming rod-shaped bacteria that ferment lactose with acid and gas formation within 48 hours at 35°C). Coliforms are normal inhabitants of human intestines, and their presence is considered an indication of human wastes entering the water. *Escherichia coli* is the predominant **fecal coliform.** Some coliforms, common soil and plant inhabitants, are **nonfecal coliforms** and are less important as indicators.

    There are two basic methods for the detection and enumeration of coliforms. These are summarized in Figure 27.10 in the text. The first is the **multiple-tube fermentation technique.** Fifteen tubes of lactose broth are inoculated by a specified dilution series of water samples. A positive presumptive test such as this can be confirmed (confirmed test) by streaking samples from the test onto differential media such as eosin methylene blue (EMB) agar (which contains lactose). Distinctive colonies with dark centers or a metallic sheen are formed by coliforms.

    Coliforms may also be detected by the **membrane filter method.** Bacteria are retained on the surface of the membrane filter and will form colonies of a distinctive appearance when the membrane is transferred to a pad soaked with a suitable nutrient medium.

A newer method of detecting *E. coli* makes use of the fact that it normally produces the enzyme β-glucuronidase. When a sample containing *E. coli* is added to a medium containing a substrate, MUG, the β-glucuronidase forms a product that is visible under ultraviolet light. Coliforms other than *E. coli* produce only the enzyme β-galactosidase, which forms a yellow color in media containing the substrate ONPG.

Water Treatment

If turbid, water can be held in a reservoir to allow some particulate matter to settle out. The water then undergoes a **flocculation treatment,** by addition of aluminum potassium sulfate (alum), for example. This chemical forms a floc, which carries suspended colloidal material with it as it settles. The next step is passage through beds of sand **(sand filtration),** which removes about 99% of the bacteria and viruses remaining at this stage. Microorganisms are trapped mostly by surface adsorption onto the sand grains. Activated charcoal, which absorbs chemical pollutants, may supplement sand filtration. Water is then **chlorinated** to kill microorganisms before distribution. **Ozone** is considered a possible replacement for chlorination. Ozone (O_3) is generated by electricity at the site of treatment. Arrays of **ultraviolet** tube lamps can also be used to disinfect water.

Sewage Treatment

After water is used, it becomes sewage, a term that includes household washing wastes, industrial wastes, and rainwater from street drains. Sewage is mostly water with as little as 0.03% particulate matter.

Primary Treatment. The usual first step in sewage treatment is **primary treatment.** Settling chambers remove sand and so on; skimmers remove floating oil and grease; floating debris is ground. Following this, solid matter is allowed to settle out in sedimentation tanks. Sewage solids collecting on the bottom (40% to 60% of suspended solids are removed here) are called **sludge.** Sludge is removed periodically, and the effluent (liquid flowing out) may undergo secondary treatment.

Biochemical Oxygen Demand. **Biochemical oxygen demand (BOD)** is a measure of the biologically degradable organic matter in water, as determined by measuring the amount of oxygen required by bacteria to metabolize it. A classic method is to follow for five days the lowering of oxygen levels in water samples in a sealed bottle. The more oxygen used up as the bacteria utilize the organic matter in the sample, the greater the BOD.

Secondary Treatment. Primary treatment removes 25% to 35% of the BOD; the remainder is largely in the form of dissolved organic matter. **Secondary treatment** involves high aeration to encourage aerobic microorganisms to oxidize dissolved organic matter to carbon dioxide and water. In the **activated sludge system,** air is added to aeration tanks. This encourages the growth of organisms that rapidly oxidize organic matter and causes formation of a suspended flocculant sludge. After several hours, the sludge is passed to a tank, where it settles to the bottom. Some of the sludge is recycled to the activated sludge tanks as a "starter" for the next sewage batch (this is the basis for the term *activated sludge*). If the sludge floats, the phenomenon is called **bulking,** a common problem. The growth of filamentous bacteria such as *Sphaerotilus natans* is a possible cause of bulking. Activated sludge systems remove 75% to 90% of the BOD. **Trickling filters** are a secondary treatment method whereby sewage is sprayed over a bed of rocks or formed plastic about 6 feet deep. A slime layer composed of microorganisms that aerobically decompose dissolved organic matter in the passing sewage soon forms on the rocks. Trickling filters remove only 80% to 85% of the BOD, but they are less troublesome to operate and less sensitive to toxic sewage and overloads.

Sludge Digestion.   Sludge accumulating from primary and secondary treatment can be treated in **anaerobic sludge digesters.** The first stage of activity in this process is a fermentation forming organic acids and carbon dioxide. Microorganisms then produce hydrogen and carbon dioxide from these organic acids. The hydrogen and carbon dioxide are used by anaerobic methane-producing bacteria to form methane. Most methane is derived from the energy-yielding reduction of carbon dioxide by hydrogen gas:

$$CO_2 + 4H_2 \longrightarrow CH_4 + 2H_2O$$

Other methane-producers split acetic acid to yield carbon dioxide and methane:

$$CH_3COOH \longrightarrow CH_4 + CO_2$$

Much of the sewage is, therefore, converted to methane and carbon dioxide. Residual solids are pumped to shallow drying beds or to filters to reduce the volume. This sludge is disposed of in several ways, including use as a soil conditioner.

Septic Tanks.   **Septic tanks** settle out suspended solids, which are periodically pumped out, in a holding tank. The effluent flows through a system of perforated piping into a soil drainage field, where it enters the soil and is decomposed by soil microorganisms.

Oxidation Ponds.   Many small communities and many industries use **oxidation ponds** (also called **lagoons** or **stabilization ponds**). Many incorporate a two-stage system. The first stage is deep and anaerobic and settles out sludge. Effluent is then pumped into other ponds that are shallow enough to be aerated by wave action. The growth of algae is encouraged in order to produce oxygen to help maintain aerobic conditions in such high organic matter loads. Bacterial action in decomposing organic matter generates the carbon dioxide used by the algae in photosynthesis in the shallow waters.

Tertiary Treatment.   Primary and secondary treatments do not remove all of the biodegradable organic matter. Secondary treatment effluent contains about 50% of the original nitrogen and 70% of the phosphorus. These proportions can have a great impact on a lake's ecosystem (see earlier discussion of eutrophication). **Tertiary treatment** removes almost all of the BOD, nitrogen, and phosphorus. Nitrogen is converted to ammonia and evaporated into the air or converted to nitrogen gas by denitrifying bacteria. Tertiary treatment is very expensive and is used at this time only in highly developed lake areas where eutrophication must be avoided, even at high costs.

Solid Municipal Waste

Solid municipal waste (garbage) is often placed into compacted **landfills.** Materials tend to degrade very slowly. Methane, a product of microbial activity, can be economically recovered. Many communities lessen the amount of organic material entering landfills by **composting.** Microbial action, much of it at thermophilic temperatures, converts much of the organic material to stable material that resembles humus and is a useful soil supplement.

Self-Tests

In the matching section, there is only one answer to each question; however, the lettered options (a, b, c, etc.) may be used more than once or not at all.

I. Matching

1. A gas that gives soil its characteristic odor.

2. Heavy growth of algae in a body of water.

3. An accumulation of dark organic matter that is relatively resistant to further degradation.

a. Humus

b. Topsoil

c. Geosmin

d. Blooms

e. Biomass

II. Matching

1. *Rhizobium* species important in this process.

2. *Pseudomonas* species important in this process.

3. *Azotobacter* species important in this process.

4. *Nitrobacter* species important in this process.

5. Forms a mantle over small roots.

6. Forms arbuscules inside plant root cells.

7. Truffles.

a. Symbiotic nitrogen fixation

b. Nonsymbiotic nitrogen fixation

c. Nitrification

d. Denitrification

e. Ammonification

f. Endomycorrhiza

g. Ectomycorrhiza

III. Matching (Aquatic Zones)

1. The zone in the region along the shore with rooted vegetation and light that penetrates to the bottom.

2. The deeper water under the limnetic zone.

3. Bottom sediments.

4. The water surface away from shore.

5. Where the primary producers are mainly located.

6. Where the methane-producing bacteria are found.

a. Limnetic zone

b. Benthic zone

c. Littoral zone

d. Profundal zone

IV. Matching (Sewage Treatments)

1. Removes almost all BOD, nitrogen, and phosphorus, not all by biological means.

2. Methane-producing bacteria are an important element.

3. Mainly designed to encourage growth of aerobic bacteria to oxidize dissolved organic matter to carbon dioxide and water.

4. *Zoogloea* bacteria form floc in one method.

a. Primary

b. Secondary

c. Anaerobic sludge digestion

d. Tertiary

V. Matching

1. A morphological form of *Rhizobium*.

2. A water plant that may have a symbiosis with nitrogen-fixing cyanobacteria.

3. A possible substitute for chlorination treatment of water.

4. A shrimplike creature found in the ocean.

5. The filamentous organism *Sphaerotilus natans* may be a factor in this.

a. Krill

b. Bacteroids

c. Lichen

d. *Azolla*

e. Phytoplankton

f. Ozone

g. Bulking

VI. Matching

1. A human and animal pathogen found in soil.

2. An insect pathogen found in soil.

3. Fecal coliform.

4. Conversion of ammonia to nitrite.

5. Nonsymbiotic nitrogen fixation involves this bacterium.

a. *Bacillus thuringiensis*

b. *Bacillus anthracis*

c. *Nitrosomonas*

d. *Beijerinckia*

e. *Escherichia coli*

VII. Fill-in-the-Blanks

1. The anaerobic environment of swamps and bogs leads to slower decomposition and therefore a higher organic matter content, about _____ %.

2. The most numerous organisms in soil are _____ .

3. Many of our antibiotics are produced by filamentous soil organisms of the genus _____ .

4. Estimates of the biomass of _____ in the soil indicate that this microbial group probably equals that of bacteria and actinomycetes combined.

5. Carbon dioxide is about _____ % of the atmospheric gases.

6. Molecular nitrogen is about _____ % of the atmospheric gases.

7. The form of nitrogen preferred by most plants as a nitrogen source is _____ .

8. The removal of amino groups from amino acids to form ammonia is called _____ .

9. Denitrification is a process by which nitrates are used as an electron acceptor in place of _____ .

10. The *Cyanobacteria* protect their nitrogenase systems from oxygen in _____ .

11. *Rhizobium* bacteria infect a plant by attaching to a root hair and causing formation of an _____ .

12. _____ are a mutualistic association between a fungus and an alga or cyanobacterium.

13. Pesticides resistant to biodegradation are described as being _____ .

14. The symbiont of the nitrogen-fixing alder tree is an _____ .

15. Many cyanobacteria fix nitrogen, and for them the only requirement for heavy growth in most lake waters is the addition of small amounts of _____ .

16. Streaking out the results of a presumptive test for coliforms onto EMB agar is part of the _____ test.

17. Addition of alum to water in purification is called _____ treatment.

18. In sand filtration, the microorganisms are trapped mostly by _____ onto the sand grains.

19. Sewage solids collecting on the bottom are called _____ .

20. Biochemical oxygen demand is a measure of the biologically degradable _____ in water, as determined by the amount of _____ required by bacteria to metabolize it.

21. When the sludge fails to settle out properly in activated sludge processing, it is called _____ .

22. In oxidation ponds, the growth of algae is often encouraged in order to produce _____ .

VIII. Label the Art

CARBON CYCLE

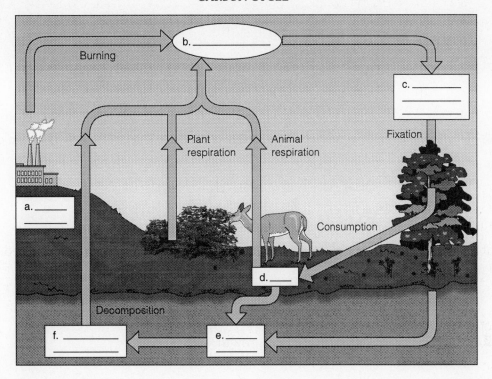

IX. Critical Thinking

1. Explain the benefits of the filamentous growth habit exhibited by actinomycetes.

2. What role do soil microorganisms play in the carbon cycle?

3. Define ammonification, nitrification, denitrification, and nitrogen fixation.

4. Discuss the role of algae in freshwater microbial populations. In what zone of the aquatic environment are algae primarily found?

5. Explain the mechanism that results in the bioluminescence of some saltwater bacteria.

6.   What problems are associated with the biodegradable detergents that were introduced in 1964?

7.   What are indicator organisms, and how are they used in the evaluation of water quality?

8.   Compare and contrast two methods of secondary treatment used in sewage treatment.

9.   What are oxidation ponds? What are the advantages and disadvantages associated with oxidation ponds?

10.   How do mycorrhizae contribute to plant growth?

CHAPTER 28

Applied and Industrial Microbiology

Learning Objectives

After completing this chapter, you should be able to:

- Provide a brief history of the development of food preservation.
- Describe thermophilic anaerobic spoilage and flat sour spoilage by mesophilic bacteria.
- Describe how canning, pasteurization, low temperature, aseptic packaging, and radiation are used to preserve foods.
- Provide examples of chemical food preservatives, and explain why they are used.
- Outline at least four beneficial activities of microorganisms in food production.
- Differentiate between primary and secondary metabolites.
- Describe the role of microorganisms in the production of alternative energy sources, industrial chemicals, and pharmaceuticals.
- Define industrial fermentation, bioreactor, and biotechnology.

FOOD MICROBIOLOGY

Food Preservation and Spoilage

Modern civilization and the support of large populations could not exist if it weren't for effective methods of preserving food. Drying food, or adding salt or sugar, lowers the available moisture and prevents spoilage. The acidity found in the natural fermentation of milk or vegetable juices prevents growth of many spoilage bacteria. Comparatively, preservation of foods by heat (**canning**) and refrigeration requires considerable technical sophistication. Canned foods today undergo what is called **commercial sterilization**, the minimum processing necessary to destroy the endospore-forming pathogen *Clostridium botulinum*. To ensure this result, the **12D treatment** is applied, by which a theoretical population of botulism bacteria would be decreased by 12 logarithmic cycles (10^{12} endospores would be reduced to only one).

Canning

Thermophilic anaerobic spoilage is possible in low-acid canned foods. The can is usually swollen from gas production, and the contents have a lowered pH and a sour odor. The cause is thermophilic clostridia, which survived commercial sterilization because of unusually resistant endospores. In **flat sour spoilage,** caused by thermophilic organisms such as *Bacillus stearothermophilus,* the can is not swollen by gas production. Both of these types of spoilage occur with storage at higher-than-normal temperatures. The thermophilic organisms involved will not grow at normal storage temperatures. Canned-food spoilage by **mesophilic bacteria** is due to can leakage or underprocessing. The former is more likely to result in spoilage by nonendospore-formers, and the latter spoilage by endospore-formers. Contamination by leakage can be caused by sucking cooling water past the sealant in the crimped lid as the can cools. These types of spoilage often result in odors of putrefaction in high-protein foods. Botulism is a possible consideration in mesophilic spoilage.

Some acidic foods, such as tomatoes and fruits, are preserved by heating at temperatures below the boiling point. This type of preservation is possible because organisms like molds, yeasts, and occasional species of acid-tolerant nonendospore-forming bacteria (the only organisms likely to grow in such foods) are easily killed. One problem with such acidic foods can be the heat-resistant **sclerotia** (specialized resistant bodies) of certain molds, which can survive temperatures of 80°C for a few minutes. Other problems are **heat-resistant ascospores** of the mold *Byssochlamys fulva,* and the endospore-forming *Bacillus coagulans,* a bacterium capable of growth at a pH of almost 4.0.

Pasteurization is not sterilization, but mild heating intended to eliminate certain spoilage or disease-causing bacteria without spoiling the quality of the food.

Aseptic Packaging

The container used in **aseptic packaging** is usually made of laminated paper or plastic that cannot tolerate conventional heat treatment. Rolls of packaging materials are fed into a machine that sterilizes the material in hot hydrogen peroxide. The containers are formed while still in the sterile environment and filled with conventionally heat-sterilized foods.

Low-Temperature Preservation

Low temperatures slow the reproduction time of microorganisms; however, some molds and bacteria can grow at a significant rate even below the freezing point of water. A properly set refrigerator maintains a temperature of 0° to 5°C. Although many microorganisms will grow slowly at these temperatures and eventually alter the taste and appearance of foods, pathogenic bacteria will not. Exceptions are the clostridia causing type E botulism, *Listeria monocytogenes,* and *Yersinia enterocolitica.* Freezing does not kill bacteria outright in significant numbers, but frozen bacterial populations become dormant and decline slowly with time. Some parasites, such as the trichinosis roundworm, are killed by several days of freezing.

Radiation and Food Preservation

The use of **ionizing radiation** to preserve food is theoretically possible. Actual sterilization by radiation probably causes too much alteration in the taste of most foods for commercial use. The use of radiation to treat spices, eliminate insects in some foods, and kill trichinosis worms in pork has recently been approved. **Microwaves** kill bacteria only by their heating effects on food.

Chemical Preservatives

Sorbic acid and **sodium benzoate** prevent mold growth in certain acidic foods, such as cheese and soft drinks, which are susceptible to mold-type spoilage. **Calcium propionate** is a fungistat used in breads. It prevents the growth of surface molds and the *Bacillus* species of bacteria that causes ropy bread. These organic acids do not work by any pH effect in inhibiting mold growth, but rather interfere with mold metabolism or the integrity of the cellular membrane. They are readily metabolized by the body.

    Sodium nitrate and **sodium nitrite** are found in meat products such as ham, bacon, wieners, and sausage. The active ingredient is sodium nitrite. Sodium nitrate is a reservoir that some bacteria use as a substitute for oxygen, producing nitrites in the process. The nitrite preserves the red color of meat by reacting with blood components and prevents the germination and growth of any botulism endospores. There has been some concern that reaction of nitrites with amino acids forms carcinogenic **nitrosamines,** and that this result might be a reason for removing nitrates from foods. It has been concluded, however, that the total removal of nitrates would not be justified because their value balances the low risks involved; however, nitrate amounts in foods have generally been reduced. The antibiotic **nisin,** active against endospore-forming bacteria, has no clinical uses and is the most widely used antibiotic in food. The antifungal antibiotic **natamycin** is occasionally used in dairy products.

Foodborne Infections and Microbial Intoxications

Some foodborne diseases are caused by a pathogen carried in contaminated food to the host, where it grows and produces damaging toxins. Such foodborne **infections** are exemplified by gastroenteritis, typhoid fever, and dysentery. In **microbial intoxications,** the toxin is formed by microbial growth in the food and then ingested with the food. Examples of such diseases are botulism and staphylococcal food poisoning.

    Dairy products, often consumed without cooking, are particularly likely to transmit diseases. Standards are therefore strict. **Pasteurized Grade A milk** is required to have a standard plate count of less than 20,000 bacteria per milliliter and not more than 10 coliforms per milliliter. **Cultured** products such as buttermilk and cultured sour cream are required to contain no more than 10 coliforms per milliliter. **Grade A dry milk** products must have a standard plate count of less than 30,000 bacteria per gram or 10 coliforms per gram. **Certified raw milk** standards specify no more than 10,000 bacteria per milliliter or 10 coliforms per milliliter.

The Role of Microorganisms in Food Production

Cheese

Cheeses come in many types, all requiring the formation of a **curd,** which can then be separated from the liquid fraction, or **whey.** Except for unripened cheeses such as ricotta and cottage, the curd undergoes a microbial ripening process. The curd is milk protein, **casein,** and is usually formed by action of the enzyme, **rennin** (chymosin), aided by acid conditions that are provided by inoculation with lactic acid-producing bacteria. These lactic bacteria provide the flavor and aroma.

Cheeses generally are classified by hardness. Romano and parmesan cheeses are *very hard* cheeses; cheddar and Swiss are *hard;* limburger, blue, and roquefort cheeses are *semisoft;* and camembert is a *soft* cheese. Hard cheddar and Swiss cheeses are ripened by lactic bacteria growing anaerobically in the interior. The longer the incubation time, the more acidity and the sharper the taste. (The holes in Swiss are from carbon dioxide produced by a *Propionibacterium* species of bacteria.) Semisoft cheeses are ripened by bacteria and other organisms growing on the surface. Blue and roquefort cheeses are ripened by *Penicillium* molds inoculated into the cheese. The open texture allows adequate oxygen to reach the molds. Camembert cheese is made in small packets so that the ripening enzymes of surface molds will diffuse into the cheese.

Other Dairy Products

Butter is made by churning cream until the fat phase forms globules of butter separated from the liquid buttermilk fraction. Lactic acid bacteria are allowed to grow in the cream and provide the flavor from the **diacetyls** that give the typical butter flavor and aroma. **Buttermilk** usually is made by inoculating skim milk with bacteria that form lactic acid and diacetyls.

Yogurt is made from low-fat milk by first evaporating much of the water in a vacuum pan. This process is followed by inoculation with a species of lactic acid-producing *Streptococcus* that grows at elevated temperatures. Stabilizers are added to aid in the formation of the thick texture. **Kefir** and **kumiss** are beverages of Eastern Europe. In these, acid-producing bacteria are supplemented with a lactose-fermenting yeast to give the drinks a small alcoholic content.

Nondairy Fermentations

Yeasts are used in baking; the sugars in bread dough are fermented to ethanol and carbon dioxide. The carbon dioxide makes the bubbles of leavened bread, and the ethanol evaporates in baking. Fermentation also is used in making sauerkraut, pickles, and olives. In Asia, soy sauce is a popular fermented food.

Beer and **ale** are products of fermentation of grain starches by yeast. The starch is first converted to glucose and maltose by a process called **malting.** Barley is allowed to sprout, then dried and ground. This results in malt, which contains amylases to degrade starch. **Distilled spirits,** such as whiskey, vodka, and rum, are made by fermentation of carbohydrates from grains, potatoes, and molasses, respectively. The alcohol is then distilled off.

Wines are typically made from the yeast fermentation of grapes, which usually need no additional sugar, although more can be added to make more alcohol. Malic acid production makes some wines too acid, and bacteria are used to convert the malic acid to the less acidic lactic acid **(malolactic fermentation).** Wine can be spoiled by aerobic bacteria that convert alcohol into acetic acid. This process is used to make **vinegar.** Ethanol is made by the anaerobic fermentation of carbohydrates by yeasts. The ethanol is then aerobically oxidized to acetic acid by *Acetobacter* and *Gluconobacter* bacteria.

Microorganisms as a Food Source

In terms of world nutrition, protein is in particularly short supply. Microorganisms may become more important to the food supply because they can convert simple substrates into **single-cell protein (SCP)** very rapidly. Cellulose, methanol, petroleum hydrocarbons, and many industrial wastes might serve as substrates. Because photosynthesis is an inexhaustible energy supply, algae are attractive sources of SCP but are somewhat unpalatable. Bacteria are the most likely organisms from which to make SCP; they have a faster reproductive rate than fungi and can use more substrates than yeasts. High nucleic acid contents may be a problem in SCP because they may precipitate ailments such as gout. The cyanobacterium *Spirulina* has a long history as a natural food.

INDUSTRIAL MICROBIOLOGY

Industrial microbiology—the use of microbes to make industrial products—has centered on the lactic acid and ethanol fermentations. In the future, as petroleum becomes more scarce, we may see a return to many microbial fermentations. In the coming years there will be a revolution in the application of genetically engineered microorganisms in industry—**biotechnology.**

Fermentation Technology

Vessels for industrial fermentations are called **bioreactors.** They are designed to control aeration, pH, and temperatures. Microbes in industrial fermentation produce either primary metabolites, such as ethanol, or secondary metabolites, such as penicillin. A **primary metabolite** is formed at the same time as the cells, and their production curves are almost parallel. **Secondary metabolites** are not produced until the growth phase **(tropho-**

phase) has entered the stationary phase of the growth cycle. The following period, during which most of the secondary metabolite is produced, is the **idiophase.** To improve strains of industrial microbes, radiation is often used to create useful mutants.

Immobilized Enzymes and Organisms

Some industrial processes make use of enzymes immobilized on surfaces while a continuous flow of substrate passes over them. Live or dead cells are also immobilized on such surfaces as small spheres or fibers, to make industrial products.

Industrial Products

Amino Acids

Certain **amino acids,** such as **lysine,** are especially important as nutritional supplements because animals cannot synthesize them. They are present only in low levels in vegetable proteins and are therefore a valuable product. One bacterium that is used to produce lysine is *Corynebacterium glutamicum,* which normally produces both threonine and lysine. A mutant form of the bacterium that lacks the enzymes necessary to make threonine is used to produce lysine in large amounts. Enough threonine is added to the medium to allow this mutant organism to grow, but not enough is added to trigger feedback inhibition, which stops the production of lysine. The mutant produces lysine continuously because lysine alone does not cause feedback inhibition. When this same bacterial species is provided with only a minimal amount of the vitamin biotin, it can be induced to excrete **glutamic acid,** which is used to make the flavor enhancer monosodium glutamate. The microbially synthesized amino acids **phenylalanine** and **aspartic acid** are ingredients in the artificial sweetener NutraSweet®.

Citric Acid

Citric acid is produced industrially by the mold *Aspergillus niger* from molasses. This happens when the mold is provided with only a limited supply of iron and manganese.

Enzymes

Enzymes are valuable industrial products: **amylases** for syrups and paper sizing, **glucose isomerase** to convert glucose into fructose, **proteases** for baking, **proteolytic** enzymes for meat tenderizers and additives to detergents.

Vitamins

Microbes produce several commercial vitamins. **Vitamin B$_{12}$** and **riboflavin** are two examples.

Pharmaceuticals

Antibiotics are an important industrial product. Many are still produced by microbial fermentations. Work continues on selection of more productive mutants and manipulations to increase efficiency. **Vaccines** are also an important industrial product. **Steroids,** such as cortisone, and estrogens and progesterone used in birth control pills, are products of microbial conversions (see Figure 28.1).

Uranium and Copper

The recovery of otherwise unprofitable grades of **uranium** and **copper ore** is made possible by the activity of the bacterium *Thiobacillus ferrooxidans*. Water is sprinkled over a pile of the ore, and the bacteria change insoluble copper compounds to an oxidized, soluble form that moves out of the ore piles and can be reclaimed.

Microorganisms

Baker's yeast is a good example of microbes themselves as an industrial product. Others are the symbiotic nitrogen-fixing bacteria *Rhizobium* and *Bradyrhizobium,* which are used to inoculate leguminous plants. The insect pathogen *Bacillus thuringiensis* is commonly sold in preparations applied to garden plants and trees.

Alternative Energy Sources Using Microorganisms

Conversion of biomass (various organic wastes) into alternative fuel sources is called **bioconversion. Methane** is an important source of such a process. It is produced even today in commercial amounts, from large landfills and cattle-feeding lots. **Ethanol** represents an important additive to motor fuels in many parts of the world.

FIGURE 28.1 Production of steroids. Conversion of a precursor compound such as a sterol into a steroid by *Streptomyces.* The addition of a hydroxyl group to carbon number 11 might require more than 30 steps by chemical means, but the microorganism can add it in only one step.

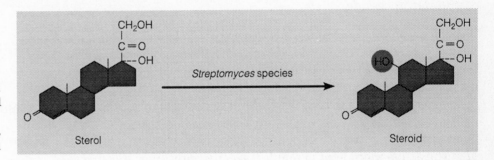

Self-Tests

In the matching section, there is only one answer to each question; however, the lettered options (a, b, c, etc.) may be used more than once or not at all.

I. Matching

1. Involved in making a less-acidic wine.

2. A possible carcinogen in meat products.

3. An enzyme that forms milk curd.

4. Responsible for the flavors and aroma of dairy products.

5. Resistant bodies found on a number of molds.

6. An essential amino acid.

7. A product made by *Aspergillus niger* from molasses.

a. Sclerotia

b. Malolactic fermentation

c. Nitrosamines

d. Rennin

e. Diacetyls

f. Citric acid

g. Lysine

II. Matching (Canned Food Spoilage Types)

1. Caused by a thermophilic *Bacillus* species.

2. Swollen can; contents would have low pH and a sour odor.

3. Botulism would be a possible consideration in this type.

4. Probable cause would be can leakage or underprocessing.

a. Flat sour

b. Mesophilic

c. Thermophilic anaerobic

III. Matching (Food Chemicals)

1. A fungistat used in breads to inhibit molds and ropy bread spoilage.

2. The active ingredient in many meats that preserves the red color and inhibits germination and growth of botulism organisms.

3. Used to prevent mold growth in certain acidic foods.

a. Sodium benzoate

b. Sodium nitrite

c. Sodium nitrate

d. Calcium propionate

IV. Matching

1. Typhoid fever.

2. Staphylococcal food poisoning.

3. Botulism.

a. Foodborne intoxication

b. Foodborne infection

V. Matching (Milk Microbiological Standards Per Milliliter or Gram)

1. Grade A pasteurized.

2. Dry milk.

3. Certified raw milk.

4. Cultured sour cream.

a. 20,000 bacteria/10 coliforms

b. 10,000 bacteria/10 coliforms

c. 10 coliforms

d. 30,000 bacteria/10 coliforms

VI. Fill-in-the-Blanks

1. Cottage cheese is classified as a(n) _____ cheese.

2. Cheddar cheese is classified by hardness as a(n) _____ cheese.

3. Blue and roquefort cheeses are ripened by _____ molds.

4. The first step in making _____ is to evaporate much of the water from low-fat milk in a vacuum pan.

5. To make vinegar, ethanol is first made by an anaerobic fermentation of carbohydrates by _____ . The ethanol is then aerobically oxidized to _____ acid by _____ bacteria.

6. One potential problem in the use of single-cell protein is high _____ content, precipitating ailments such as gout.

7. Glutamic acid is a microbial product used to make the flavor enhancer _____ .

8. Vessels for industrial fermentations are called _____ .

9. Steroids can be made by microbial alteration of available _____ .

10. Canned food undergoes the minimum processing needed to destroy any *Clostridium botulinum* endospores; this is called _____ .

Differences:

1. AIDS may also be transmitted by sharing of needles, in breast milk, and via blood.
2. Syphilis is effectively treated with antibiotics; several difficulties complicate the development of antiviral drugs.

6. Procaryotes have a single, circular chromosome that is not enclosed by a nuclear envelope. This differs from eucaryotic organisms that have a true, membrane-bound nucleus.

   Procaryotes lack true (membrane-bound) organelles.

   Most procaryotes have peptidoglycan in their cell walls.

   Binary fission (rather than mitosis) is a common form of reproduction in procaryotic organisms.

7. Once DNA was established as the hereditary material and its structure and replication were understood, other research led to discoveries about the regulation of gene function in bacteria. Then, in the late 1960s, Paul Berg showed that fragments of human or animal DNA could be attached to bacterial DNA. These new techniques enabled the easy and economical production of large quantities of proteins such as insulin, human growth hormone, and interferon.

8. The relationship between normal microbiota and the human host is mainly one of balance. Normal microbiota usually benefit their host, for example by producing some essential B vitamins and vitamin K. Health is maintained when human mechanisms of defense are able to overcome the disease-producing effects of microorganisms. The same microorganisms can cause disease when host resistance is not sufficient. When this occurs, antimicrobial drugs may be necessary to help reestablish this delicate balance.

9. Although Pasteur disproved spontaneous generation and established that yeasts play a role in fermentation, the germ theory was still difficult for people to accept. Many people still believed that disease was either God's punishment for unacceptable behavior or caused by foul odors from sewage or swamps. They couldn't accept that tiny microbes could travel through the air, or be transmitted by bedding or clothing from person to person.

   Bassi and Pasteur both worked on silkworm diseases. Bassi (1835) demonstrated that an invisible microbe, a fungus, caused disease in silkworms. Several years later (1865) Pasteur established that yet another microbe, this time a protozoa, also caused infections in silkworms. Both of these discoveries established a link between the "dubious" microbes and disease.

10. Sourdough is 8–10 times more acidic than conventional bread due to the fermentation of maltose into lactic and acetic acids by the bacterium *Lactobacillus sanfrancisco*. *Saccharomyces exiguus*, rather than *S. cerevisiae*, is used in sourdough for its ability to thrive in this acidic environment.

Chapter 2

Matching

I.   1.h 2.f 3.a 4.e 5.c 6.h 7.i 8.g
II. 1.c 2.j 3.f 4.h 5.e 6.g 7.d 8.a 9.k
III. 1.g 2.d 3.c 4.e 5.b
IV. 1.d 2.e 3.e 4.a 5.f 6.g 7.j 8.h

Fill-in-the-Blanks

1. chemical element 2. reactants; product 3. four 4. isotopes 5. peptide bonds 6. deoxyribose
7. thymine 8. uracil 9. adenosine triphosphate (ATP) 10. single- 11. primary 12. nucleotides
13. pyrimidines 14. adenine, guanine, cytosine, thymine, uracil 15. tertiary 16. 20 17. stereoisomers
18. boiling point; buffer 19. compound 20. less 21. activation energy 22. carbohydrates 23. di-
24. lipids 25. electrons 26. exergonic

Label the Art

2. N 7 14 7 7 7   3. Cl 17 35 17 18 17   4. O 8 16 8 8 8 5. O 8 17 8 9 8

Critical Thinking

1. Catalysts lower the activation energy of a chemical reaction by increasing the rate of the reaction and by orienting the reactant molecules toward each other. Both of these mechanisms increase the probability of a reaction occurring. If the molecules move faster, they will collide more often, increasing the number of reactions. Catalysts speed up the reaction without raising the temperature. Orienting the molecules toward each other further increases the probability that the reaction will occur. Catalysts are important in living systems because they accelerate biological reactions without raising the temperature. Increased temperature would result in the denaturing of protein molecules and the death of the cell.

2. Carbon is important because of the condition of its outer electron shell. In this shell there are four electrons and four vacancies. This allows carbon to react with many other atoms and with other carbon atoms to form a variety of chains and rings. These chains and rings are the basis of many organic compounds in living cells.

3. Lipids are a major component of membranes, in terms of both structure and function. Membranes serve as a boundary between the cell and the environment. Lipids function in energy storage and are also a component of some bacterial cell walls.

4. Waste products excreted by bacteria alter the pH of the medium. pH buffers—such as potassium phosphate, monobasic—are added to media to prevent drastic changes in pH. Without the addition of buffers, the media would become acidic enough to inactivate bacterial enzymes and kill the organism.

5. Advantages:

   1. Bacteria can extract pollutants that have combined with soil and water and would otherwise be difficult to separate.

   2. Bacteria may chemically alter a harmful substance into one that is less harmful or even beneficial. For example, *Pseudomonas* can alter mercury into its relatively harmless elemental form.

   Disadvantages:

   1. Bacteria work too slowly in the cleanup of toxic spills.

   2. Natural numbers of "clean-up" bacteria are not sufficient in cases of large-scale contamination.

   Scientists are trying to improve efficiency of pollution-fighting bacteria through genetic engineering. The number of oil-degrading bacteria can be augmented by bioremediation, the addition of nitrogen and phosphorus plant fertilizers to the beach.

6. Products:

$$\begin{array}{cccccc} & H & H & & CH_3 & H \\ & | & | & & | & / \\ HO-C-&C-&N-&C-&C-&N-H \ + \ H_2O \\ & || & | & & || & | \\ & O & H & & O & H \end{array}$$

7. There are four possible levels of protein structure: primary, secondary, tertiary, and quaternary.

   The primary structure is the unique sequence of amino acids making up the protein molecule. The sequence is genetically determined.

   The secondary structure is the twisting or folding of the polypeptide chain. This occurs due to hydrogen bonding joining the atoms of peptide bonds at different locations along the polypeptide chain.

   Tertiary structure results from additional folding of the secondary structure. This folding is due to interactions between various amino-acid side groups in the polypeptide chain.

Quaternary structure is observed in some proteins. It results when two or more tertiary proteins form an aggregate.

Protein function is related to its ability to recognize and bind to other molecules—for example, to a substrate.

8. The molecular weight of methane is 16 [12 + (4 × 1)], so one mole of methane weighs 16 grams.

The number of moles of methane in 240 grams is thus 240 ÷ 16 = 15.

9. ATP is important because it is the principal energy-carrying molecule of the cell. It stores the chemical energy released by some reactions and releases energy for reactions that require energy. ATP releases large amounts of usable energy when it loses its terminal phosphate group in the following reaction:

ATP → ADP + P + energy

The energy released from decomposition reactions can be stored by attaching a terminal phosphate to ADP, the reverse of the previous reaction.

10.

DNA	RNA
Double-stranded molecule	Single-stranded molecule
A, T, G, and C	A, U, G, and C
Codes for RNA	Codes for protein
Deoxyribose sugar	Ribose sugar
	3 kinds of RNA: mRNA, tRNA, rRNA

Chapter 3

Matching

I. 1.d 2.a 3.c 4.f 5.e 6.b
II. 1.e 2.g 3.a 4.c 5.d
III. 1.a 2.d 3.e 4.f 5.d 6.f 7.c

Fill-in-the-Blanks

1. billion 2. 2000 3. glass 4. ultraviolet light 5. 1/100,000; better 6. negative 7. scanning
8. smear 9. fixing 10. 100,000× 11. mordant 12. *Mycobacterium; Nocardia* 13. tuberculosis or leprosy 14. simple 15. mordant 16. acidic 17. darkfield 18. interference contrast

Critical Thinking

1. The limit of resolution of the compound light microscope is about 0.2 μm. This limit of resolution is due to the wavelength of white light, which is used in compound light microscopes.

2. Immersion oil has the same refractive index as the glass of the objective lens. By using immersion oil, light rays are prevented from refracting from the slide as they enter the air. This has the same effect as increasing the diameter of the lens and improves the resolving power.

3. The term *refractive index* refers to a measure of the relative velocity at which light passes through a material. In order to view a specimen with a microscope, the specimen must contrast with the medium in which it is viewed. This can be achieved by changing the refractive index of the specimen by staining.

4. a. Fluorescence microscopy

   b. Electron microscopy

   c. Brightfield microscopy

   d. Darkfield microscopy

   e. Phase-contrast microscopy

5. Basic dyes are salts composed of a positive ion and a negative ion. The positive ion contains the color in basic dyes. Three basic dyes used to stain bacteria are crystal violet, methylene blue, and safranin. Basic dyes are used to stain bacteria because bacteria are slightly negatively charged and so attract the positive, colored ion portion of basic stains.

6. A mordant is a chemical substance that is used to intensify the stain. A mordant may be used to increase the affinity of a stain for the specimen. It may also be used to coat a structure (such as a flagellum) to make it thicker and easier to view after it is stained.

7. a. A negative stain using India ink or nigrosin and a simple stain such as safranin. The India ink (or nigrosin) stains the background but doesn't penetrate the capsule. The safranin is also repelled by the capsule but does stain the cell. The capsule shows up as a halo surrounding the cell against a dark background.

   b. For this purpose a simple stain such as safranin or methylene blue will work fine.

   c. Acid-fast staining would be appropriate to diagnose infections of *Mycobacterium* and *Nocardia*. The red dye, carbolfuchsin, binds strongly to a waxy substance in the cell wall of these organisms but not to other "nonacid-fast" bacteria.

   d. The Gram staining reaction is helpful information when choosing an antimicrobial drug.

8. When iodine is added to a smear after previous staining with crystal violet, they combine to form a complex (CV-I complex) that is larger than the crystal violet molecule that initially entered the cells. The CV-I complex is too large to be washed out of the intact peptidoglycan layer of gram-positive cells. When decolorizing gram-negative cells, the alcohol washes away the outer lipoprotein layer and the crystal violet from the thin layer of peptidoglycan.

9. Light can be adjusted with the diaphragm and the condenser. Increased contrast is achieved by reducing the amount of light. This can be done by lowering the condenser or closing the diaphragm.

10. Total magnification = magnification of ocular × magnification of objective lens

    Low power = 100X

    High dry = 400X (450 on some microscopes)

    Oil immersion = 1000X

    Some compound microscopes can achieve 2000x with the oil immersion lens.

Chapter 4

Matching

I. 1.h 2.c 3.a 4.d 5.e 6.f 7.j
II. 1.a 2.a 3.b 4.a 5.b 6.a
III. 1.i 2.g 3.d 4.c 5.a 6.f
IV. 1.a 2.j 3.g 4.h 5.f 6.c
V. 1.f 2.g 3.e 4.d 5.b 6.c
VI. 1.d 2.a 3.c

Fill-in-the-Blanks

1. polysaccharide 2. phagocytosis 3. plasmids 4. grana 5. pores 6. endospores 7. cisternae
8. pseudomurein 9. septum; forespore 10. matrix 11. centrioles 12. golgi 13. hypotonic
14. simple diffusion; osmosis; facilitated diffusion 15. negative 16. basal 17. flagellin 18. pilin
19. pleomorphic 20. glycocalyx

Critical Thinking

1. The glycocalyx is a sticky, viscous, gelatinous polymer that surrounds some bacterial cells. It may be composed of polysaccharide, polypeptide, or a combination of these two substances. Depending on how the material is arranged and attached to the cell, it may be referred to as a slime layer or a capsule. The glycocalyx is associated with bacterial virulence because it helps protect the bacterium from phagocytosis by white blood cells and helps the bacterium to adhere to and colonize a host.

2. The difference between flagellum attachment in gram-positive and gram-negative cells is in the makeup of the basal body. Gram-positive cells have only the inner pair of rings (see figure below), which are attached to the plasma membrane. Gram-negative cells have both pairs of rings; the outer pair are attached to the cell wall, the inner pair to the plasma membrane.

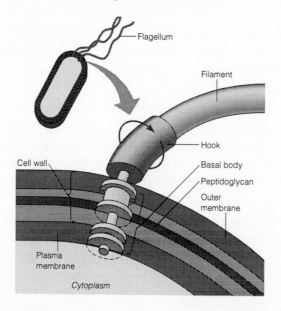

Structure of a procaryotic flagellum. Parts and attachment of a flagellum of a gram-negative bacterium.

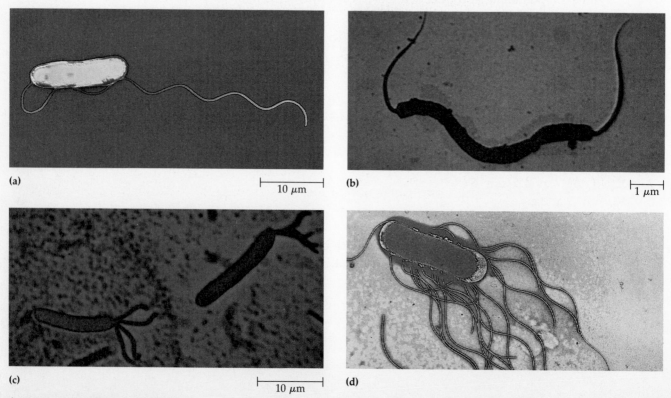

Arrangement of bacterial flagella. Four basic types of flagellar arrangements.

3. Teichoic acids are a component of the gram-positive cell wall and consist primarily of an alcohol and phosphate. The role of teichoic acids is not entirely clear, but because they are negatively charged it is believed that they may bind and regulate the movement of cations into and out of the cell. They may also play a role in cell growth by regulating the activity of autolysins, allowing for the insertion of additional cell wall subunits while preventing excessive cell wall breakdown. Finally, teichoic acids provide much of the antigenicity of the cell wall, providing a means to identify bacteria using serologic methods.

4. Substances that dissolve easily in lipids can most easily cross the plasma membrane. These include oxygen, carbon dioxide, and nonpolar organic molecules. Also, small molecules such as water are able to cross the plasma membrane easily.

5. There will be no change in a bacterial cell in an isotonic solution; water leaves and enters the cell at the same rate.

   A bacterial cell placed in a hypotonic solution will undergo osmotic lysis because more water will enter the cell than the cell wall can contain.

   A hypertonic solution will cause a bacterial cell to undergo plasmolysis, the osmotic loss of water due to increased solutes outside of the cell.

6. a. Endospore formation begins after replication of the chromosome. A chromosome and a small quantity of cytoplasm are separated from the rest of the cell by an infolding of the plasma membrane called a spore septum. The spore septum becomes double-layered and surrounds the chromosome and cytoplasm. The newly formed structure is referred to as a forespore. Thick layers of peptidoglycan are laid down between the two membranes, and a thick, protein spore coat forms around the outside of the membrane.

   b. Six genera of gram-positive bacteria have the ability to form endospores.

   c. Genera of bacteria that can form endospores include *Bacillus*, *Sporolactobacillus*, *Clostridium*, *Desulfotomaculum*, *Sporosarcina*, and *Oscillospira*.

   d. Endospore-forming bacteria may be able to survive periods when water and/or nutrients are not available. Also, due to the resistant nature of endospores, these organisms may be able to survive exposure to high temperatures and to some toxic substances.

   e. Sporulation: The process of spore and endospore formation.

   Forespore: A structure consisting of a chromosome, cytoplasm, and endospore membrane inside a bacterial cell.

   Germination: The process of starting to grow from a spore or endospore.

7. Peptidoglycan is a substance that is found in varying quantities in most procaryotic cells. Peptidoglycan is unique to procaryotic cells and is never found in eucaryotic cells. Antibiotics such as the penicillins and the cephalosporins act specifically against peptidoglycan and therefore have low toxicity in humans. These drugs prevent the formation of the peptide cross bridges of peptidoglycan, preventing synthesis of a functional cell wall.

8. The endosymbiont hypothesis addresses the evolution of eucaryotic cells, specifically the origin of their mitochondria and chloroplasts. This hypothesis proposes that eucaryotic cells evolved from symbiotic relationships between procaryotic cells having different metabolic abilities. Evidence supporting the endosymbiont theory is seen in mitochondria and chloroplasts. Both organelles have 70S ribosomes, the type seen in procaryotic cells, rather than the typical 80S eucaryotic ribosomes. Also, mitochondria and chloroplasts multiply and grow within eucaryotic cells.

9. The ER is a membranous system that runs throughout the eucaryotic cell. It is continuous with the plasma and nuclear membranes. Functions include protein and lipid synthesis and molecule transport and storage. The sacs of the Golgi complex are sometimes attached to the ER. The Golgi complex packages and secretes the products of the ER.

10. The semirigid cell wall of the procaryotic cells supports the cell and determines the shape (morphology) of the cell.

Eucaryotic-cell support and shape are provided by the cytoskeleton, a network of microtubules, microfilaments, and intermediate filament.

Chapter 5

Matching

I. 1.a 2.f 3.e 4.c 5.d 6.h 7.g
II. 1.f 2.e 3.e 4.c 5.b 6.a
III. 1.c 2.f 3.c 4.e 5.a 6.b
IV. 1.a 2.c
V. 1.a 2.d 3.e 4.b

Fill-in-the-Blanks

1. saprophyte 2. anoxygenic 3. saturated 4. noncompetitive 5. competitive 6. oxidation 7. substrate-level 8. oxygen 9. 38 10. 2 11. deamination 12. decarboxylation 13. substrate 14. pantothenic 15. chemoautotrophic 16. biochemical 17. glycolysis 18. Embden–Meyerhof 19. CoA 20. nucleotides

Label the Art

a. Pyruvic acid b. NADH c. CO_2 d. Acetyl CoA e. NADH f. CO_2 g. NADH h. CO_2 i. ATP j. $FADH_2$ k. NADH

Critical Thinking

1. Catabolic and anabolic reactions are referred to as coupled reactions because catabolic reactions furnish the energy necessary to drive anabolic reactions.

2. Competitive enzyme inhibitors bind to and fill the active site of an enzyme. They compete with the substrate for the active site of the enzyme. The inhibitor does not undergo any reaction to form a product. This binding may or may not be reversible.

   Noncompetitive inhibitors interact with some other part of the enzyme, a process that is referred to as allosteric inhibition. The binding of enzyme and inhibitor results in a change in the active site of the enzyme. This prevents binding of the substrate so the reaction cannot occur.

3. In cyclic photophosphorylation the electrons liberated from chlorophyll pass through the electron transport chain and eventually return to the chlorophyll.

   In noncyclic photophosphorylation, electrons released from chlorophyll pass through the electron transport chain to the electron acceptor, $NADP^+$. Electrons are replaced in chlorophyll from the splitting of water.

4. Provides a means for the breakdown of pentose sugars.

   Produces intermediates that are precursors in the synthesis of nucleic acids, some amino acids, glucose from CO_2 in photosynthesis.

   The process is an important producer of the coenzyme NADPH from $NADP^+$.

5. Glycolysis occurs in most cells. This process splits glucose into two three-carbon sugars. These sugars are oxidized; they release energy and their atoms are rearranged to form two molecules of pyruvic acid. Oxygen is not required.

   Pentose phosphate pathway produces important intermediate pentoses that act as precursors in the synthesis of nucleic acids, some amino acids, glucose from CO_2 in photosynthesis, and some amino acids. It also produces reduced NADPH.

Entner–Doudoroff pathway produces two molecules of NADPH and one molecule of ATP from each molecule of glucose. Bacteria that can use this pathway are able to metabolize glucose without glycolysis or the pentose phosphate pathway.

6. 1. An electron donor (organic or inorganic)

   2. A final electron acceptor

   3. A system of electron carriers

7.

Process	Electron acceptor
Aerobic respiration	O_2
Anaerobic respiration	NO_3^-, SO_4^{2-}, CO_3^{2-}
Fermentation	Some organic molecule

8.

Nutritional category	Energy source	Carbon source
Photoautotroph	Light	Carbon dioxide
Photoheterotroph	Light	Organic compounds
Chemoautotroph	Inorganic compounds	Carbon dioxide
Chemoheterotroph	Organic compounds	Organic compounds

9. Bacteriochlorophylls—The photosynthetic pigment found in green sulfur and purple sulfur bacteria.

   Intracytoplasmic membranes—Invaginations of the plasma membrane in purple sulfur bacteria.

   Chlorosomes—Plasma membrane folds in green sulfur bacteria containing bacteriochlorophylls.

10. Flavoproteins—Protein molecules containing flavin that can perform alternating oxidations and reductions.

    Cytochromes—Proteins with an iron-containing group that can exist alternately in a reduced or oxidized form. Several different cytochrome molecules found in electron transport chains.

    Ubiquinones—Small, nonprotein carrier molecules.

Chapter 6

Matching

I. 1.g 2.d 3.c 4.a 5.f 6.e
II. 1.e 2.a 3.b 4.f 5.d 6.c 7.g
III. 1.b 2.i 3.f 4.g 5.j
IV. 1.d 2.b 3.a
V. 1.d 2.a 3.d 4.c
VI. 1.b 2.a 3.f 4.d

Fill-in-the-Blanks

1. polysaccharide 2. *Cyanobacteria* 3. mesophiles 4. number 5. broth 6. deep-freezing
7. binary 8. absorbance (also optical density) 9. dry weight measurement 10. facultative
11. facultative 12. phosphate; amino acids 13. culture medium 14. 40°C 15. pour plate method
16. peptones 17. living host cells 18. reduction 19. optimum

Label the Art

I. 1. Cell elongates and DNA divides 2. Cell wall and plasma membrane begin to divide 3. Cross wall forms completely around divided DNA 4. Cells separate II. a. Lag phase b. Log, or exponential growth, phase c. Stationary phase d. Death, or logarithmic decline, phase

Critical Thinking

1. Temperature—Microorganisms vary in their preferred range of temperature for growth. They can be divided into three main groups: the psycrophiles (cold-loving), the mesophiles (moderate-temperature-loving), and the thermophiles (heat-loving). Within their temperature range, microorganisms have minimum, optimum, and maximum growth temperatures. Generally, low temperatures slow growth. Excessively high temperatures inactivate enzymatic systems of the cell.

   pH—Microorganisms also vary in their preference for pH. Although most bacteria prefer a neutral pH (7), some species can tolerate extremely acidic or alkaline environments.

   Osmotic pressure—High osmotic pressure caused by the addition of sugar, salt, or other solutes results in the osmotic loss of water from most bacterial cells. This process is referred to as plasmolysis. Halophilic bacteria can tolerate high concentrations of salt. Low osmotic pressure may result in the lysis of some bacterial cells when excessive water enters the cells and causes them to swell.

2. Carbon—Carbon is necessary for the synthesis of the organic components of the cell. Bacteria may obtain carbon from organic substances like carbohydrates and lipids or from inorganic sources such as carbon dioxide.

   Nitrogen, sulfur, and phosphorus—These and other elements are necessary for the synthesis of cellular materials such as DNA, RNA, protein, and ATP. Bacteria rely on a variety of methods for obtaining these elements.

   Trace elements such as zinc, iron, and molybdenum necessary for the function of various enzymes are usually obtained from contaminants in water.

   Oxygen—Oxygen may or may not be needed depending on the microorganisms' relationship to oxygen (aerobic, anaerobic, and so on).

   Organic growth factors—These are organic compounds that microorganisms need to synthesize vitamins. Others include amino acids, purines, and pyrimidines.

3. a. High osmotic pressure (sugar).
   b. High osmotic pressure (salt) and acidic pH.
   c. High osmotic pressure (salt) and low water content.
   d. Acidic pH.
   e. Heat and pressure.
   f. Acidic pH.

4. a. Molds.
   b. Molds or acidophilic bacteria.
   c. Extremely halophilic bacteria.
   d. Molds or acidophilic bacteria.
   e. Endospore-forming thermophilic bacteria.
   f. Molds or acidophilic bacteria.

5. Nitrogen-fixing microorganisms absorb gaseous nitrogen (N_2) and convert it into a form that plants can utilize. Some nitrogen-fixing bacteria live symbiotically in the roots of certain plants such as legumes. The fixed nitrogen is used by both the bacterium and the plant. The process increases soil fertility and benefits agriculture.

6.

Relation to oxygen	Where in the tube does growth occur?	Why?
Obligate aerobe	Near the top of the medium	Dissolved oxygen diffused through surface of the medium
Facultative anaerobe	Best near surface but throughout tube	They can survive without oxygen but grow better in the presence of oxygen
Obligate anaerobe	At the bottom	Oxygen is poisonous to these organisms
Aerotolerant anaerobe	Evenly thoughout tube	These organisms don't use oxygen
Microaerophile	In a narrow band of the medium	They will grow at the depth having the optimum oxygen concentration

7. Hydrogen gas generated when water mixed with the contents of the chemical packet reacts with the atmospheric oxygen trapped in the jar on the surface of the palladium catalyst to form more water. This process removes the oxygen, creating a suitable environment for the cultivation of anaerobic bacteria.

8. a. Blood agar would be used to diagnose strep throat. The causative agent of strep throat, *Streptococcus pyogenes*, lyses red blood cells, a process referred to as hemolysis. Hemolysis of blood cells in blood agar produces a clear area surrounding the colonies. This reaction, considered with the patient's symptoms, allows for diagnosis of strep throat.

   b. A complex medium such as nutrient agar (or broth) meets the requirements for growth for most heterotrophic bacteria and fungi.

   c. Mannitol salt agar (MSA) would be a good choice for the isolation of *Staphylococcus aureus*. MSA contains 7.5% NaCl, which discourages the growth of competing organisms. The pH indicator allows *S. aureus* to be differentiated from similar organisms based on the ability to ferment mannitol.

   d. Because soil bacteria may be present in small numbers, enrichment media would be appropriate. It could also act as a selective medium favoring the growth of the desired microbe to detectable levels while discouraging the growth of others.

9. See figure on next page.

   Lag phase—The period immediately following inoculation to fresh media in which little or no growth occurs. A time of intense metabolic activity as the cells gear up for reproduction.

   Log phase—The period of growth in which cellular reproduction is most active. Generation time is at a minimum.

   Stationary phase—The number of new cells being produced equals the number of cell deaths; the period of equilibrium.

   Death phase—The number of dead cells exceeds the number of living cells until only a small portion of the population exists or the population dies out completely.

   Exposure to antibiotics during stationary phase would cause the most adverse effects on the bacterial population. This is because antibiotics are most effective against growing cells.

10. Advantages are that the plate count method reflects the number of living cells in the sample and is easy to perform.

    A disadvantage is that it takes 24 or more hours for results to be available.

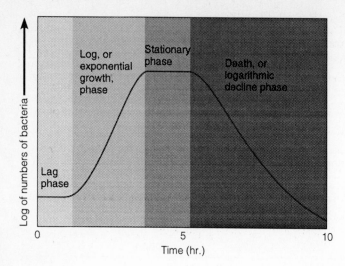

Chapter 7

Matching

I. 1.g 2.b 3.d 4.e 5.c 6.f

II. 1.c 2.b 3.e 4.d 5.f

III. 1.i 2.b 3.d 4.d 5.e 6.f 7.g 8.j 9.a 10.c 11.j

Fill-in-the-Blanks

1. nonionizing 2. singlet 3. X rays, gamma rays, high-energy electrons 4. tincture 5. iodophore 6. ammonia 7. 70% 8. formalin 9. ozone 10. glutaraldehyde 11. -stat 12. under pressure (as in autoclave) 13. autoclaves 14. DNA 15. porins 16. use-dilution 17. 260 18. fungi, such as molds and yeasts 19. disulfide 20. glutaraldehyde (Cidex®) 21. triclocarban

Critical Thinking

1. Heavy metals such as silver, copper, and mercury have value as antimicrobial agents. Metal ions affect bacterial proteins, resulting in denaturing of bacterial proteins.

   Silver nitrate ointment used to treat the eyes of newborns is an example of an application. The ointment helps prevent infection of the eyes by gonorrhea and chlamydia.

   Oligodynamic action refers to the ability of small amounts of a heavy metal compound to exert antimicrobial activity.

2. Many antimicrobial agents are effective against limited types of microbes. For example, many antiseptics and disinfectants are more effective against gram-positive than gram-negative bacteria. The reason for this is not clearly understood. Pseudomonads and mycobacteria are especially resistant to chemical agents. It is also useful to know the Gram designation of the pathogen when prescribing antibiotics. For example, penicillin is more effective against gram-positive bacteria.

3. a. Autoclaving at 121°C, 15 psi for 15 minutes will kill all organisms and their endospores.

   b. The milk should be sterilized by ultra-high-temperature (UHT) treatment.

   c. Vaccines are heat-sensitive and must be filter-sterilized.

   d. Most media can be safely autoclaved. Heat-sensitive media can be filter-sterilized.

4.

Chemical agent	Injures cell membrane	Denatures/inactivates protein	Dissolves lipid	Inactivates enzymes
Ethylene oxide		X		
Phenolics	X	X		X
Halogens	X	X		
Chlorhexidine	X	X		
Alcohols	X		X	

5. a. An oxidizing agent such as hydrogen peroxide would be a good choice. Oxidizing agents are especially effective against anaerobic bacteria.

   b. Chlorhexidine is useful for surgical scrubbing because it is bacteriocidal against both gram-positive and gram-negative organisms.

   c. Ethylene oxide would be appropriate because it is 100% effective and can penetrate the wrapping material covering the bandage.

   d. The addition of a compound such as methylparaben would inhibit mold growth.

6. Sterilization refers to the destruction or the removal of *all* microbial life, including endospores. There are many ways to achieve sterilization, including the use of heat, chemical agents, or filtration. Sanitation is the *reduction* of pathogens on inanimate objects (such as eating utensils) to "safe" levels. This may be achieved by mechanical cleaning or with chemical agents.

7. Phenol coefficient expresses the effectiveness of an antiseptic or disinfectant when compared to phenol under standard test conditions. A phenol coefficient less than 1 indicates that the test chemical is less effective than phenol. A phenol coefficient greater than 1 indicates the test chemical is more effective than phenol under the same conditions.

8. a. The use of high concentrations of salt or sugar creates a hypertonic environment that results in the osmotic loss of water from microbial cells. Advantages are that this is a simple way to preserve meat and fruit. Applications include jams and jellies. Disadvantages are that molds may grow on foods prepared this way and that it isn't a practical way to preserve many foods.

   b. Desiccation involves drying food (for example, meat and fruit). The lack of water retards the growth and reproduction of microbes. The advantage is that it is an easy way to preserve some foods. The disadvantages are that many microorganisms are able to survive desiccation for long periods of time and are revived upon the addition of moisture. Applications are beef jerky and sun-dried tomatoes.

   c. Refrigeration is a simple and relatively effective way to retard the spoilage of food. Although many bacteria can survive and even reproduce at refrigerator temperatures, the rate of chemical reactions is slowed.

   d. Filtration is the passage of gas or liquid through a screenlike material with pores small enough to retain microbes. There are many applications of filtration, such as sterilizing heat-sensitive materials. It is difficult to filter-sterilize viscous materials such as some media.

9. Nonionizing radiation (for example, UV light) damages the DNA of exposed cells and is used to control microbes in air and to sterilize vaccines, serums, and toxins. A serious disadvantage of nonionizing radiation is that because of its relatively low energy content, it penetrates poorly. Organisms protected by practically anything are not affected.

10. Germicide—A chemical agent that rapidly kills microbes but not necessarily their endospores.

    Bacteriostasis—A condition in which bacterial growth and multiplication are inhibited, but bacteria are not killed.

    Degerming—The removal of transient microbes from the skin by cleaning.

Chapter 8

Matching

I. 1.f 2.b 3.d 4.c 5.a 6.e 7.g
II. 1.c 2.d 3.a 4.b
III. 1.a 2.a 3.b 4.d
IV. 1.g 2.b 3.e 4.d
V. 1.b 2.a 3.a 4.a 5.a
VI. 1.a 2.b

Fill-in-the-Blanks

1. ultraviolet light 2. auxotroph 3. replica plating 4. gene 5. histidine 6. replication fork
7. polyribosome 8. degeneracy 9. genotype 10. competent 11. Hfr 12. gamete 13. meiosis 14.
constitutive 15. repressor 16. operator 17. resistance transfer factor 18. phage 19. thymine 20.
uracil 21. transposons 22. transposon 23. introns

Label the Art

a. Enzymes b. Parental double helix c. Proteins d. Parental DNA e. Leading strand f. DNA
polymerase g. Lagging strand h. RNA polymerase i. RNA primer j. DNA polymerase k. DNA
polymerase l. RNA primer m. DNA n. DNA ligase o. Lagging strand p. DNA polymerase
q. Replication fork r. Parental strand s. RNA polymerase t. RNA primer u. DNA polymerase
v. DNA polymerase w. DNA ligase x. left arrow

Critical Thinking

1. Hydrogen bonds between the nitrogenous bases hold the complementary strands of DNA together.
   Although individual hydrogen bonds are weak, the numerous hydrogen bonds in a DNA molecule make
   it a relatively weak structure.

2. The two complementary strands making up a DNA molecule have different chemical directions. This
   means that as a replication fork moves along a parental DNA molecule, the new strands are synthesized
   in opposite directions. The leading strand is synthesized continuously in the 5′ → 3′ direction. The
   lagging strand is produced in the opposite direction. It is synthesized in fragments that are joined
   together later.

3. When conditions are optimum—for example, in log phase—bacterial cells can initiate multiple
   replication forks on the origin of the chromosome. A new pair begins before the original pair finishes.

4. Transcription produces a strand of mRNA complementary to the DNA strand. This mRNA carries the
   coded information for making proteins to ribosomes, where proteins are synthesized. tRNA plays a role
   in the translation of mRNA into functional protein. tRNA molecules add amino acids to the growing
   polypeptide chain in the order specified by the sequence of nucleotides in the mRNA molecule.

5. Base substitutions, also called point mutations, are common mutations in which a single base of a DNA
   molecule is replaced with a different base. This may not cause a problem due to the redundancy in the
   genetic code, or it may result in missense or a nonsense mutation.

   Frameshift mutations involve the insertion or deletion of one or more nucleotide pairs and can shift the
   reading frame (the three-by-three groupings) of nucleotides. This usually results in a long stretch of
   missense and production of a nonfunctional protein. Translation is likely to be terminated when a
   nonsense codon is encountered.

6. A mechanism referred to as global regulation helps bacteria respond to environmental stresses by
   turning on or off large sets of genes. This allows for the production of proteins that help the cell survive
   elevated temperatures or depleted quantities of such essential substances as oxygen, nitrogen, carbon,
   and phosphorus.

7. *Genotype* refers to the genetic makeup of an organism. It represents the potential properties of an organism. *Phenotype* refers to the actual, expressed characteristics and is a manifestation of genotype.

8. Ionizing radiation, such as X rays and gamma rays, penetrate well and have high energy contents. They ionize molecules, making them very reactive. Some of the affected ions combine with bases in DNA, resulting in replication errors. Ionizing radiation may even cause breakage of the covalent bonds in the sugar-phosphate backbone of DNA, resulting in physical breaks in the chromosome. Bacteria do not possess a mechanism to repair this very serious damage.

   Nonionizing radiation such as UV light doesn't carry as much energy or penetrate as effectively. UV light will, however, cause thymine dimers to form in the DNA strand. If the dimers are not repaired, then proper transcription and replication of DNA can't occur.

   Bacteria (and other organisms) have photoreactivating enzymes that can repair damage caused by UV light by splitting the dimers. Occasionally this repair process will result in errors and is yet another source of mutation.

9. Jumping genes, or transposons, are small segments of DNA that can move from one region of a DNA molecule to another, to another DNA molecule, or to a plasmid. This is not a common occurrence. They may contain genes that allow bacteria to resist toxins or antibiotics. Because transposons can be carried between cells and even species via viruses or plasmids, they can play a significant role in evolution.

10. Some bacterial cells have a type of plasmid referred to as an R factor. R factors carry genes that provide resistance to one or more antibiotics and can be passed from cell to cell and even to other species through the process of conjugation. When antibiotics are used as a supplement in animal feeds, bacteria able to resist these drugs are selected by preferentially diluting the effectiveness of antibiotics.

Chapter 9

Matching

I. 1.a 2.d 3.b 4.c

Fill-in-the-Blanks

1. recombinant 2. complementary 3. base 4. plasmids 5. gene library 6. probes 7. Southern blotting 8. polymerase chain 9. Ti 10. protoplast

Label the Art

I. a. Restriction enzyme b. Recognition sites c. "Sticky" ends d. Another source e. Perhaps a plasmid f. Restriction enzyme g. Base-pairing h. A plasmid i. DNA ligase j. Recombinant DNA k. Recognition sites l. "Sticky" end II. a. Target DNA strands b. Nucleotides c. DNA polymerase d. Primers e. DNA polymerase f. Target DNA g. Target DNA sequence h. Primer

Critical Thinking

1. Police could use recombinant DNA techniques to amplify the DNA from the skin sample or from the white cells in the blood sample. This material would be compared to the suspect's DNA using the Southern blot method to do a DNA fingerprint. If the samples match, then the police have positive identification tying the suspect to the crime—that is, unless the suspect has a twin!

2. Restriction enzymes are a special class of DNA-cutting enzymes found in many bacteria. Genetic engineers use purified forms of restriction enzymes to make recombinant DNA. Restriction enzymes recognize and cut (or digest) only one particular sequence of nucleotide bases in DNA; many different restriction enzymes exist. Some restriction enzymes make staggered cuts in DNA that are referred to as "sticky ends." If two fragments of DNA from different sources are cut with the same restriction enzyme, they will have identical "sticky ends." They may be spliced and then rejoined with the enzyme DNA ligase.

3. Genetic engineering produces human hormones that are less expensive, less allergenic, available in large quantities, and pose no risk of transmitting disease.

4. a. Animal cells may be made "competent" to take up external DNA by soaking them in a solution of calcium chloride. After this treatment many of the animal cells will take up the DNA; recombination will occur in some of the cells. Direct introduction of foreign DNA into animal cells can be achieved by microinjection using a glass micropipette.

   b. Plant cells may have foreign DNA introduced by first enzymatically removing the cell wall to create protoplasts. Protoplasts are then fused by adding polyethylene glycol to form a hybrid cell. The DNA in the hybrid cell may then undergo recombination naturally. Another method utilizes a "gene gun" that shoots DNA-coated tungsten or gold particles through the cell wall. Some cells will incorporate and express the new DNA.

   c. Yeast cells may also be made competent by soaking them in a calcium chloride solution or through the process of electroporation. This involves using an electrical current to form microscopic pores in the cell membrane. DNA is able to enter through these pores.

5. cDNA, or complementary DNA, is an artificial gene made from an mRNA template of eucaryotic DNA and that contains only exons. This is significant because bacterial cells lack the ability to process and remove introns from eucaryotic DNA.

6. R plasmids having antibiotic-resistance genes against ampicillin and tetracycline are used as vectors to introduce new genes into bacterial cells. The restriction site used to insert the new genes into the plasmid is located in the middle of the tetracycline gene. This means that cells that receive the new DNA will be resistant to ampicillin but not to tetracycline. After treating bacterial cells with R plasmid vectors, the cells can be identified by plating them on media containing nutrients and either ampicillin or tetracycline. Transformed cells will grow on the ampicillin but not on the tetracycline.

7. A gram-positive bacterium would be the best choice for two reasons. First, gram-positive bacteria lack a cell-wall component called endotoxin that is found in gram-negative bacteria. Endotoxin causes fever and shock in animals and would pose a serious problem if present in gene products. Second, gram-negative cells such as *E. coli* don't usually secrete protein products, and the process used to harvest them is too expensive for industrial applications.

8. Subunit vaccines consist of a protein portion of a pathogen produced by a genetically engineered organism—for example, a yeast. Because the protein is made by a yeast rather than a treated pathogen or a portion of the actual pathogen, there is no chance that the disease will be transmitted.

9. a. Gene therapy may help some people with cells having mutant or defective genes. Some of these cells are removed; the gene in question is "snipped out" with restriction enzymes and is replaced with the normal gene. The transformed cells are replaced in the body and should function normally.

   b. DNA fingerprinting involves the Southern blotting method and is useful in forensic medicine to compare a sample of hair, blood, or semen found at the crime scene with a sample from a suspect.

   c. Genetic engineering may also be used to develop new strains of agricultural crops. Using the Ti plasmid from the bacterium *Agrobacterium tumefaciens* as a vector to insert new genes into many plants, herbicide- and pesticide-resistant plants have been developed.

10. One of the concerns about genetic engineering is that the engineered microbes could become human pathogens. For this reason, some organisms that are intended for use in the environment contain inserted "suicide genes." These genes eventually turn on to produce a toxin that will kill the microbes, effectively limiting their survival after accomplishing their task.

Chapter 10

Matching

I. 1.a 2.c 3.d 4.e 5.e 6.a

II. 1.a 2.d 3.b 4.b 5.e

III. 1.f 2.b 3.a 4.c 5.e 6.g

Fill-in-the-Blanks

1. taxonomy 2. binomial nomenclature 3. taxa 4. *Bergey's* 5. phage 6. phylogenetic 7. antisera
8. G + C ratio 9. genera 10. peptidoglycan

Label the Art

1. a. Methanogens b. Halophiles c. Thermoacidophiles 2. a. Actinomycetes b. Lactobacillus
c. Bacillaceae d. Micrococcaceae e. Other gram-positive bacteria 3. a. Cyanobacteria
b. Spirochetes c. Green sulfur bacteria d. Pseudomonadaceae e. Enterobacteriaceae f. Other
gram-negative bacteria

Critical Thinking

1. a. Taxonomy classifies organisms, thereby establishing relationships between them.

   b. Taxonomy serves as a reference for identification of known organisms.

   c. Taxonomy (and nomenclature) serve as a universal language, avoiding the confusion associated with the use of common names.

2. The endosymbiont theory attempts to explain the evolution of eucaryotic cells. The theory suggests that eucaryotic cells evolved from procaryotic cells living inside one another as endosymbionts. Evidence to support the theory includes the following:

   a. Similarities between procaryotes and eucaryotic organelles such as mitochondria and chloroplasts. All have 70S ribosomes.

   b. Procaryotic cells, mitochondria, and chloroplasts all multiply by binary fission.

   c. Fossil evidence shows that procaryotes evolved first, approximately 3.5 billion years ago, whereas eucaryotic cells evolved only 1.4 billion years ago.

   d. Modern day examples of endosymbionts exist—for example, the protozoa *Myxotrichia paradoxa*.

3. Kingdom Monera—Extremely diverse nutritional abilities; heterotrophic or autotrophic.

   Kingdom Fungi—Absorb nutrients through the cuticle; saprophytic or parasitic heterotrophs.

   Kingdom Protista—Primarily heterotrophs.

   Kingdom Plantae—Photosynthetic (autotrophic) nutrition.

   Kingdom Animalia—Heterotrophic; ingestion through some kind of mouth.

4. a. Archaeobacteria lack peptidoglycan in their cell walls, a substance that is found in the cell wall of most eubacteria.

   b. Archaeobacteria can live in more extreme environments than most eubacteria can tolerate.

   c. Even more metabolic diversity is seen within the archaeobacteria than in the eubacteria.

5. Eucaryotic species refers to a group of closely related organisms that are capable of interbreeding. A bacterial species is a population of cells with similar characteristics indistinguishable from each other. Strains are groups of cells derived from a single cell.

6. *Bergey's Manual of Determinative Bacteriology* organizes bacteria for identification purposes using criteria such as cell-wall type, morphology, staining reactions, oxygen requirements, and biochemical testing. *Bergey's Manual of Systematic Bacteriology* divides procaryotic organisms into four volumes reflecting evolutionary (systematic) relationships.

7. All three are serological methods used to identify bacteria. In the slide agglutination test, samples of an unknown bacterium are placed in a drop of saline on each of several slides. A different known antiserum is added to each sample. Bacteria will clump (agglutinate) when mixed with antibodies specific for that strain. ELISAs are read by a computer scanner. Known antibodies are placed in wells of a microplate along with the unknown bacterium. Reaction between the bacterium and an antibody provide identification. The Western blot identifies bacterial antigens in a patient's blood using electrophoresis.

8. a. Forensics

   b. Epidemiology—tracing the source of infection

   c. To determine relatedness of bacteria by comparison of restriction fragments

9. First, DNA fragments of the bacteria that you wish to identify are produced with restriction enzymes. These fragments should react with the bacteria of interest but not with closely related organisms. After cloning sufficient numbers of the fragments, they are radioactively or fluorescently tagged. These probes are mixed with single-stranded DNA from the sample to be tested. If the organism in question is present, the DNA probes will hybridize with its DNA and can be detected by the radioactivity or fluorescence of the probes.

10. Viruses are fundamentally different from the members of the five kingdoms in that they are not composed of cells. This means that viruses are neither procaryotic or eucaryotic. One hypothesis concerning the evolution of viruses suggests that they arose from independently replicating nucleic acids, such as plasmids. The other hypothesis says that they evolved from degenerative cells that eventually lost their ability to survive independently.

Chapter 11

Matching

I.   1.a 2.f 3.b 4.d 5.e 6.a
II. 1.h 2.f 3.d 4.c 5.g 6.e 7.b 8.f
III. 1.b 2.g 3.c 4.e 5.f 6.a 7.h
IV. 1.b 2.e 3.f 4.d 5.a 6.g 7.i 8.h
V.   1.e 2.c 3.a 4.b 5.d

Fill-in-the-Blanks

1. flagella 2. serovars (serotypes) 3. red 4. *Erwinia* 5. heterocysts 6. *Fusobacterium*
7. *Staphylococcus* 8. Alpha- 9. Beta- 10. *Clostridium* 11. *Nocardia* 12. reticulate body
13. geosmin 14. *Coxiella burnetii* 15. *Streptomyces* 16. *Caulobacter* 17. *Leptospira*
18. *Pseudomonas* 19. *Streptococcus* 20. vibrios

Critical Thinking

1. Because some ulcers are caused by the bacterium *Helicobacter pylori*.

2. The bacterium was *Shigella*. The cantaloupes may have been contaminated by pickers in the field. Very often running water for hand washing is not available for the pickers.

3. *Vibrio parahaemolyticus*, an organism that may contaminate shellfish.

4. The types of criteria used to classify plants and animals are not apparent in bacteria. Classifying bacteria using a taxonomic approach is more difficult and can result in confusion. A few bacteria are actually included in two sections of *Bergey's Manual*. For example, *Gardnerella vaginalis* is included in two sections of *Bergey's*: the facultatively anaerobic gram-negative rods because its thin wall stains gram-negative, and the irregular, nonspore-forming gram-positive rods because its cell wall is structurally similar to gram-positive cell walls.

5. a. *Hemophilus*.

   b. *Legionella*.

   c. *Mycoplasma*.

6. Lactobacilli are aerotolerant bacteria that lack a cytochrome system. They are unable to use oxygen as a final electron acceptor; they produce lactic acid from simple carbohydrates. The acidic by-products of their metabolism create an environment that inhibits competing bacteria, giving *Lactobacillus* an advantage.

7. Chemoautotrophic bacteria are significant in agriculture because they are able to use inorganic chemicals as energy sources. For example, nitrifying bacteria such as *Nitrobacter* and *Nitrosomonas* use reduced nitrogenous compounds, including ammonium or nitrite ions, for energy. These compounds are converted to nitrate, which is used by plants for a nitrogen source.

8. a. An extreme halophile such as *Halobacterium* or *Halococcus*.

   b. *Rhizobium*.

   c. *Desulfovibrio*.

9. a. Members of both of these groups are gram-negative. The spirochetes are coil-shaped and actively motile by means of axial filaments. The spirochetes include several human pathogens. Some, like *Borrelia*, are transmitted by ticks or lice. The facultatively anaerobic gram-negative rods are a large group displaying much diversity. Some members are motile by means of flagella, whereas others are nonmotile. Many significant pathogens are included in this group.

   b. Rickettsias and chlamydias are included in the same group; both are tiny, obligate intracellular parasites. Rickettsias are rod-shaped or are coccobacilli. There are several pathogenic rickettsias, many of which are transmitted by insects and ticks. There are only three species of chlamydias; all are coccoid and pathogenic. These organisms are transmitted by contact or by airborne respiratory routes. Both rickettsias and chlamydias are cultivated in laboratory animals, in cell cultures, or in the yolk sacs of chicken embryos.

   c. The mycoplasmas and the mycobacteria have few similarities beyond their similar names and their tendencies to produce filaments that resemble fungi. Mycoplasmas are small bacteria that lack cell walls and are highly pleomorphic. They will grow with difficulty on artificial media with added sterols but are usually grown by cell culture. Mycobacteria are rod-shaped organisms with cell walls that stain acid-fast.

10. *Bergey's Manual of Systematic Bacteriology* uses a systematic or taxonomic approach to organize bacteria. Organisms are separated into divisions on the basis of cell-wall characteristics. They are divided into sections according to gram stain reaction, cell morphology and arrangement, oxygen requirements, motility, and nutritional and metabolic properties.

Chapter 12

Matching

I. 1.g 2.a 3.b 4.d 5.h 6.j 7.i
II. 1.c 2.b 3.c 4.c 5.c 6.a
III. 1.i 2.f 3.e 4.a 5.b 6.d
IV. 1.d 2.d 3.a 4.d
V. 1.f 2.c 3.d 4.h 5.b 6.e 7.a 8.g

Fill-in-the-Blanks

1. merozoites; sporozoites 2. Zygomycota 3. mycosis 4. Basidiomycota 5. Dermatophytes 6. dimorphic 7. candidiasis 8. Deuteromycota 9. thallus 10. six 11. cercaria 12. definitive; intermediate 13. hermaphroditic 14. trophozoite 15. hyphae; mycelium 16. yeasts 17. coenocytic 18. ascus 19. subcutaneous 20. thrush 21. vacuoles 22. schizogony 23. conjugation 24. cercaria 25. stipes 26. mutualistic 27. bladders

Label the Art

1. a. Food vacuole b. Pseudopods c. Nucleus 2. a. 1st–3rd flagella b. 4th flagellum c. Nucleus d. Oral groove e. Food vacuoles 3. a. Nuclei 4. a. Cyst wall 5. a. Pellicle b. Cytostome c. Cytopharynx d. Food vacuoles e. Cilia f. Contractile vacuole g. Macronucleus h. Micronucleus i. Anal pore

Critical Thinking

1. Fungi are classified into phyla according to the type of sexual spore they produce. The reason for using sexual spore type is that each fungus can produce only one type of sexual spore but may be able to form more than one type of asexual spore.

Phylum	Asexual spores	Representative genus
Deuteromycota	Conidiospores, chlamydospores, arthrospores	*Epidermophyton*
Ascomycota	Conidiospores	*Aspergillus*
Zygomycota	Sporangiospores	*Rhizopus*
Basidiomycota	Conidiospores	*Coprinus*

2. *Pneumocystis* is difficult to place taxonomically due to its lack of identifiable structures. Until recently *Pneumocystis* was classified as a protozoa, but comparison of its rRNA to other forms of life suggests that it may be a fungus.

3. Mycoses are classified according to the level of tissue infected and by transmission.

Category of mycosis	How transmitted	Level of tissue
Systemic	Inhalation of spores	Deep within body
Subcutaneous	Implantation of spores or hyphae	Beneath skin
Cutaneous	Contact	Hair, skin, nails
Opportunistic	Opportunistic organism, compromised host	Varies
Superficial	Contact or fomite	Hair shafts, superficial epidermal cells

4. a. Algin extracted from the cell walls of algae is used to thicken foods such as ice cream and cake decorations.

   b. Algin is also used in nonfood products such as rubber tires and hand lotions.

   c. Agar is extracted from a red algae and is used to solidify microbiological media.

   d. Diatoms are used as diatomaceous earth.

5. Lichens are a combination of green algae (or cyanobacteria) and a fungus and exist in a mutualistic relationship. The algae or cyanobacteria are photosynthetic and provide nutrients to the relationship. The fungal member provides attachment and protection from desiccation in the harsh, rocky environment in which they live.

6. Some protozoa are able to form a protective cyst under adverse conditions. These conditions include a lack of food, moisture, or oxygen, or when toxic substances are present. The cyst serves to allow the organism to survive outside of the host and be transmitted to a suitable new host.

7. The protozoa is *Entamoeba histolytica*. It is transmitted by ingestion of the cyst in contaminated food or water.

8. An *Anopheles* mosquito carrying sporozoites bites a human. The sporozoites are carried to the liver by the blood. Schizogony occurs in liver cells and produces thousands of merozoites. Merozoites enter the blood stream and infect RBCs. Young, ring-shaped trophozoites develop. The RBCs eventually rupture, releasing more merozoites. This process causes chills and fever in the patient. Some merozoites develop into male and female gametocytes that can be picked up by another *Anopheles* mosquito. Within the mosquito the gametocytes unite to form a zygote. The zygote forms an oocyst from which sporozoites are formed.

9. a. Advise infected people not to defecate in the water.

   b. Eliminate the snail host.

   c. Cook crayfish before eating them.

10. a. *Enterobius vermicularis* (the pinworm) and *Ascaris lumbricoides*.

    b. *Necator americanus* and *Trichinella spiralis*.

    c. *Echinococcus granulosis*.

Chapter 13

Matching

I. 1.a 2.c 3.b 4.e 5.d
II. 1.g 2.h 3.c 4.d 5.e 6.a 7.l 8.i 9.j
III. 1.c 2.g 3.d 4.e 5.b
IV. 1.a 2.e 3.b 4.c

Fill-in-the-Blanks

1. uncoating 2. temperate 3. leukemias 4. embryonated 5. latent 6. plaque-forming
7. translocation 8. host range 9. tumor 10. transform 11. retrovirus 12. transplantation
13. eclipse 14. tail 15. lysozyme 16. prophage 17. lysogeny 18. contact inhibition 19. D
20. R 21. benign

Label the Art

a. Attachment b. Penetration c. Biosynthesis d. Maturation e. Release f. Capsid g. Tail sheath
h. Tail fiber i. Baseplate j. Pin k. Cell wall l. Plasma membrane m. Tail n. Tail core o. Tail
p. DNA q. Capsid r. Tail fibers s. Viron

Critical Thinking

1. The problem results from the fact that viruses take over the reproductive machinery of host cells to replicate. This means that drugs that inhibit viral replication will also affect reproduction of the host's cells.

2. When viruses infect a host, the host's immune system reacts by producing specific antibodies that act against that virus. Some viruses are able to escape antibodies because proteins on their surface or on their spikes mutate. This means that the antibodies that were originally formed will no longer react with the virus, making them ineffective.

3. Viral classification is still in its infancy. The International Committee of Taxonomy of Viruses has not yet established higher taxa (order through kingdom). Families have been established, and these names end in -viridae; genus names end in -virus. Viruses are grouped by nucleic acid type, morphology, the presence or absence of an envelope, and reproductive strategy.

   Viral species are defined as a group of viruses sharing the same genetic information and ecological niche.

4. AIDS research has been slowed down by the lack of animals to use for experimentation of drugs against the virus. Simian and feline AIDS are caused by a closely related virus, and the diseases they cause develop in a few months. This provides researchers with a model to study viral growth in various tissues.

5. The final stage of the lytic cycle involves release of the virions from the host cell. This is accomplished when lysozyme is synthesized within the cell. This enzyme breaks down the cell wall, resulting in lysis and release of the virions. In the lysogenic cycle the phage remains latent, incorporating its nucleic acid into that of the host. The lytic cycle may be induced by some spontaneous event such as exposure to UV light. Lysogeny also results in the following:

   a. Lysogenic cells are immune to reinfection by the same phage.

   b. The infected host cell may have new properties.

   c. Lysogeny makes specialized transduction possible.

6. In latent viral infections the virus may remain in equilibrium with the host and not produce the disease for a long period, sometimes years. Some individuals have subclinical infections. As with the chickenpox virus, some viruses may remain latent in host cells for years. Examples of latent viral infections are chickenpox and cold sores.

   Slow viral infections are conditions that progress gradually, in some cases over a period of years, and are believed to be caused by a virus. These infections are usually fatal. An example of a slow viral infection is subacute sclerosing panencephalitis.

7. Retroviruses induce tumors because some of them contain promoters that turn on oncogenes; others actually contain oncogenes. Also the fact that the double-stranded DNA of these viruses (produced by reverse transcription of the viral RNA) is incorporated into the DNA of the host and introduces new material to the host's genome can in itself cause problems.

8. Evidence to support the hypothesis includes the fact that CJD runs in families (although it is also transmitted by infected nerve tissue). A major portion of the prion has been identified on chromosome 20 and is designated PrP. An abnormal form of the gene is found in the brains of animals with scrapie.

9. A method referred to as PCR was used to amplify RNA from autopsy specimens and eventually helped researchers to identify Hantavirus as the cause of the mysterious deaths.

10. a. Attachment (absorption) of the virus to a complementary receptor on the bacterial cell.

    b. Penetration occurs when the phage injects its nucleic acid into the bacterium. The capsid remains outside of the bacterium.

    c. Biosynthesis of viral nucleic acid and proteins occurs in the cytoplasm of the host cell.

    d. Maturation of the virus involves assembly of the viral components.

    e. Release is accomplished when lysozyme is produced, resulting in lysis of the host cell and release of the newly formed virions.

Chapter 14

Matching

I.   1.a 2.d 3.c 4.b
II. 1.e 2.a 3.c 4.d
III. 1.a 2.b 3.c
IV. 1.a 2.c
V.   1.g 2.c 3.d 4.e 5.f 6.a

Fill-in-the-Blanks

1. nosocomial 2. carriers 3. mechanical 4. zoonoses 5. incidence 6. acute 7. sporadic
8. bacteremia 9. secondary 10. symptoms 11. epidemiology 12. Centers for Disease Control and
Prevention 13. local 14. systemic 15. subclinical 16. prevalence 17. subacute 18. pandemic
19. Isolate the pathogen from the animal and show that it is the same as the original organism.
20. pathology

Label the Art

a. Eyes b. Nose and throat c. Mouth d. Skin e. Large intestine f. Urinary and genital systems

Critical Thinking

1. The relationship between normal microbiota and the host may be commensalism or mutualism.
   Examples of contributions made by normal microbiota include the synthesis of K and B vitamins by
   E. coli, and microbial antagonism in a healthy host.

2. In most pregnancies the fetus will remain germ-free until birth. The first microbiota that the infant will
   encounter will be *Lactobacilli* from the mother's vagina. They will become the predominant organisms in
   the newborn's intestines until the introduction of *E. coli* and other bacteria. Various other microbiota will
   become established as the infant comes in contact with its environment.

3. Microbial antagonism refers to the competition that exists among a host's normal microbiota for space
   and nutrients. This process protects the host from colonization with potentially pathogenic organisms.
   Four examples of microbial antagonism include the following:

   —*Lactobacilli* create an acidic environment that discourages the growth of *Candida albicans*, a common
   cause of vaginitis.

   —Streptococci living in the mouth prevent the growth of other gram-positive organisms.

   —Bacteriocins produced by *E. coli* inhibit the growth of *Shigella* and *Salmonella*.

   —Normal microbiota of the intestines inhibit growth of the pathogen *Clostridium difficile*.

4. —Urinary catheters.

   —Intravenous catheters.

   —Respiratory therapy equipment.

   —Needles.

   —Surgical dressings.

5. The infection control nurse (or epidemiologist) identifies problems and possible sources of infection in
   the hospital. The infection control nurse does so by looking for antibiotic-resistant strains of bacteria and
   by inspecting equipment, the cafeteria, the facility, and so on.

6. a. The period of incubation occurs between the initial infection and the first appearance of any signs and
      symptoms.

   b. The prodromal period is the brief period following incubation, characterized by early, mild symptoms
      of the disease.

   c. The period of illness is that period in which the disease is most acute.

d. During the period of decline the signs and symptoms subside.

e. The period of convalescence is when the patient regains strength and health.

7. Synergism means that the effect of two microbes acting together is greater than the effect of either acting alone. This definition is demonstrated by the cooperation that exists between oral streptococci and the pathogens that cause periodontal disease and gingivitis, and that between *Mycoplasma* and HIV.

8. Koch's postulates are most easily applied to microbes that can be grown on artificial media. To help overcome this problem, alternate methods of culturing and detecting microbes have been developed. Detection of these organisms might involve culturing in a guinea pig and in the yolk sacs of chick embryos.

9. Emerging infectious diseases are new or changing diseases such as AIDS and tuberculosis. Several factors contribute to the development of emerging infectious diseases. Factors contributing to their development include:

a. Pathogens moving from one host to another due to changes in the pathogen or in host susceptibility.

b. The increasing human population, producing changing disease patterns.

c. Ecological changes affecting both the host and the pathogen.

d. Air travel, which contributes to the rapid distribution of disease.

10. Both mechanical and biological transmission of disease involve arthropods. Mechanical transmission refers to the passive transport of the pathogen by an insect's feet or another body part. This most often involves contamination of food. In biological transmission the insect bites an infected person and in turn becomes infected. The pathogen reproduces in the insect and is transmitted to an uninfected person by another insect bite.

Chapter 15

Matching

I. 1.e 2.a 3.f 4.c 5.c 6.k 7.g 8.j
II. 1.a 2.a 3.b 4.b 5.b
III. 1.d 2.c 3.b
IV. 1.d 2.b 3.e

Fill-in-the-Blanks

1. parenteral 2. animals 3. phagocytosis 4. collagenase 5. hyaluronic 6. endo 7. cytocidal
8. lysosomes 9. contact inhibition 10. syphilis 11. M 12. hypothermic 13. keratins 14. portal of entry 15. ID_{50} (infectious dose) 16. sclerotia 17. interleukin-1 18. interferon

Label the Art

A. 1. Mucous membranes a. Respiratory tract b. Gastrointestinal tract c. Genitourinary tract
d. conjunctiva 2. Skin 3. Parenteral route **B.** 1. Capsules 2. Cell wall components 3. Enzymes
C. 1. Direct damage 2. Toxins a. Exotoxins b. Endotoxins 3. Hypersensitivity

Critical Thinking

1. Adhesins are surface molecules found on microbial pathogens that bind to specific receptor molecules on the appropriate host cells. Because adhesins bind only to complementary receptors, the infection process is very specific.

Examples of adhesins include the following:

Neisseria gonorrhoeae—adhesin containing fimbriae

E. coli—adhesin containing fimbriae

Streptococcus mutans—glycocalyx

Actinomyces—fimbriae that attach to the glycocalyx of *S. mutans*

2. The waxy cell wall of *M. tuberculosis* helps it avoid phagocytosis by cells of the immune system, allowing the microorganism to reproduce and become established in the host.

3. a. *Streptococcus pyogenes.*

   b. The respiratory system.

   c. Airborne transmission of cough-produced droplets or vehicle transmission, such as transport of the streptococci on dust particles.

4. Tetanus toxin is produced by *Clostridium tetani*. It is a neurotoxin that binds to nerve cells that control the contractions of some skeletal muscles. Tetanus toxin blocks relaxation of these muscles, resulting in spasmodic contractions in the patient. The condition is referred to as "lockjaw."

   C. botulinum produces botulinum toxin, a neurotoxin that prevents transmission of nerve impulses to muscle cells. This results in a symptom referred to as "flaccid paralysis." The patient may require respiratory and/or cardiac assistance. The botulinum toxin is produced only by phage-infected *C. botulinum* cells.

5. Hemolysins and coagulases are exoenzymes, produced by some gram-positive bacteria, that are associated with virulence. Hemolysins attack red blood cells and, in the case of streptolysins, white blood cells too. Coagulases coagulate blood by converting fibrinogen into fibrin, serving to protect the bacteria from host defensive mechanisms.

6. a. Some species have capsules that help them to avoid phagocytosis by host cells.

   b. M protein in the cell wall of *Streptococcus pyogenes* helps it avoid phagocytosis.

   c. Leukocidins are enzymes that destroy neutrophils and macrophages.

   d. Streptolysins are hemolysins produced by streps that lyse red and white blood cells.

   e. Kinases, like streptokinase, break down fibrin and dissolve clots formed by the body to isolate the infection.

   f. Some species have resistance plasmids, enabling them to resist the effects of some antibiotics.

   g. Hyaluronidase is an enzyme that helps streps spread from the initial site of infection.

7. a. Gram-negative.

   b. Endotoxin from the cell wall of lysed cells caused the symptoms. The endotoxin was released as cells were killed by the antibiotics.

   c. The release of prostaglandins by the hypothalamus in response to endotoxin resulted in the fever. Aspirin inhibits synthesis of prostaglandins and diminishes the fever.

   d. By the second day of antibiotic therapy the endotoxin had broken down.

8. Septic shock seen in some gram-negative infections is due to endotoxins produced by the pathogen. The endotoxin causes macrophages to secrete a substance called tumor necrosis factor (TNF) or cachectin. TNF binds to and alters the metabolism of certain tissues in the body. For example, TNF damages and increases permeability of capillaries, causing the loss of large amounts of fluids. Blood pressure drops and results in shock and other serious effects.

9. When infection with HIV occurs, the body responds by producing antibodies specific for receptors on the surface of the virus. HIV is able to avoid the effects of these antibodies because of the architecture of its surface. The receptors are located in the valleys or folds that cover the surface of the virus. The antibodies formed against HIV are too large to make contact with the receptor sites. These folds are also the correct shape for the long, slender CD4 molecules found on the surface of the host cells, so infection can take place.

10. Both ergot and aflatoxin are toxins produced by fungi. Ergot is a product of *Claviceps purpurea*, whereas aflatoxin is made by *Aspergillus flavus*. Ingestion of ergot causes hallucinations and constricts capillaries, resulting in gangrene of the extremities. Aflatoxin may be altered into a carcinogenic compound by the body after ingestion.

Chapter 16

Matching

I. 1.e 2.b 3.f 4.c 5.a
II. 1.b 2.c 3.e 4.f 5.g 6.h 7.d
III. 1.b 2.c 3.a
IV. 1.f 2.a 3.b 4.g

Fill-in-the-Blanks

1. cilia 2. sweat 3. neutrophils 4. neutrophils 5. cascade 6. differential white blood cell count
7. transmembrane 8. mononuclear phagocytic (reticuloendothelial) 9. margination 10. hypothalamus
11. complement 12. basophils 13. granules 14. adherence 15. opsonization 16. inflammation
17. diapedesis 18. kinins, prostaglandins, or leukotrienes 19. stroma 20. alternate 21. eosin
22. eosinophils 23. histamine 24. transferrin

Label the Art

a. Tonsil b. Right lymphatic duct c. Right subclavian vein d. Thymus gland e. Lymphatic vessel
f. Bone marrow g. Lymphatic vessel h. Left subclavian vein i. Thoracic duct j. Spleen k. Small
intestine l. Peyer's patch m. Lymph node

Critical Thinking

1. —Mucus membranes lining the respiratory tract help prevent the penetration of pathogens.

   —Mucus-coated hairs of the nose trap microbes, dust, and so on.

   —The ciliary escalator helps remove microbes from the lower respiratory tract.

   —The epiglottis covers the larynx during swallowing, preventing microbes from entering the lower respiratory tract.

2. Sebum is an oily substance, produced by the sebaceous glands of the skin, that coats hair and skin. Unsaturated fatty acids are a component of sebum, and along with lactic acid create a low-pH environment that inhibits the growth of many bacteria and fungi. Some species of normal microbiota are resistant to sebum and even metabolize it, causing an inflammatory response that leads to acne.

3. Bacteria such as *Clostridium botulinum* and *Staphylococcus aureus* don't actually survive stomach acids. Instead, they produce acid-resistant toxins that produce the symptoms of the diseases that they cause. Other microorganisms are protected from the effects of acid by food particles. Finally, the bacterium *Helicobacter pylori* is able to neutralize the acid and grow in the stomach lining.

4. Phagocytosis refers to the ingestion of microorganisms or particulate matter by a cell. Phagocytosis is a means by which the body counters infection.

   —Lipids in the cell wall of *Mycobacterium* help it to avoid phagocytosis.

   —Capsules help bacterial cells avoid phagocytosis.

   —Toxins produced by staphylococci and *Actinobacillus* kill phagocytes.

   —Some cells evade the immune system by entering and reproducing inside phagocytes.

   —Some microorganisms prevent fusion of the phagosome with a lysosome, preventing digestion.

   —M proteins of streps inhibit attachment of phagocytes.

5. Wandering macrophages and fixed macrophages make up the mononuclear phagocytic system. Wandering macrophages develop from monocytes that migrate to the infected area. They leave the blood and migrate to the infected area, where they scavenge and phagocytize bacteria and debris. Fixed macrophages are located in certain tissues such as the liver, lungs, and nervous system. Fixed macrophages remove microorganisms from blood or lymph as these substances pass through the organs.

6. Opsonization refers to the coating of microorganisms with certain plasma proteins to promote attachment and phagocytosis.

7. Both histamines and kinins increase permeability of blood vessels and cause vasodilation. Histamine is found in several kinds of cells, including mast cells, basophils, and blood platelets, and is released when these cells are damaged. Histamine is also released in response to stimulation by certain components of the complement system. Kinins are chemicals present in blood plasma that attract granulocytes to the infected site when activated.

8. Diapedesis (cell walking) is the amoeboid movement of phagocytes that allows them to reach the damaged area. Phagocytes are then able to eliminate invading microorganisms. After the majority of microorganisms and damaged tissue has been removed from the infected site, phagocytes die. The dead cells and body fluids are referred to as pus. The pus pushes to the surface of the body or into an internal cavity for dispersal. Alternately, pus may be absorbed by the body.

9. Stroma is supporting connective tissue—for example, the capsule that surrounds the liver. Parenchyma is the functioning part of the tissue—for example, the liver cells. Both types of cells may be involved in tissue repair. If only parenchyma cells are active in repair, then perfect or near-perfect repair of the tissue is likely. If stroma repair cells are active, then it is likely that scar tissue will form.

10. When C3 is cleaved into two fragments, C3a and C3b, the fragments can induce cytolysis, inflammation, and opsonization. C3a initiates a sequence of reactions involving C5, C6, C7, C8, and C9 (known as the membrane attack complex) invading the cell membrane of the attacking pathogen. C3a can also lead to inflammation. C3b acts as an opsonin by coating microorganisms and promoting phagocytosis.

Chapter 17

Matching

I. 1.a 2.b 3.b 4.b 5.a 6.b
II. 1.a 2.c 3.c 4.b 5.d
III. 1.a 2.b 3.c 4.d 5.e
IV. 1.e 2.b 3.c 4.d 5.a
V. 1.b 2.c 3.e 4.a 5.d
VI. 1.d 2.b 3.c

Fill-in-the-Blanks

1. innate 2. antigenic determinant 3. Fc 4. J-chain 5. titer 6. colostrum 7. secretory 8. serum
9. natural killer 10. interleukin-1 11. CD4 12. haptens 13. plasma cell 14. T-independent
15. secondary, or anamnestic 16. chimeric 17. serology 18. affinity 19. immunotoxin 20. perforin

Label the Art

I. a. B cells b. Antigens c. Memory cells d. Plasma cells e. Antibodies f. Clone of B cells
g. B cells h. Antigen i. B cell j. Antigen k. Proliferating l. B cells m. Memory cells n. B cells
o. Plasma cells p. Plasma cells q. Antibodies II. a. Antigen–antibody complexes b. Neutralization
c. Agglutination d. Precipitation e. Complement f. Phagocytosis g. Inflammation h. Cell lysis

Critical Thinking

1. Innate resistance refers to the resistance that an individual has to diseases that affect other species and other individuals of the same species. Natural selection contributes to variation within the same species. Acquired immunity refers to immunity that an individual develops against certain types of microbes or foreign substances. The process may be natural or artificial, or active or passive.

2. The infant should definitely be immunized. Although antibodies against diphtheria were passed to the fetus through a process known as transplacental transfer, they are only temporarily effective. These maternal antibodies provide immunity until his or her immune system matures.

3. Haptens are small compounds that attach to foreign substances too small to elicit an immune response on their own. This complex will cause antibodies to be formed. Once they have been formed, the antibodies will react with the hapten independent of the carrier molecule. Some people will develop an allergic response to the hapten.

4. The hinge region is a flexible area in the antibody molecule that allows for changes in conformation to fit different-sized antigens.

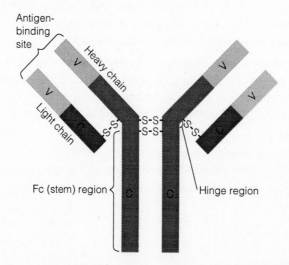

5. Within the same class, antibody molecules differ in the amino-acid sequence of their variable portions. The variable portion of one light chain and one heavy chain make up what is known as the antigen-binding site. The specific sequence of the amino acids in the variable region is determined by the antigen with which it will bind. All antibodies of the same class have the same constant region.

   Antibodies from different classes have different constant regions. There are five major categories of constant regions.

6.

Class of Ig	% in serum	Transplacental transfer?	Complement system?	Example of function
IgG	80	Yes	Yes	Increases phagocytosis
IgM	5–10	No	Yes	Aggregates antigens
IgA	10–15	No	No	Inhibits attachment
IgD	0.2	No	No	Little known
IgE	0.002	No	No	Causes inflammatory reactions

7. IgM antibodies are most effective at cross-linking foreign cells because they have numerous binding sites, more so than are found on antibodies of other classes.

8. Myelomas are tumors formed by cancerous B cells. Cells from myelomas (and other cancerous cells) are referred to as immortal because they can be propagated in culture indefinitely. When fused to a noncancerous antibody-producing plasma cell and cloned, a hybridoma is formed. These antibody-secreting cells can be maintained in culture and are used to produce monoclonal antibodies.

9. The "T cell" system classifies cells of the cell-mediated immune system by their function. According to this system, there are four categories of T cells, including T_H (helper T cells), T_C (cytotoxic T cells), T_D (delayed hypersensitivity T cells), and T_S (suppressor T cells).

   The other system of classification is based on a type of cell surface receptor referred to as a CD marker (cluster of differentiation). There are two types of receptors, CD4 and CD8. Helper T cells have CD4 markers. Cytotoxic and suppressor cells have CD8 markers. T_D cells may have either CD marker.

10. Cytokines are soluble chemical messengers produced by different cells such as lymphocytes and macrophages. Cytokines allow cells of the immune system to communicate and work more effectively.

    Interleukin-1 stimulates T_H cells in the presence of antigens and attracts phagocytes.

    Interleukin-2 is involved in proliferation of helper T cells and B cells. It activates cytotoxic T cells and natural killer cells.

    Gamma-interferon inhibits intracellular viral reproduction.

    Tumor necrosis factor-beta is cytotoxic to tumor cells; enhances phagocytic cells.

    Granulocyte-macrophage-colony-stimulating factor stimulates the formation of red and white blood cells from stem cells.

Chapter 18

Matching

I. 1.g 2.b 3.d 4.a 5.c 6.f
II. 1.a 2.c 3.b

Fill-in-the-Blanks

1. variolation 2. titer 3. indirect 4. attenuated 5. titer 6. herd 7. conjugate

Label the Art

a. Antigen b. Antibody c. Zone of antibody excess d. Zone of equivalence e. Zone of antigen excess

Critical Thinking

1. Herd immunity refers to a situation in which most individuals of a population are immune to a disease. If an outbreak of the disease occurs, there will not be enough susceptible individuals to support an epidemic.

2. Attenuated and inactivated vaccines are both examples of whole-agent vaccines. The process for making the vaccines and the results achieved by using the vaccines differ. Attenuated vaccines use weakened microorganisms to confer immunity to the recipient. Very often viruses used for this type of vaccine are derived from viruses that have mutated during long-term cell culture. Many attenuated vaccines provide lifelong immunity and may be 95% effective. This is due to the fact that the attenuated virus replicates in the body, increasing the original dose. Inactivated vaccines use killed viruses. The viruses are killed by treatment with formalin or other chemicals. Even though the viruses are dead, they still stimulate an immune response. Inactivated vaccines are usually used in situations in which live vaccines are considered too risky.

3. Recombinant vaccines, or subunit vaccines, are produced by genetically altered organisms such as a yeast. The genetically engineered yeast produces a portion of the virus, such as a viral coat protein. Subunit vaccines are safer than whole-agent vaccines because they cannot cause infection in the recipient under any circumstances.

4. The vaccinia virus is a good choice for the production of subunit vaccines because it is harmless and has a relatively large genome. This means that there is room to insert several genes so that the vaccinia virus might be used as a vaccine for several diseases simultaneously.

5. The relatedness of antigens can be determined (at least a rough determination can be made) by the Ouchterlony test. Relatedness of the antigens is indicated by the formation of a visible precipitate in the agar between the wells on the Ouchterlony plate.

6. Titer refers to the concentration of serum antibody and is the lowest dilution of the antibody that will result in agglutination. In general, the higher the serum antibody titer, the greater the immunity to the disease. This means that a titer of 1:312 shows greater immunity to a disease than a titer of 1:94.

7. The procedure used to diagnose strep throat involves coating particles with antibodies to detect strep antigens. Advantages to this indirect agglutination method over traditional methods include speed (this method takes 10 minutes as opposed to 24 hours) and specificity. Culturing on blood agar for the demonstration of beta hemolysins allows for a presumptive diagnosis of strep throat.

8.

Procedure	Advantages	Disadvantages
a. Complement fixation	Can detect small amounts of Ab	Great care and good controls necessary
b. Radioimmunoassay	Can detect many substances like drugs and Abs	Uses radioactive substances
c. ELISA	Can detect Ab or Ag; easy to read results	None to speak of!

9. The helper T cells can be identified and sorted using a fluorescence-activated cell sorter. This involves labelling the CD4 marker on the cells with a fluorescent dye.

10. Genetic immunization would involve injecting the DNA that codes for a protein antigen, rather than the antigen itself. The animal's body produces the protein, recognizes it as foreign, and then produces antibodies specific for the protein. This procedure would eliminate the risk involved with inactivated and attenuated vaccines.

Chapter 19

Matching

I. 1.d 2.b 3.f 4.c 5.a 6.e
II. 1.d 2.d 3.e 4.a 5.g
III. 1.d 2.a 3.c 4.e
IV. 1.a 2.d 3.c 4.b 5.e
V. 1.d 2.a 3.c 4.d 5.b 6.a
VI. 1.a 2.c 3.c

Fill-in-the-Blanks

1. tumor necrosis 2. systemic 3. O 4. isograft 5. major histocompatibility complex 6. human leukocyte antigens 7. antigen 8. versus host 9. Graves' 10. epinephrine 11. cell-mediated

12. anti-Rh⁺ antibodies 13. myasthenia gravis 14. hives 15. clonal deletion 16. privileged site
17. cyclosporine, FK 506, or rapamycin 18. privileged tissue 19. positive 20. agranulocytosis

Label the Art

I. a. gp 120 b. RNA c. Cylindrical core d. Envelope e. Reverse transcriptase f. Capsomere
g. gp 41 h. gp 160 II. a. HIV b. T cell c. CD4 receptor d. DNA

Critical Thinking

1. The first exposure to poison ivy sensitized Susie's T_D cells. The second exposure resulted in a cell-mediated hypersensitivity reaction causing T_D cells to release cytokines, the primary cause of the inflammatory reaction. The substances in poison ivy causing the immune response are catechols, which act as haptens, combining with skin proteins to provoke an immune response.

2. Superantigens are antigens that cause a drastic immune response. They act as nonspecific antigens, indiscriminately activating many T-cell receptors at once. This causes the release of large amounts of cytokines and in turn the production of a flood of T cells. Enterotoxins produced by some staphylococci act as superantigens.

3. When an antigen such as pollen contacts two adjacent antibodies of the same specificity, it will bind to each one, bridging the space between them. The formation of this bridge triggers mast cells and basophils to degranulate, releasing a variety of mediators responsible for the symptoms of anaphylactic reactions.

4. a. Histamines increase the dilation and permeability of blood capillaries, resulting in edema, erythema, runny nose, and difficulty in breathing.

   b. Prostaglandins affect the smooth muscles of the respiratory system and cause increased mucus secretion.

   c. Leukotrienes usually cause prolonged contractions of certain smooth muscles, contributing to spasms of the bronchial tubes associated with asthma attacks.

5. It is difficult to know if the patient is allergic to penicillin. A skin rash without hives (in response to penicillin) is not always significant in children, but without a reliable method to test for penicillin allergy, it might be wise to err on the side of caution and avoid the use of penicillin in the future.

6. People with type A blood have anti-B antibodies in their plasma, so they will react to type B blood on the first exposure. Rh⁻ people do not have anti-Rh antibodies in their plasma, so they won't react to Rh⁺ blood until the second exposure.

7. Serum sickness refers to inflammatory reactions that occur upon repeated exposure to antibodies—for example, the antibodies in antitoxins produced in other animals.

8. Clonal deletion and clonal anergy are processes that occur during fetal development or shortly after birth. Clonal deletion refers to the elimination of T cells that will target host cells, whereas clonal anergy is the inactivation of these cells. These processes are essential for self-tolerance.

9. a. Type II autoimmune reactions involve antibody reactions to surface cell antigens but don't result in destruction of the cell. An example is Graves' disease, a condition in which antibodies attach to the thyroid, blocking receptors for thyroid-stimulating hormone. This results in the overproduction of this hormone. Another example is myasthenia.

   b. Type III autoimmune reactions are not completely understood; however, individuals produce antibodies directed at components of their own cells. Examples include systemic lupus and rheumatoid arthritis.

   c. Type IV autoimmune reactions involve the destruction of cells, primarily by T cells. Examples include Hashimoto's thyroiditis and insulin-dependent diabetes.

10. Some of the obstacles that must be overcome to produce an AIDS vaccine are:
    —The lack of a suitable animal model to study replication of the virus and its response to experimental drugs.

—Not being able to use whole-agent vaccines is a disadvantage because they are generally more effective than subunit vaccines.

—Antigenic variants of HIV pose yet another obstacle in the development of a vaccine.

Chapter 20

Matching

I. 1.e 2.a 3.b
II. 1.c 2.b 3.f 4.a 5.e 6.d
III. 1.a 2.c 3.e 4.b 5.d
IV. 1.e 2.b 3.d 4.f 5.c 6.a
V. 1.b 2.a 3.c 4.d 5.e
VI. 1.a 2.f 3.e 4.c 5.g 6.b 7.h

Fill-in-the-Blanks

1. peptidoglycan 2. chemotherapy 3. sterols 4. DNA 5. cytosine 6. salvarsan 7. penicillinase 8. selective toxicity 9. aplastic anemia 10. rifampin 11. streptomycin 12. β-lactam ring 13. aminoglycoside 14. vancomycin 15. minimum inhibitory concentration (MIC) 16. minimum bactericidal concentration (MBC) 17. imidazole 18. ribosomes 19. superinfection 20. monobactams 21. AZT

Critical Thinking

1. a. Selective toxicity—The drug should be toxic to the pathogen but not to the host.

   b. The drug should not produce hypersensitivity in most hosts.

   c. The drug must be soluble in body fluids so that it can rapidly penetrate body tissues. It must also remain in the body long enough to be effective.

   d. Microorganisms shouldn't become readily resistant to the drug.

2. Antibiotics were first added to animal feed to lower the incidence of infection in closely penned animals. Another reason that antibiotics are still added to animal feed is that they accelerate the growth of the animal. The practice has been linked with *Salmonella* infections in humans from meat and milk. FDA testing has shown that most milk and meat has little or no detectable antibiotics, but the practice is still seen as unfavorable by many. Continued use of antibiotics in animal feed will result in the development of antibiotic-resistant strains of bacteria.

3. Advantages:

   Broad-spectrum activity means that the identity of the pathogen need not necessarily be known; this saves valuable time.

   Many important pathogens and opportunistic organisms are eliminated by broad-spectrum antibiotics.
   Disadvantages:

   Normal microbiota are killed, allowing opportunistic microbiota (for example, yeasts) to proliferate.

   Resistant strains of bacteria develop with the indiscriminant use of broad-spectrum drugs.

4. Chloramphenicol binds to the 50S portion of the procaryotic ribosome and inhibits formation of peptide bonds.

   Tetracycline prevents the attachment of tRNA to mRNA.

   Erythromycin also binds to the 50S unit but it prevents translocation.

   Streptomycin changes the shape of the 30S portion of the procaryotic ribosome, resulting in the misreading of mRNA.

5. Drugs like amphotericin can be toxic to humans due to the cholesterol in the membranes of their cells. It is more toxic to fungal cells because they have mostly ergosterol, against which the drug is most effective.

6. Ethambutol—Advantages: helps avoid resistance in *Mycobacterium*; disadvantages: weak antitubercular drug.

   Sulfonamides—Advantages: several valuable applications; disadvantages: they cause allergic reactions in many people.

   Carbapenems—Advantages: very broad-spectrum activity; disadvantages: none (serious) so far!

   Nitrofurans—Advantages: broad-spectrum activity; disadvantages: usually reaches effective concentrations in urine only.

7. Nystatin is used successfully to treat local vaginal infections without serious problems because it has poor solubility. This prevents the drug from entering body tissue in toxic amounts.

8. Acyclovir is an analog of a guanine-containing nucleoside. The drug terminates the synthesis of viral DNA but has little effect on host-cell DNA.

9. The main difficulty associated with antiprotozoan drugs is that these organisms are eucaryotic, so drugs that affect them are more likely to affect the host. Specific problems include DNA damage, carcinogenic action, and nerve damage. Examples of antiprotozoan drugs include the following:

   Metronidazole—inhibits nucleic acid synthesis; used to treat *Trichomonas* and *Giardia* infections.

   Quinacrine—blocks DNA synthesis; used to treat *Giardia* infections.

10. Synergism refers to the use of two or more antimicrobial drugs simultaneously. It has been found that their combined effect is greater than the effect of either given alone.

    Combinations of antimicrobial drugs should be given for the following purposes:

    a. To prevent or minimize development of resistant strains.

    b. To take advantage of the synergistic effect.

    c. To provide optimal therapy in life-threatening, time-critical situations.

    d. To lessen drug toxicity by reducing the necessary drug concentration.

Chapter 21

Matching

I. 1.c 2.d 3.a 4.b 5.e
II. 1.b 2.d 3.c 4.a
III. 1.c 2.a 3.e 4.b 5.d
IV. 1.b 2.d 3.d 4.c 5.e 6.d
V. 1.a 2.a 3.a 4.d 5.b
VI. 1.c 2.d 3.a 4.b
VII. 1.c 2.b 3.a 4.e 5.d

Fill-in-the-Blanks

1. lysozyme 2. diphtheroids 3. Reye's 4. carbuncle 5. beta 6. pyocyanin 7. acne 8. measles (rubeola) 9. mycosis 10. viremia 11. trachoma 12. 1 13. sty 14. antiphagocytic 15. sebum 16. smallpox 17. trimester 18. capitis 19. coagulase 20. erythrogenic 21. pedis 22. *Acanthamoeba*

Label the Art

a. Blood vessel b. Oil gland c. Duct of sweat gland d. Sweat gland e. Hair follicle f. Adipose tissue g. Nerve h. Sweat pore i. Stratum corneum j. Hair shaft k. Hair erector muscle l. Epidermis m. Dermis n. Subcutaneous layer

Critical Thinking

1. The etiologic agent is probably *Staphylococcus aureus*. It is a common cause of nosocomial infections because it is a part of the normal microbiota of most people. That means hospital staff, visitors, and even the patient are possible sources of infection.

2. a. *Pseudomonas* dermatitis. Transmitted in swimming pools, hot tubs, and so on. Contributing factors include the diminished effects of chlorine due to heavy usage; hot water also facilitates entry of bacteria into hair follicles.

   b. Respiratory infections. Airborne or fomite transmission; compromised hosts are at risk.

   c. Otitis externa (swimmer's ear) transmitted in swimming pools. People who spend a lot of time in pools are at risk.

3. *Propionibacterium acne* is part of the normal microbiota of the skin. This organism is able to metabolize sebum and produces acid by-products that discourage the growth of potential pathogens.

   When sebum channels become blocked and rupture, bacteria—especially *Propionibacterium acne*—become involved. The bacterium metabolizes sebum into free fatty acids that cause an inflammatory response. This leads to tissue damage and possible scarring.

4. The symptoms of shingles are different from those of chickenpox because shingles is a different expression of the virus. After having had chickenpox, the patient has partial immunity to the virus.

5. Herpes simplex viruses, type I and type II, are both double-stranded DNA viruses that infect up to 90% of the population of the U.S.A. Herpes simplex type I usually causes oral infections, and the virus may be stimulated into renewed activity by sunburn, hormonal changes, or emotional upset. Herpes simplex type II infections are often of the genitals. The viruses are very similar.

6. When a new measles vaccine was introduced in 1963, the number of cases per year dropped drastically. 1983 saw a record low number of cases. Since then, measles cases have risen slowly. Factors contributing to this rise include the fact that not everyone was immunized and that the vaccine was only about 95% effective.

7. *Candida albicans*, the cause of vaginal yeast infections, is part of the normal microbiota of many women. Anything that upsets the delicate balance of normal microbiota can contribute to development of a yeast infection. The use of antibiotics eliminates most bacterial microbiota, allowing *Candida* to overgrow. Immunocompromised patients are also at risk.

8. The physician suspects *Acanthamoeba* keratitis and is looking for trophozoites or cysts in the corneal scraping.

9. The efforts to eliminate smallpox were successful primarily because of the lack of animal reservoirs. Once an effective vaccine was available, a concerted effort was coordinated by the World Health Organization that resulted in elimination of the disease.

10. a. The acidic metabolic by-products of *Propionibacterium* may result in acne.

    b. Normal microbiota that metabolize sweat contribute to body odor.

    c. The yeast *Pityrosporum* may be responsible for dandruff.

Chapter 22

Matching

I. 1.a 2.d 3.b 4.f 5.c
II. 1.e 2.a 3.d
III. 1.a 2.a 3.a 4.a 5.d 6.b 7.g
IV. 1.a 2.f 3.d 4.e 5.c
V. 1.b 2.b

VI. 1.a 2.b 3.d 4.b
VII. 1.a 2.c 3.c

Fill-in-the-Blanks

1. encephalitis 2. central 3. meningitis 4. blood–brain barrier 5. peripheral 6. Sabin 7. leprosy 8. Eastern equine encephalitis 9. tetanus 10. nervous 11. sausage 12. cerebrospinal 13. human diploid 14. nitrates 15. lepromatous (progressive)

Critical Thinking

1. The signs and symptoms of tetanus are caused by a potent neurotoxin produced by *Clostridium tetani* and include severe muscle contractions (spastic paralysis). Several factors contribute to the transmission of tetanus. They include improperly cleaned deep puncture wounds, inadequate immunity (vaccinations not current), and lessened immunity as seen in the elderly.

2. The toxin causing botulism is extremely potent and causes the disease in minute quantities. These quantities are too small to elicit an immune response. Antibiotics are not indicated in the treatment of botulism because the toxin is preformed and the bacteria may not even be present.

3. Infants under 1 year of age should not be fed honey because their intestinal microbiota is not well established, making them susceptible to infant botulism.

4. Infection with polio virus usually occurs by ingestion of feces-contaminated water. Primary areas of viral multiplication are the throat and small intestine, resulting in a sore throat and nausea.

5. The rabies virus has an unusually long incubation period. This makes it possible for a patient to respond to vaccination and develop immunity to the virus during the incubation period.

6. The major problem in developing a vaccine against trypanosomiasis is the ability of the trypanosome to change protein coats. This ability also allows the organism to avoid the host's immune system.

7. a. Sheep scrapie—The animal rubs against fences and walls, gradually loses motor control, and dies.

    b. Creutzfeldt–Jakob disease (CJD)—This disease is similar to scrapie but often occurs in families.

    c. Kuru—Infection with kuru is associated with cannibalistic practice and is very rare.

    d. Bovine spongiform encephalopathy—Bovines infected with this disease become unmanageable. The disease is thought to be caused by a scrapie-like pathogen.

8. Hansen's disease is another name for leprosy, which is caused by *Mycobacterium leprae*. Transmission is not completely understood, but it has been observed that infected people shed the bacterium in nasal secretions and from their lesions. The disease is diagnosed by the detection of acid-fast bacteria.

9. The Salk vaccine uses formalin-inactivated virus and requires a series of injections. Booster shots are necessary every few years to maintain full immunity.

    The Sabin vaccine contains three strains of live, attenuated viruses and is administered orally. Immunity is effective but in rare cases infection caused by the vaccine does occur.

10. *Cryptococcus neoformans* is commonly found in the soil and is associated with areas contaminated with pigeon droppings. Infection occurs when *Cryptococcus* is inhaled and is often subclinical. The organism can spread through the bloodstream to other parts of the body, including the brain and meninges. This disease is usually expressed as chronic meningitis and is fatal if untreated. Immunocompromised people and those taking steroids are especially at risk of progressive cryptococcosis.

Chapter 23

Matching

I. 1.e 2.a 3.b
II. 1.c 2.h 3.g 4.f 5.b

III. 1.b 2.f 3.d 4.d 5.e 6.a
IV. 1.g 2.f 3.e 4.a 5.c 6.b 7.b 8.g 9.g
V. 1.b 2.k 3.d 4.c 5.e 6.i 7.j

Fill-in-the-Blanks

1. debridement 2. plasma 3. leukocyte 4. endotoxins 5. pericardium 6. *Staphylococcus aureus*
7. *Streptococcus* 8. peritonitis 9. *Streptococcus pyogenes* 10. tularemia 11. swine 12. anthrax
13. rheumatic fever 14. infectious mononucleosis 15. hyperbaric 16. tularemia 17. *Bacteroides;*
Clostridium 18. tularemia 19. puerperal sepsis 20. tularemia; brucellosis 21. sporozoite
22. *falciparum* 23. schistosomiasis 24. malaria

Label the Art

a. Superior vena cava (main upper vein) b. Lung c. Capillaries in lung d. Inferior vena cava (main lower vein) e. Liver f. Intestine g. Aorta (main artery) h. Heart i. Spleen j. Stomach k. Kidneys l. Capillaries in intestines

Critical Thinking

1. Gram-negative rods—for example, *E. coli* and *Enterobacter aerogenes*—are the organisms most often associated with septicemia. The symptoms of septicemia are the result of endotoxins that are released as the bacterial cells are lysed.

2. The causative agent was probably *Staphylococcus aureus, Streptococcus pyogenes*, or another organism associated with dental infections. Microorganisms that are released into the blood during tooth infections can find their way to the heart, resulting in endocarditis.

3. Untreated cases of strep throat may lead to rheumatic fever. Approximately 3% of untreated cases of strep throat lead to rheumatic fever.

4. *Brucella abortus* is transmitted in milk and causes a mild and self-limiting disease.

   Brucella melitensis is also transmitted in milk. Animal reservoirs for this species include sheep and camels. The disease caused by this species is serious and may result in disability or even death.

   Brucella suis is transmitted by swine and results in the formation of destructive abscesses.

5. A hyperbaric chamber is a chamber that contains a pressurized, oxygen-rich atmosphere. The oxygen saturates the infected tissues and prevents the growth of obligate anaerobes such as *Clostridium*. This procedure is used to treat abdominal gas gangrene.

6. The etiologic agent is *Bacillus anthracis*; the diagnosis is pulmonary anthrax. The patient probably worked with sheep or goats or with their hides, a common cause of anthrax in the Middle East.

7. The etiologic agent is any of several species of *Borrelia*. This organism is most often transmitted by the bite of an infected soft tick. The fever returned because *Borrelia* changed its antigenic type to avoid the effects of the host's immune system.

8. The suspected etiologic agent is *Borrelia burgdorferi*; the diagnosis is Lyme disease. The physician instructed the patient to finish the antibiotics because a negative test for Lyme disease is not conclusive due to the lack of reliable tests for this condition. Under these circumstances it makes sense for the patient to finish the antibiotics.

9. EBV was established as the cause of infectious mononucleosis by accident. During a study of Burkitt's lymphoma, a technician previously negative for EBV contracted the virus and developed infectious mononucleosis.

10. Congenital infections with *Toxoplasma* may result in spontaneous abortion or a stillborn fetus. Reactivated toxoplasmosis in immunocompromised people may also be fatal.

Chapter 24

Matching

I. 1.c 2.b 3.d 4.a 5.e 6.a

II. 1.e 2.d 3.b 4.c 5.d 6.b 7.e 8.a

III. 1.a 2.c 3.c 4.b 5.d 6.d 7.d 8.d 9.e

Fill-in-the-Blanks

1. influenza 2. alveoli 3. pleura 4. epiglottitis 5. AIDS 6. IgA 7. miliary 8. BCG
9. pneumococcal pneumonia 10. cutaneous diphtheria 11. tuberculosis 12. macrophages
13. drift 14. shift 15. sterile 16. pertussis (whooping cough) 17. pertussis (whooping cough)
18. hemagglutinin, neuraminidase 19. hemolysis 20. diphtheria 21. *Pneumocystis* pneumonia
22. Q fever 23. tuberculosis

Label the Art

a. N spike b. Protein layer c. H spike d. Lipid bilayer e. 8 RNA segments

Critical Thinking

1. Because there are so many agents that can cause colds, it would be difficult or even impossible to make a vaccine effective against them all. Research has shown that most of these organisms use as few as two receptors to attach to the host's mucosa. For this reason researchers have considered developing a vaccine that will elicit production of an antibody to block those receptors.

2. Although *Bordetella* does not invade tissue, it will grow on cilia of the trachea, impeding their action and resulting in loss of the ciliated cells and an accumulation of mucus. The patient coughs desperately to eliminate the mucus. In small children the coughing may be so forceful as to break the ribs.

3. If a person yields a positive skin test for tuberculosis, the next step would be to do a chest X ray. If the X ray is positive, then it is necessary to attempt to isolate the organism and to do an acid-fast stain on a sputum sample.

4. People of European descent have a certain level of innate immunity to tuberculosis because many generations have been exposed to the disease. Native Americans, Asians, and Hispanics lack this innate immunity.

5.

Causative agent	Transmission	Drug of choice
Streptococcus pneumoniae	Cough-produced droplets of normal microbiota	Penicillin
Klebsiella pneumoniae	Associated with malnutrition and alcoholism	Cephalosporins, gentamicin
Mycoplasma pneumoniae	Airborne droplets	Tetracycline, erythromycin
Legionella pneumophila	Aerosol formed by contaminated water	Erythromycin and maybe rifampin

6. The suspected etiologic agent is *Chlamydia psittaci*, a bacterium whose transmission is associated with birds. The pet-shop owner was probably infected by inhalation of the bacterium on dried particles from contaminated bird droppings. The drug of choice is tetracycline.

7. Approximately 10% of people with Q fever will develop endocarditis. It can take as long as 5–10 years for endocarditis to develop. Researchers speculate that the organism resides in the liver during that time.

8. H spikes and N spikes are projections located on the outside of an influenza virus. H spikes (hemagglutinin) allow the virus to recognize and attach to host cells before infecting them. N spikes (neuraminidase) help the virus to separate from the host cell after intracellular reproduction.

9. Earthquakes produce dust that may harbor these organisms. This situation is especially true in years with an especially wet winter and an especially dry summer.

10. Atypical pneumonia is difficult to diagnose because *Mycoplasma pneumoniae* is difficult to isolate and to grow on artificial media. Atypical pneumonia is often self-limiting, and the patient will have recovered before the organism is isolated.

Chapter 25

Matching

I. 1.e 2.a 3.b 4.c 5.d 6.f 7.h 8.g
II. 1.g 2.f 3.d 4.g 5.c 6.e 7.b 8.g
III. 1.c 2.a 3.d 4.f 5.g 6.e 7.b
IV. 1.b 2.f 3.e 4.a 5.h 6.i

Fill-in-the-Blanks

1. *solium* 2. 0:1 3. jaundice 4. cysticeri 5. fish 6. *solium* 7. ergot 8. *sonnei* 9. *Vibrio* 10. aflatoxin 11. IgA 12. dextran 13. periodontal 14. A 15. mannitol 16. pharynx; esophagus 17. small; large 18. hepatitis B 19. trichinosis 20. metronidazole

Label the Art

a. Enamel b. Dentin c. Pulp d. Bone e. Root f. Decay g. Healthy tooth with plaque h. Decay in enamel i. Advanced decay j. Decay in dentin k. Decay in pulp

Critical Thinking

1. Food contaminated with *S. aureus* is not safe to eat after reheating. This is due to the heat-stable enterotoxin produced by *S. aureus*. The toxin causes the symptoms of staphylococcal food poisoning.

   Salmonella-contaminated food is safe to eat if properly cooked. The symptoms of salmonella food poisoning are due to colonization of the organism in the GI tract of the host and not due to a toxin.

2. The incubation period after infection with *Salmonella* can vary from 12 hours to 2 weeks. The symptoms of salmonellosis are due to colonization of the host's GI tract with *Salmonella*. The length of the incubation period is related to the number of cells initially ingested.

3. Both of these infections are caused by species of the genus *Vibrio*.

   Cholera is endemic in Asia; parahaemolyticus gastroenteritis is common in many parts of the world, especially in Japan.

   V. parahaemolyticus gastroenteritis is associated with salt water; the etiologic agent requires 2% salt to grow.

   Both infections are associated with contaminated water and can be transmitted by contaminated seafood.

   Both cause the loss of large amounts of water through diarrhea.

   Cholera has a high mortality rate; *V. parahaemolyticus* gastroenteritis has a low mortality rate.

4. Food poisoning with *C. perfringens* is associated with meat contaminated by the butchered animal's intestinal contents. The organism thrives on the nutrients provided by the meat, and cooking lowers the levels of oxygen, further encouraging growth. Meat should not be allowed to cool slowly so as to discourage growth of *Clostridium*.

5. Complications associated with CMV occur upon maternal infection with the virus and in the immunosuppressed. Maternal infection can seriously harm the fetus, even resulting in death. Immunocompromised individuals may develop life-threatening pneumonia.

6. Rotavirus is the most common cause of viral gastroenteritis, causing disease mostly in small children. After a 2–3 day incubation period, the following symptoms develop: low-grade fever, diarrhea, and vomiting. These symptoms may last for 5–8 days.

   The virus referred to as the Norwalk virus is also a common cause of viral gastroenteritis. The incubation period is 2 days, and symptoms include nausea, abdominal cramps, diarrhea, and vomiting. They last for approximately 2 days.

7. Infection with *Giardia* is relatively common because there are many animal reservoirs that harbor the protozoa.

8. The cyst of *E. histolytica* allows the organism to survive the highly acidic environment found in the stomach. Stomach acids break down the cyst wall so that the trophozoite is released into the intestines.

9. The HDV antigen isn't capable of causing infection without first being covered by the HBV envelope.

10. The four pathogenic species of *Shigella* are *S. sonnei*, *S. dysenteriae*, *S. flexneri*, and *S. boydii*. *S. dysenteriae* has the highest mortality rate, approximately 20%. *S. sonnei* is most common in the U.S.A.

Chapter 26

Matching

I. 1.b 2.d 3.e 4.a 5.b 6.d 7.d 8.e 9.c
II. 1.d 2.b 3.a 4.c
III. 1.e 2.d 3.c 4.a 5.b
IV. 1.d 2.a 3.a 4.b

Fill-in-the-Blanks

1. ovulation 2. urethritis 3. ureteritis 4. salpingitis 5. opthalmia neonatorum 6. chlamydia 7. reagin 8. chancroid 9. 2 10. fallopian 11. ectopic 12. kidneys 13. leptospirosis 14. arthritis 15. fimbriae 16. congenital 17. vulvovaginitis candidiasis (vaginitis) 18. gummas 19. *Escherichia coli* 20. soft chancre 21. acyclovir

Label the Art

I. a. Urinary bladder b. Pubic bone c. Ductus (vas) deferens d. Urethra e. Penis f. Urethral opening g. Scrotum h. Ureter i. Rectum j. Seminal vesicle k. Ejaculatory duct l. Prostate gland m. Anus n. Epididymis o. Testis II. a. Uterine (fallopian) tube b. Ovary c. Uterus d. Pubic bone e. Urinary bladder f. Urethra g. Clitoris h. Labium majus i. Rectum j. Cervix k. Vagina l. Anus m. Labium minus III. a. Uterine tube b. Ovary c. Endometrium d. Cervix e. Vagina f. Ovary g. Uterus

Critical Thinking

1. Within a few weeks of birth lactobacilli will dominate normal microbiota of the female genital system. This is due to the influence of maternal estrogens.

   By a few weeks later, other bacteria will become established, including corynebacteria and a variety of cocci and bacilli.

   Lactobacilli will again dominate during puberty and maturity as estrogen levels increase.

2. Women have a shorter urethra than do men; the urethra in women is closer to the anus than it is in men; the female urethra has an abundance of normal microbiota.

3. Kidney damage associated with glomerulonephritis is due to antigen–antibody complexes that interact with complement. These complexes are deposited in the glomeruli, where they cause inflammation and kidney damage.

4. PID stands for pelvic inflammatory disease. It is a generalized term that refers to any extensive bacterial infection of the pelvic organs. One cause of PID is gonorrhea. Any other generalized infection of the pelvic organs can result in PID—for example, a contaminated IUD.

5. Oral contraceptives have contributed to the increased incidence of gonorrhea and other STDs for two major reasons. First, oral contraceptives increase the moisture content and raise the pH of the vagina. This increases the susceptibility of mucosal cells to infection. Second, oral contraceptives have in many cases replaced condoms and spermicides, both of which help prevent disease transmission.

6. a. Venereal Disease Research Laboratory (VDRL) is a slide flocculation test and is probably the most widely used serological screening test.

   b. The rapid plasma reagin (RPR) card test is similar to the VDRL test and is also in common use. Both of these tests are nonspecific and test for reagin-type antibodies.

   c. Fluorescent treponemal antibody absorption test (FTA-ABS) is an indirect immunofluorescence test in which an avirulent strain of *Treponema* is allowed to react with a sample of the patient's serum. Fluorescence indicates a positive test.

7. Clue cells are associated with *Gardnerella* vaginosis and are sloughed-off vaginal epithelial cells covered with bacteria, primarily *Gardnerella vaginalis*. Signs include a frothy, foul smelling vaginal discharge and clue cells in the discharge. The drug of choice is metronidazole.

8. Three diseases caused by *Chlamydia trachomatis* include the following:

   Lymphogranuloma venereum is a disease seen most often in tropical countries. After 7–12 days, a small lesion appears at the site of infection, ruptures, and heals. One week to 2 months later the organisms invade the lymphatic system, cause scarring, and may result in enlargement of the genitals in males or rectal narrowing in females.

   Trachoma is the single greatest cause of blindness and is common in the arid regions of Africa and Asia and in the southwestern U.S.A. It is transmitted by hands, fomites, and flies.

   Nongonococcal urethritis (NGU) refers to any inflammation of the urethra not caused by gonorrhea. The most common cause is *Chlamydia*. Signs and symptoms include painful urination and a watery discharge.

9. Neonatal herpes may result in spontaneous abortion or serious damage to the fetus, such as mental retardation and defective sight and hearing.

10. Trichomoniasis is more common in females than in males because *Trichomonas vaginalis* may be part of the normal microbiota of women.

Chapter 27

Matching

I. 1.c 2.d 3.a
II. 1.a 2.d 3.b 4.c 5.g 6.f 7.g
III. 1.c 2.d 3.b 4.a 5.c 6.b
IV. 1.d 2.c 3.b 4.b
V. 1.b 2.d 3.f 4.a 5.g
VI. 1.b 2.a 3.e 4.c 5.d

Fill-in-the-Blanks

1. 95 2. bacteria 3. *Streptomyces* 4. fungi 5. 0.03 6. 80 7. nitrate 8. ammonification 9. oxygen
10. heterocysts 11. infection thread 12. lichens 13. recalcitrant 14. actinomycete (*Frankia*)
15. phosphorus 16. confirmed or confirmatory 17. flocculation 18. adsorption 19. sludge
20. organic matter; oxygen 21. bulking 22. oxygen

Label the Art

a. Wood and fossil fuels b. CO in atmosphere c. Plants, algae, cyanobacteria d. Animals e. Dead organisms f. Soil and water microbes

Critical Thinking

1. Actinomycetes are common inhabitants of the soil. Their filamentous growth habits allow them to bridge the gap between soil particles in dry conditions and maximizes surface area, giving Actinomycetes a nutritional advantage over other soil-dwelling microbes.

2. Bacteria and fungi decompose organic (plant and animal) material, releasing carbon dioxide into the atmosphere and making it available to start the cycle over again.

3. Ammonification—the removal of amino groups from amino acids to form ammonia.

   Nitrification—refers to the oxidation of nitrogen from ammonia to nitrites or nitrates.

   Denitrification—the reduction of nitrates to nitrites or nitrogen gas.

   Nitrogen fixation—the conversion of nitrogen into ammonia.

4. Algae are mostly found in the limnetic zone of a freshwater habitat. Algae serve as the primary producers, supporting all other life in the habitat.

5. Bioluminescent bacteria have an enzyme called luciferase. This enzyme picks up electrons from flavoproteins in the electron transport chain and emits some of the electron's energy as a photon of light.

6. The high phosphate content of the new biodegradable detergents was unchanged by most sewage treatment processes and so led to eutrophication of streams and lakes. This is an overabundance of nutrients that causes an overgrowth of algae and cyanobacteria and the death of other organisms in the aquatic environment.

7. Indicator organisms are organisms that are consistently found in human feces in sufficient numbers to be detected in water when fecal contamination occurs. The indicator organism must also grow at least as well as would pathogenic organisms in feces. In the U.S.A. coliform organisms are used as indicator organisms. Their detection is an indication of fecal contamination of water.

8. The activated sludge system involves the addition of air or pure oxygen to the effluent from primary treatment. This encourages the growth of aerobic organisms that oxidize much of the effluent's organic matter.

   Trickling filters involve the spraying of sewage over a bed of rocks or molded plastic. Aerobic bacteria form a gelatinous film on these materials and act very much the same way as was just discussed.

9. Oxidation ponds, or stabilization ponds, are used in some small communities for water treatment. The advantage of oxidation ponds is that they are inexpensive to build. The disadvantage is that they require considerable quantities of land.

10. There are two types of mycorrhizae: endomycorrhizae (vesicular-arbuscular mycorrhizae) and ectomycorrhizae. Both function as root hairs of plants. They extend the surface area through which plants can absorb nutrients, especially phosphorus.

Chapter 28

Matching

I.   1.b 2.c 3.d 4.e 5.a 6.g 7.f
II. 1.a 2.c 3.b 4.b
III. 1.d 2.b 3.a
IV. 1.b 2.a 3.a
V.   1.a 2.d 3.b 4.c

Fill-in-the-Blanks

1. unripened 2. hard 3. *Penicillium* 4. yogurt 5. yeasts; acetic; *Acetobacter* or *Gluconobacter* bacteria 6. nucleic acid 7. monosodium glutamate 8. bioreactors 9. sterols 10. commercial sterilization 11. 12 12. *Clostridium botulinum* Type E, *Yersinia enterocolitica*, or *Listeria monocytogenes* 13. trichinosis 14. pasteurization 15. whey 16. *Propionibacterium* 17. yogurt 18. hydrogen peroxide 19. heating effects 20. secondary 21. idiophase

Label the Art

I. a. Primary metabolite b. Ethanol produced II. a. Secondary metabolite b. Trophophase c. Idiophase d. Penicillin produced

Critical Thinking

1. True sterilization kills or eliminates all of the microbes from food. Commercial sterilization is not as rigorous; it is designed to kill *Clostridium botulinum* endospores. If this is achieved, then any other bacteria that cause food spoilage or are pathogenic will be killed as well. The lower temperatures used in commercial sterilization result in a better-tasting food product.

2. Advantages are that irradiation of food with gamma rays is very effective at eliminating microorganisms, worms, and insect pests; and penetration of the food is excellent due to the high energy content of gamma radiation. Disadvantages are that the process takes several hours and must be conducted behind protective walls, and public apprehension concerning radiation.

   Advantages are that high-energy electron accelerators are faster than gamma-ray irradiation; it takes only seconds to treat food. Disadvantages are that this method doesn't penetrate well, so it is effective only for thin food products such as sliced meat.

3. a. Sorbic acid is used to prevent the growth of molds in acidic foods such as cheese and soft drinks.

   b. Sodium nitrate and sodium nitrite are added to some meat products to keep the meat red and to prevent the germination and growth of botulism organisms.

   c. Calcium propionate is used in bread to inhibit the growth of fungi and *Bacillus*.

4. The standards are more stringent for raw milk because it may contain pathogenic microorganisms. Raw milk has been linked to several outbreaks of salmonellosis.

5. Flavor is determined by the ripening process; for example, the flavor of Roquefort cheese is due to *Penicillium* that is injected into the cheese before ripening.

   Hardness is dependent on the moisture content of the cheese; the less moisture, the harder the cheese.

6. To make soy sauce, the mold *Aspergillus oryzae* is grown on wheat bran and then added to cooked soybean and crushed wheat along with lactic acid bacteria. This process produces fermentable carbohydrates. The mixture is then allowed to undergo prolonged fermentation, resulting in soy sauce.

7. *Spirulina* could be grown and used as a food source for the colonists.

8. Primary metabolites are products that are formed at essentially the same time as the bacterial cells. Ethanol is an example of a primary metabolite.

   Secondary metabolites are not produced until the microbes have nearly completed their growth phase. An example of a secondary metabolite is penicillin.

9. Lysine is produced by *Corynebacterium glutamicum* and is used as a food supplement. Glutamic acid is also produced by *C. glutamicum* and is used to make the flavor enhancer MSG.

10. Bioconversion refers to the conversion of biomass into alternate fuel sources such as methane. This practice would help meet our ever-increasing energy needs while helping to dispose of solid waste.